COMMUNICATIONS SATELLITE DEVELOPMENTS: TECHNOLOGY

Edited by
William G. Schmidt
CML Satellite Corporation
Washington, D. C.

Gilbert E. LaVean
Defense Communications Agency
Reston, Virginia

Volume 42
PROGRESS IN
ASTRONAUTICS AND AERONAUTICS

Martin Summerfield, Series Editor-in-Chief
Princeton University, Princeton, New Jersey

Technical papers selected from the AIAA 5th Communications Satellite Systems Conference, April 1974, subsequently revised for this volume.

Published by the American Institute of Aeronautics and Astronautics in cooperation with the MIT Press

American Institute of Aeronautics and Astronautics
New York, New York

The MIT Press
Cambridge, Massachusetts and London, England

Library of Congress Cataloging in Publication Data
Main entry under title:

AIAA Communications Satellite Systems Conference, 5th,
   Los Angeles, 1974.
   Communications satellite developments, technology.

   (Progress in astronautics and aeronautics; v. 42)
   "Technical papers selected from the AIAA 5th Communications Satellite Systems Conference, April 1974, subsequently revised for this volume."
   Includes bibliographies.
   1. Artificial satellites in telecommunication——Congresses.
I. Schmidt, William G. II. LaVean, Gilbert E. III. Title. IV. Series.
TL507.P75 vol. 42 [TK5104]   629.1'08s [621.38'0422]   75-45243
ISBN: 0-915928-06-X

Copyright © 1976 by
American Institute of Aeronautics and Astronautics

All rights reserved. No part of this book may be reproduced in any form or by any means, electronic or mechanical, including photocopying, recording, or by any information storage and retrieval system, without permission in writing from the publisher.

Table of Contents

Preface — xi

I Spacecraft Technology — 1

Efficient High-Capacity Communications Satellites — 3
J. H. HOCKENBERRY, D. W. KREJCI, W. A. KOENIG, AND J. P. KLOCKSIEM

A Constrained Lens Antenna for Multiple-Beam Satellites — 25
H. S. LU, W. G. SCOTT, T. SMITH AND A. SMOLL

ATS-6 Interferometer: A Precision, Wide-field-of-View Attitude Sensor — 35
M. A. TEICHMAN, F. L. MAREK, J. J. BROWNING, AND A. K. PARR

Development and Air Bearing Test of a Double-Gimballed Momentum Wheel Attitude-Control System — 49
J. J. KALLEY JR. AND H. L. MORK

Solar vs Nuclear Power: Is There a Choice? — 83
BERNARD RAAB AND JAY J. KARLIN

II Terminal Technology — 113

Earth Station for the U. S.–USSR Direct Communications Link — 115
W. P. HOOPER, W. M. ROGERS, AND J. G. WHITMAN JR.

Economic Considerations for Low-Capacity SHF Satellite Communications Earth Terminals — 145
G. P. PETRICK AND C. M. ABRAHAMSON

Operational Experience with Small Unattended Television Receive Earth Stations — 167
A. D. D. MILLER

III Control Technology — 181

Telesat Satellite Control System — 183
H. KOWALIK

Design of a Ground Control System to Operate Domestic and Maritime Satellites — 201
A. J. E. VAN HOVER AND W. J. GRIBBIN

Satcom System Control Concepts for Increased Link Availability — 217
A. N. INCE, D. W. BROWN, AND J. A. MIDGLEY

IV Multiple Access, Modulation, and Coding — 240

Single Channel Per Carrier Voice Transmission Via Communications Satellite — 243
CHARLES C. SANDERSON AND LLOYD G. LUDWIG

Comparison of Two Basic Doppler
Compensation Methods 263
P. A. KULLSTAM

Methods of Alleviation of Ionospheric Scintillation
Effects on Digital Communications 279
JAMES L. MASSEY

V   **Transmission Path Effects**   289

Impact of Rain Attenuation on 18/30–GHz
Satellite Systems 291
D. JARETT AND L. D. SPILMAN

Morphology of Ionospheric Scintillation 311
R. K. CRANE

Geophysical Properties of the Ionospheric
Irregularities Responsible for Radio Scintillation 347
J. P. MCCLURE

Modeling and Prediction of Ionospheric
Scintillation 367
EDWARD J. FREMOUW

Impact of Scintillation on Transionospheric
Communications 397
HOWARD A. BLANK AND THOMAS S. GOLDEN

Index to Contributors to Volume 42   419

Progress in
Astronautics and Aeronautics

Martin Summerfield,
Series Editor
PRINCETON UNIVERSITY

VOLUMES
1. Solid Propellant Rocket Research. 1960

EDITORS
Martin Summerfield
PRINCETON UNIVERSITY

2. Liquid Rockets and Propellants. 1960

Loren E. Bollinger
THE OHIO STATE UNIVERSITY

Martin Goldsmith
THE RAND CORPORATION

Alexis W. Lemmon Jr.
BATTELLE MEMORIAL INSTITUTE

3. Energy Conversion for Space Power. 1961

Nathan W. Snyder
INSTITUTE FOR DEFENSE ANALYSES

4. Space Power Systems. 1961

Nathan W. Snyder
INSTITUTE FOR DEFENSE ANALYSES

5. Electrostatic Propulsion. 1961

David B. Langmuir
SPACE TECHNOLOGY LABORATORIES, INC.

Ernst Stuhlinger
NASA GEORGE C. MARSHALL SPACE FLIGHT CENTER

J. M. Sellen Jr.
SPACE TECHNOLOGY LABORATORIES

6. Detonation and Two-Phase Flow. 1962

S. S. Penner
CALIFORNIA INSTITUTE OF TECHNOLOGY

F. A. Williams
HARVARD UNIVERSITY

7. Hypersonic Flow Research. 1962

Frederick R. Riddell
AVCO CORPORATION

8. Guidance and Control. 1962

Robert E. Roberson
CONSULTANT

James S. Farrior
LOCKHEED MISSILES AND SPACE COMPANY

9. Electric Propulsion Development. 1963

Ernst Stuhlinger
NASA GEORGE C. MARSHALL SPACE FLIGHT CENTER

10. Technology of Lunar Exploration. 1963	Clifford I. Cummings and Harold R. Lawrence JET PROPULSION LABORATORY
11. Power Systems for Space Flight. 1963	Morris A. Zipkin and Russell N. Edwards GENERAL ELECTRIC COMPANY
12. Ionization in High-Temperature Gases. 1963	Kurt E. Shuler, Editor NATIONAL BUREAU OF STANDARDS John B. Fenn, Associate Editor PRINCETON UNIVERSITY
13. Guidance and Control — II. 1964	Robert C. Langford GENERAL PRECISION INC. Charles J. Mundo INSTITUTE OF NAVAL STUDIES
14. Celestial Mechanics and Astrodynamics. 1964	Victor G. Szebehely YALE UNIVERSITY OBSERVATORY
15. Heterogeneous Combustion. 1964	Hans G. Wolfhard INSTITUTE FOR DEFENSE ANALYSES Irvin Glassman PRINCETON UNIVERSITY Leon Green Jr. AIR FORCE SYSTEMS COMMAND
16. Space Power Systems Engineering. 1966	George C. Szego INSTITUTE FOR DEFENSE ANALYSES J. Edward Taylor TRW INC.
17. Methods in Astrodynamics and Celestial Mechanics. 1966	Raynor L. Duncombe U.S. NAVAL OBSERVATORY Victor G. Szebehely YALE UNIVERSITY OBSERVATORY
18. Thermophysics and Temperature Control of Spacecraft and Entry Vehicles. 1966	Gerhard B. Heller NASA GEORGE C. MARSHALL SPACE FLIGHT CENTER
19. Communication Satellite Systems Technology. 1966	Richard B. Marsten RADIO CORPORATION OF AMERICA

20.	Thermophysics of Spacecraft and Planetary Bodies Radiation Properties of Solids and the Electromagnetic Radiation Environment in Space. 1967	Gerhard B. Heller NASA GEORGE C. MARSHALL SPACE FLIGHT CENTER
21.	Thermal Design Principles of Spacecraft and Entry Bodies. 1969	Jerry T. Bevans TRW SYSTEMS
22.	Stratospheric Circulation. 1969	Willis L. Webb ATMOSPHERIC SCIENCES LABORATORY, WHITE SANDS, AND UNIVERSITY OF TEXAS AT EL PASO
23.	Thermophysics: Applications to Thermal Design of Spacecraft. 1970	Jerry T. Bevans TRW SYSTEMS
24.	Heat Transfer and Spacecraft Thermal Control. 1971	John W. Lucas JET PROPULSION LABORATORY
25.	Communication Satellites for the 70's: Technology. 1971	Nathaniel E. Feldman THE RAND CORPORATION Charles M. Kelly THE AEROSPACE CORPORATION
26.	Communication Satellites for the 70's: Systems. 1971	Nathaniel E. Feldman THE RAND CORPORATION Charles M. Kelly THE AEROSPACE CORPORATION
27.	Thermospheric Circulation. 1972	Willis L. Webb ATMOSPHERIC SCIENCES LABORATORY, WHITE SANDS, AND UNIVERSITY OF TEXAS AT EL PASO
28.	Thermal Characteristics of the Moon. 1972	John W. Lucas JET PROPULSION LABORATORY
29.	Fundamentals of Spacecraft Thermal Design. 1972	John W. Lucas JET PROPULSION LABORATORY

30. Solar Activity Observations and Predictions. 1972	Patrick S. McIntosh and Murray Dryer ENVIRONMENTAL RESEARCH LABORATORIES, NATIONAL OCEANIC AND ATMOSPHERIC ADMINISTRATION
31. Thermal Control and Radiation. 1973	Chang-Lin Tien UNIVERSITY OF CALIFORNIA, BERKELEY
32. Communications Satellite Systems. 1974	P. L. Bargellini COMSAT LABORATORIES
33. Communications Satellite Technology. 1974	P. L. Bargellini COMSAT LABORATORIES
34. Instrumentation for Airbreathing Propulsion. 1974	Allen E. Fuhs NAVAL POSTGRADUATE SCHOOL Marshall Kingery ARNOLD ENGINEERING DEVELOPMENT CENTER
35. Thermophysics and Spacecraft Thermal Control. 1974	Robert G. Hering UNIVERSITY OF IOWA
36. Thermal Pollution Analysis. 1975	Joseph A. Schetz VIRGINIA POLYTECHNIC INSTITUTE
37. Aeroacoustics: Jet and Combustion Noise; Duct Acoustics. 1975	Henry T. Nagamatsu, Editor GENERAL ELECTRIC RESEARCH AND DEVELOPMENT CENTER Jack V. O'Keefe, Associate Editor THE BOEING COMPANY Ira R. Schwartz, Associate Editor NASA AMES RESEARCH CENTER
38. Aeroacoustics: Fan, STOL, and Boundary Layer Noise; Sonic Boom; Aeroacoustic Instrumentation. 1975	Henry T. Nagamatsu, Editor GENERAL ELECTRIC RESEARCH AND DEVELOPMENT CENTER Jack V. O'Keefe, Associate Editor THE BOEING COMPANY Ira R. Schwartz, Associate Editor NASA AMES RESEARCH CENTER

39. Heat Transfer with Thermal Control Applications. 1975

M. Michael Yovanovich
UNIVERSITY OF WATERLOO

40. Aerodynamics of Base Combustion. 1975

S. N. B. Murthy, Editor
PURDUE UNIVERSITY

J. R. Osborn, Associate Editor
PURDUE UNIVERSITY

A. W. Barrows and J. R. Ward, Associate Editors
BALLISTICS RESEARCH LABORATORIES

41. Communications Satellite Developments: Systems. 1976

Gilbert E. LaVean
DEFENSE COMMUNICATIONS AGENCY

William G. Schmidt
CML SATELLITE CORPORATION

42. Communications Satellite Developments: Technology. 1976.

William G. Schmidt
CML SATELLITE CORPORATION

Gilbert E. LaVean
DEFENSE COMMUNICATIONS AGENCY

(Other volumes are planned.)

PREFACE

On April 6, 1965, just a decade ago, the first commercial operational satellite was placed in synchronous orbit. By June 1965 this satellite was providing commercial satellite service to the Atlantic region of the world. It was not until July 1969 that worldwide service was made available. In a single decade, communication satellites have increased the transoceanic communications capability by more than an order of magnitude. In addition to reduced operating costs and improved performance, users have been provided with completely new capabilities such as transoceanic television and high-speed digital service.

Only recently has this revolution in international communications capability been able to affect the United States domestic scene. The Federal Communications Commission (FCC) encountered divergent points of view that slowed down the development of a policy for the operation of U.S. domestic satellite systems such as existed in Canada (TELESAT) and the U.S.S.R. (MOLNIYA). In 1972, after five years of investigation, the Office of Telecommunications Policy (OTP) and the FCC established the "Open Skies" policy for domestic satellite systems. This policy permits the development of many U.S. domestic communication satellite systems, limited only by the economic base that is generated by the demand for this class of service. Therefore, U.S. policy and existing international agreements now recognize two categories of commercial communications satellite systems–International Systems (INTELSAT and Special Service Satellites) and Domestic Systems (U.S. and foreign).

The Communications Satellite Systems Conferences have spanned the past decade and have contributed significantly to the current successes in this field. Communications satellite technology has now attained a high level of maturity and is having an ever increasing impact on our society. Therefore, the AIAA 5th Communications Satellite Systems Conference was organized with a greater emphasis on the overall system aspects of communication satellites. This emphasis resulted in introducing sessions on policy, spectrum utilization, and geopolitical/economic/national requirements, in addition to the usual sessions on technology and system applications. This was considered essential because, as the communications satellite industry continues to mature during the next decade, it must assume an even more productive and responsible role in the world community. Therefore, the professional systems engineer must develop an ever increasing awareness of the world environment, the most likely needs to be satisfied by communication satellites, and the geopolitical constraints that will determine the acceptance of this capability and the ultimate success of the technology. The papers from the Conference are organized into two volumes of the AIAA Progress in Astronautics and Aeronautics series; the first book (Volume 41) emphasizes the system aspects, and the second book (Volume 42) highlights recent technological innovations.

Communications satellite technology has developed at a remarkable rate over the past decade. This technology explosion has created an apparently paradoxical situation in which the communications satellite systems engineer has produced excellent solutions looking for problems to solve. This dilemma has resulted from the inability of the international sociological and political machinery to respond in an effective manner to reap the benefits of this new capability. The Policy section (Chapter I, Volume 41) reviews the development of communications satellite policy as it pertains to international and domestic systems. Future policy is strongly driven by technology, but in the short term the systems engineer must develop systems that conform to the constraints of existing policy or expect long delays before new concepts are incorporated into operational systems.

The tremendous capability of communications satellites, the relatively large initial investment, and the almost immediate international implications require extensive negotiation among interested parties (and there are usually many) whenever a proposed system is presented for ratification by the affected parties. As a result of this process and the ensuing strict regulations and agreements, it has been difficult to offer new or unusual services or to extend established services in a flexible manner. The difficulty in establishing interntional organizations to cope with the problems associated with the financing and operating of these systems is discussed in Chapter II, Volume 41. The role of communications satellites in future societies and in future military strategies are also major topics of this chapter.

The communication requirements stated by the user and the geopolitical constraints together define the general problem to be solved by the systems engineer. From this information it should be possible to develop a set of system performance criteria that can be used as the basis to evaluate a range of system alterntives. Frequently, however, it is difficult to define specific values for the system criteria without first performing a sensitivity analysis in relation to the specific parameter in question. This sensitivity analysis usually is accomplished by assuming a gross system representation and allowing the parameter in question to vary across a reasonable range of values while assessing the impact on cost and performance. The resulting parametric study is used to determine the desired operating point for the parameters that drive the system design. Chapter III of Volume 41 (System Analysis and Tradeoffs) includes some excellent examples of this kind of system tradeoff analysis. These analyses apply to communications satellite systems and navigational satellite systems, and to the complex problems encountered when these functions are combined.

Communications satellite technology has matured rapidly over the past decade, and it is now possible to provide new services to an ever-increasing community of users. The new services include wideband digital service, color television, secure voice, and rapid and flexible extension of service. The new users served include mobile platforms (ships and aircraft), remote terminals, and small direct-access users. This is a greatly expanded role over the large fixed-terminal point-to-point deployment that has been the mainstay of INTELSAT. Improvements in component reliability have made it feasible to increase satellite complexity and at the same time reduce the size of the crew that operates and maintains the Earth terminals. To allow a large number of users to obtain direct access to these new services, the terminals must be economical from both the initial investment and operation and maintenance (O&M) standpoints. Terminals can be made simpler and more economical by placing an increased burden on the spacecraft (i.e., increased satellite radiated power and onboard processing). This seems to be the logical trend for the future, and so the new generations of satellites now in the planning stages are being provided with narrower beam antennas and onboard processing. Chapter IV of Volume 41 (Advanced Concepts) presents new concepts that employ processing satellites to achieve antenna beam switching and navigational positioning. The impact of introducing a private line service for a large community of users is discussed therein also.

Communications satellite technology now can provide operational systems that have demonstrated high availability and superior performance. The INTELSAT system has become the principal means of international voice communication and, of course, the only means for transoceanic television broadcast and wideband digital service. The increased demand has required INTELSAT to develop improved satellites with expanded capability. The current generation of INTELSAT IV's is being supplemented by INTELSAT IV-A's, which promise greater capacity, principally through frequency reuse. At present, however, INTELSAT is no longer the only operational system. Several U.S. domestic systems (Western Union, RCA GLOBCOM, COMSAT/AT&T-GTE, American Satellite) are operational or about to become operational, and the U.S. will soon deploy fully operational military systems. Foreign domestic satellite service is developing rapidly also. Most of these systems, however, are (or will be) nationally owned and operated. The TELESAT system is already available for service in the Canadian region. Other countries, such as Algeria and Brazil, are availing themselves of the tranponder leasing service offered by INTELSAT. The interruptible service that Algeria is obtaining will give it an excellent system at a very reasonable annual cost. Also, MARISAT will soon be supplying service for maritime ships. The DoD and NASA have supported most of the technological developments that have established the basis for this rapidly expanding industry. Except for the U.S.S.R., the DoD and NASA still have the only major space launching capabilities in the world. The major operational DoD satellite communications system is the Defense Satellite Communications System (DSCS). In 1974, the new "Phase II" DSCS satellites were declared operational in the Pacific and Atlantic regions of the world. The tactical operational service of DoD is to be established by leasing a portion of the MARISAT spacccraft in early 1976. By 1978 the FLEETSAT/AFSATCOM system will provide increased capability for this class of user. For the more distant future, DoD is looking to survivable satellite techniques that will enhance the command and control capability of the strategic forces. Chapter V of Volume 41 (Communications Satellite Systems) describes several of these systems.

Chapter I of Volume 42 presents an excellent cross section of papers covering the technologies available for future satellites. The key areas of mass versus communications power and of benefit versus risk are emphasized in the evolution of the satellite concept. In the course of these discussions, most of the technologies being considered for use in the 1980 generations of satellites are exposed. There appears to be a general trend in satcom spacecraft toward multiple antenna beams because of frequency-reuse, higher EIRP, or user discrimination requirements. Such satellites impose stringent requirements upon the positioning and orientation subsystem of the satellite, as well as upon the attitude control subsystem. A three-axis stabilization concept is usually proposed, rather than the spin-stabilization typical of the current generation systems. Nuclear-powered spacecraft have long been a topic of considerable controversy within the satellite community. In general, the nuclear power advocates were at a disadvantage since the solar cell approach was clearly more attractive on the weight-limited vehicles of today. However, the advent of high performance radioisotope power systems may change all that.

There is an increasing number of applications for satellite communications to support networks involving hundreds of potential terminals. A premium must be placed by operators of these networks on the need to keep the initial cost of each terminal as low as possible and, at the same time, to have inherent in the basic terminal design such features as will minimize the recurring costs of operation and maintenance. Tradeoffs of minimizing the initial cost of the terminal against the requisite terminal availability, together with the operational experience of Telesat Canada in utilizing small unattended receive-only Earth stations, are presented in Chapter II of volume 42 (Terminal Technology). The impact of elliptical satellite orbits (i.e., Molniya) on Earth terminal design also is discussed.

The third major element of a satcom system is the control subsystem, and this is the most all-pervasive of the three major elements because it encompasses every user, every terminal, and every satellite. As such, it must be integral to the total system design and operation and be responsive in time to the demands placed upon it. From the overall point of view, there are various types of control. Systems, satellites, terminals, operational modes, links, and channels are all subject to control procedures. The section on Control Technology (Chapter III, Volume 42) deals with the complex control terminal operations and the difficulty of integrating a mix of terminal sizes and types. These papers emphasize the need to automate these operations to reduce the response time and improve system performance.

The area of technology called baseband or communications processing encompasses modulation, multiplexing, multiple access, and coding for both the transmission path and the data source itself. A form of multiple access that has become popular of late is "single channel per carrier" (SCPC) and may be defined best as a frequency division multiple-access (FDMA) method in which each carrier is limited to one voice channel, or its data equivalent, and assigned on a dedicated basis. Because of its marked similarity to FDMA demand assignment methods, SCPC is sometimes used as a stepping stone into an FDMA demand assignment (DA) system or as a dedicated subset of such a system. SCPC is normally applied to networks of small, lightly loaded terminals in relatively isolated circumstances or as an auxiliary capability of a large Earth station. These are relatively new applications of communication technology; the impacts of the propagation media and the motion of the platforms require reassessment. These problem areas and the ability of the modems and coding equipment to combat them are addressed in Chapter IV of Volume 42 (Multiple Access, Modulation, and Coding).

One of the trends of satellite communciations systems has been the greater use of portions of the frequency spectrum other than the SHF region that spawned most of the technical and operational growth of the satcom technology. The UHF and EHF segments are of particular interest, the former for its promise of low-cost mobile terminals and the latter for very wide spectral allocations that minimize interference with terrestrial communications. Each of these regions also has unique transmission path characteristics. The EHF band and the upper portion of the SHF band are impacted to a significant degree, proportional to the frequency of interest, by rain attenuation. The UHF band and the lower part of the SHF band are influenced by ionospheric scintillations. Because of the importance of information on these phenomena to satellite communciations systems engineers, five papers on these phenomena (four of them from the AIAA 12th Aerospace Sciences Meeting) have been selected for inclusion in Chapter V of Volume 42 (Transmission Path Effects).

The systematic coverage provided by this two-volume set will serve on the one hand to expose the reader new to the field to a comprehensive coverage of communications satellite systems and technology, and on the other hand to provide also a valuable reference source for the professional satellite communication systems engineer.

We wish to thank our wives, Barbara LaVean and Ruth Schmidt, for their patience and understanding.

Gilbert E. LaVean
William G. Schmidt

November 1975

CHAPTER I—SPACECRAFT TECHNOLOGY

The paper by Hocenberry et al., the first of the five papers dealing with spacecraft technology, presents an account of the evolution of a communications satellite baseline configuration and its subsequent developments. The key areas of mass versus communications power and of benefit versus risk are emphasized in the evolution of the satellite concept. In the course of these discussions, most of the technologies being considered for use in the 1980 generations of satellite are exposed.

There appears to be a general trend in satcom spacecraft toward multiple antenna beams because of frequency-reuse, higher EIRP, or user discrimination requirements. Such satellites impose stringent requirements upon the positioning and orientation subsystem of the satellite, as well as upon the attitude control subsystem. A three-axis stabilization is usually proposed, rather than the spin-stabilization typical of the current generation systems. The next three papers of this section deal with candidate technologies for each of those areas: the bootlace lens antenna for multiple beam applications (Lu et al.), the radiometric interferometer for attitude sensing (Teichman et al.), and the double-gimballed momentum wheel for three-axis attitude control (Kalley and Mork).

Nuclear-powered spacecraft have long been a topic of considerable controversy within the satellite community. In general, the nuclear power advocates were at a disadvantage since the solar cell approach was clearly more attractive on the weight-limited vehicles of today. However, the advent of high-performance radioisotope power systems may change all of that. Raab and Karlin pose the question, "Solar versus Nuclear Power: Is there a Choice?" and conclude that successful development of one or more of the advanced nuclear power systems would provide a competitive alternate to solar-powered satellites.

EFFICIENT HIGH-CAPACITY COMMUNICATIONS SATELLITES

J. H. Hockenberry,[*] D. W. Krejci,[†] W. A. Koenig,[‡] and J. P. Klocksiem[§]

Lockheed Missiles & Space Company, Inc.,
Sunnyvale, Calif.

Abstract

The historical development of communications satellite systems has been characterized by a careful evolution of capability by the cautious introduction of new technology. The application of this philosophy to a Delta-launched satellite, with a logical progression of efficiency and communications capacity, is described. The percentage of satellite mass and power available to the communications payload is derived for a baseline domestic communications satellite and three succeeding development phases. New technologies, including high-energy apogee motor, high-efficiency solar cells, flexible substrate solar arrays, nickel-hydrogen batteries, and ion propulsion, are introduced in a logical series of developments. New communications technology, including 14/12-GHz transponders, 30/20-GHz transponders, and satellite-switched time-division multiple access (SS/TDMA), provides greater communications capacity as available payload mass and power increase.

Presented as Paper 74-462 at the AIAA 5th Communications Satellite Systems Conference, Los Angeles, Calif., April 22-24, 1974.
  [*]Assistant General Manager, Space Systems Division.
  [†]Assistant to the Assistant General Manager, Space Systems Division.
  [‡]Consulting Engineer, Communications Systems, Space Systems Division.
  [§]Senior Staff Engineer, Systems Design, Space Systems Division.

I. Introduction

The development of communications satellites over the past decade has been characterized by a careful evolution of capability and communications capacity by the cautious introduction of new technology. The reasons for this are obvious: the capital investment in current equipment is enormous, and the requirement for service continuity in operational systems is vital. From time to time, radical changes have been proposed, but generally conservatism has prevailed when hard commitments had to be made. It is not anticipated that future expansions of communications satellite capabilities will be accomplished by revolutionary changes. Thus, the problem to be solved when considering advanced technologies is how they may be introduced within an evolutionary framework. This paper presents a postulated domestic communications satellite, sized for a Delta 3914 launch vehicle, which has been conceived in such a way that advanced communications and spacecraft technologies can be introduced in an evolutionary manner.

The efficiency or utilization factor of a communications satellite has been defined to be the ratio of payload-related mass (actual payload mass plus the payload portion of power subsystem) to the satellite total useful mass in orbit.[1] The efficiency of a basic satellite design can be improved in a stepwise manner over a period of years as new technologies become available and proven. Furthermore, new communications technologies can be introduced as pilot operations in conjunction with an operational payload to obtain orbital experience with low risk prior to operational deployment.

II. Technology Evolution

Figure 1 shows a domestic communications satellite that could be placed in geostationary orbit today by a Delta 3914 launch vehicle plus an apogee motor. The satellite's payload consists of four antennas that provide United States coverage and 12 fully redundant transponder channels of 80-MHz bandwidth operating in the 6- and 4-GHz frequency bands. This baseline satellite would employ such well-proved 1974 technology as hydrazine propulsion (catalytic decomposition), momentum wheel attitude control, sun-tracking solar arrays with extendible rigid (aluminum honeycomb) panels, and nickel-cadmium batteries.

The traffic growth history of international communications satellites provides an indicator of a probable early and continuing requirement for step-by-step increases in the payload mass and power of domestic communications satellites. An orderly, cautious response to this forecast requirement is outlined in Fig. 2, which shows a baseline satellite design periodically uprated to accommodate heavier, higher-power payloads. Table 1 indicates the significant impact of the various technology improvements as they are phased into the spacecraft design.

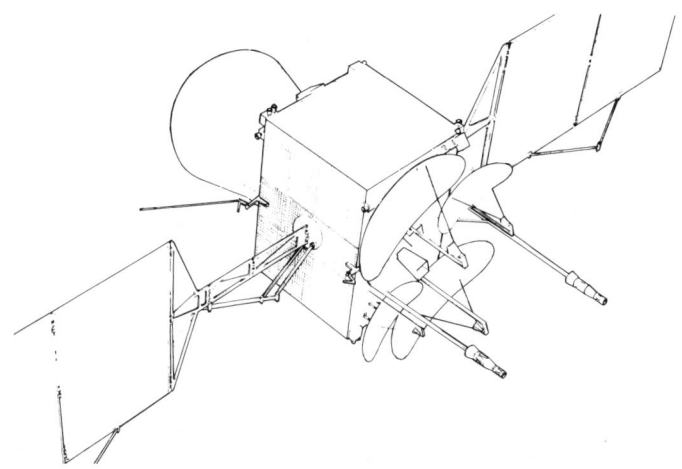

Fig. 1 Domestic communications satellite (baseline).

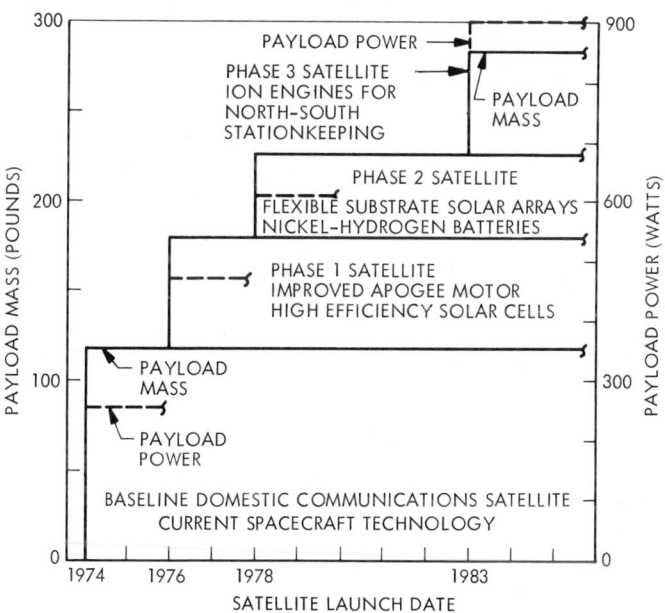

Fig. 2 Evolution of spacecraft technology.

Table 1 Spacecraft technology improvements

Phase 1 satellite	Improved apogee motor: Technology developed for the Trident Fleet ballistic missile. High-energy propellant; 100:1 expansion ratio nozzle; lightweight PRD 49-III case construction. Specific impulse is 304+ lbf-sec/lbm, and propellant mass fraction is 0.93+, compared to current values of 288 and 0.91.
	High-efficiency solar cells: 14% compared to 11% for conventional cells.
Phase 2 satellite	Flexible substrate solar arrays: 39 w/lb compared to 10 w/lb for rigid panels (both using high-efficiency solar cells).
	Nickel-hydrogen batteries: 30 w-hr/lb (total capacity) compared to 14 for nickel-cadmium batteries.
Phase 3 satellite	Ion engines for north-south stationkeeping: For this satellite mass and 7-yr life, ion engines (specific impulse 2,600 lbf-sec/lbm compared to 222 for hydrazine) result in a net addition of 96 lb of payload. With an electric power subsystem conversion factor of approximately 7.5 w/lb, this is alternatively equal to 720 w of additional payload power.

III. Baseline Satellite

The baseline satellite incorporates current technology and could be launched in 1974 with the highest confidence of a successful mission. The baseline payload and its supporting spacecraft subsystems, or platform, are described in the following paragraphs.

Baseline Satellite Communications Payload

The baseline communications subsystem is based on current technology, utilizing flight-proven components in the 6/4-GHz bands but employing two variations from current usage which greatly increase the capacity of the satellite. The first technique is the use of dual polarization with frequency reuse to double the bandwidth available. The second is the use of 80-MHz bandwidth transponder channels.

An efficient satellite transponder should operate at the crossover between a bandwidth-limited and power-limited condition for the type of modulation selected. The somewhat limited geographical area to be covered in a domestic satellite, as compared to global coverage in an international satellite, provides additional antenna gain, permitting bandwidth-limited operation with nominal rf power levels. Hence the doubling of available bandwidth via dual polarization frequency reuse provides much more effective use of d.c. power than simply increasing effective isotropically radiated power (EIRP).

It has been common practice to employ 12 transponders, each occupying 40 MHz of bandwidth, in the 500-MHz bandwidth available in the 6/4-GHz bands. Examination of the weight of components in a typical transponder using 40-MHz bandwidths discloses that approximately 75% of the mass is in the channelizing filters and traveling-wave tube amplifiers (TWTA's). One reason for using 40-MHz bandwidths is flexibility of assignment and compatibility with television signals. However, a domestic satellite logically would be operating with multiple carriers, either single channel per carrier (SCPC) or FDM/FM (where FDM is frequency-division multiplex), for much of the traffic to be handled, with all channels accessible to all Earth stations within the common beam. For this type of system, wider bandwidths with higher rf powers provide equivalent capacities with significantly less mass. Television may be handled on a multiple carrier per transponder basis. Table 2 compares the mass of a 24-channel, 40-MHz transponder and a 12-channel, 80-MHz transponder. Bandwidths wider than 80 MHz are precluded by the paucity of proven TWTA's with higher power outputs.

Table 2 Transponder mass comparison.

	40 MHz, lb	80 MHz, lb
Receivers	22	22
Input/output multiplexers	28	15
TWTA's	82 (5 w)	44 (8 w)
Miscellaneous	20	12
Total	152	93

Figure 3 shows a simplified block diagram of the transponder. The antenna arrangement for the baseline satellite is shown in Fig. 4 and the antenna beam coverage in Fig. 5. Typical link parameters are shown in Table 3. The total capacity of the baseline satellite is 20,400 50-kbit/sec channels.

Baseline Spacecraft

Significant characteristics of the baseline platform are summarized in Table 4. (Note that a 40% off-loading of orbit hydrazine and batteries brings the spacecraft mass within the capability of the Delta 2914 launch vehicle.) For the baseline spacecraft, the payload related mass (actual payload plus payload portion of power subsystem) is 193 lb, and the d.c. power is 256 w. The efficiency, defined as the ratio of payload related mass to satellite useful mass in orbit (M_p/M_s), is 25%.

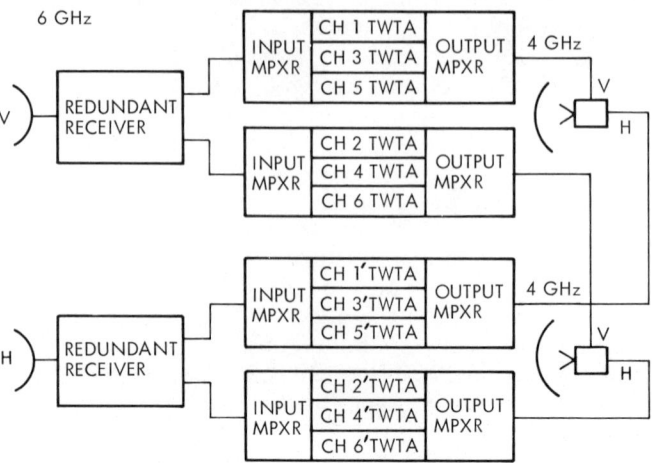

Fig. 3 Baseline 6/4-GHz transponder.

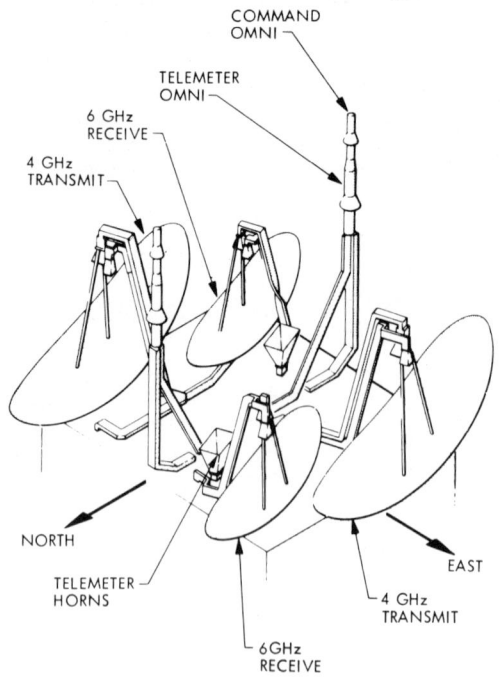

Fig. 4 Baseline satellite antenna arrangement.

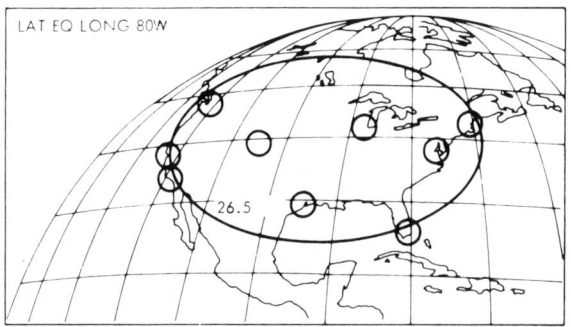

Fig. 5 Baseline antenna coverage.

Table 3 6/4-GHz link parameters

Satellite EIRP (beam edge), db-w	34.5
TWTA backoff, db	4.0
Operating margin, db	2.0
Space loss, db	196.7
Earth station G/T (33-ft diam), db/°K	29.0
Received C/T, db-w/°K	-139.2
Capacity, 4∅ phase shift keying, 3/4 coding	1700 channels (50 kbit/sec)

Table 4 Characteristics of the baseline spacecraft

Launch vehicle	Delta 3914
Useful mass in orbit, lb	771
Operational life, yr	7
Electric power	Sun-oriented, extendible rigid panel solar arrays
Eclipse capability	100% (nickel-cadmium batteries)
Attitude control	Ascent: hydrazine thrusters Orbit: double-gimballed momentum-reaction wheel Pointing accuracy, ±0.15°
Stationkeeping	±0.10° N-S and E-W using hydrazine thrusters
Growth potential	200% payload growth (see Fig. 2)

IV. Phase 1 Satellite

The phase 1 payload is described below, followed by an identification of the evolutionary changes in the phase 1 platform as compared to the baseline platform.

Phase 1 Communications Payload

The communications payload for the phase 1 satellite carries over the 12-channel, 6/4-GHz transponder from the baseline and augments it by four transponder channels operating in the 14/12-GHz bands. Antenna coverage at 14/12 GHz is provided by a shaped beam covering the contiguous United States (CONUS), with greater power directed to the Eastern U.S., where rainfall attenuation is more severe. Only one polarization is used in this modification; signals are received and transmitted in one antenna using orthogonal linear polarization. Figure 6 shows the antenna arrangement, whereas Fig. 7 shows the shaped-beam coverage at 14/12 GHz.

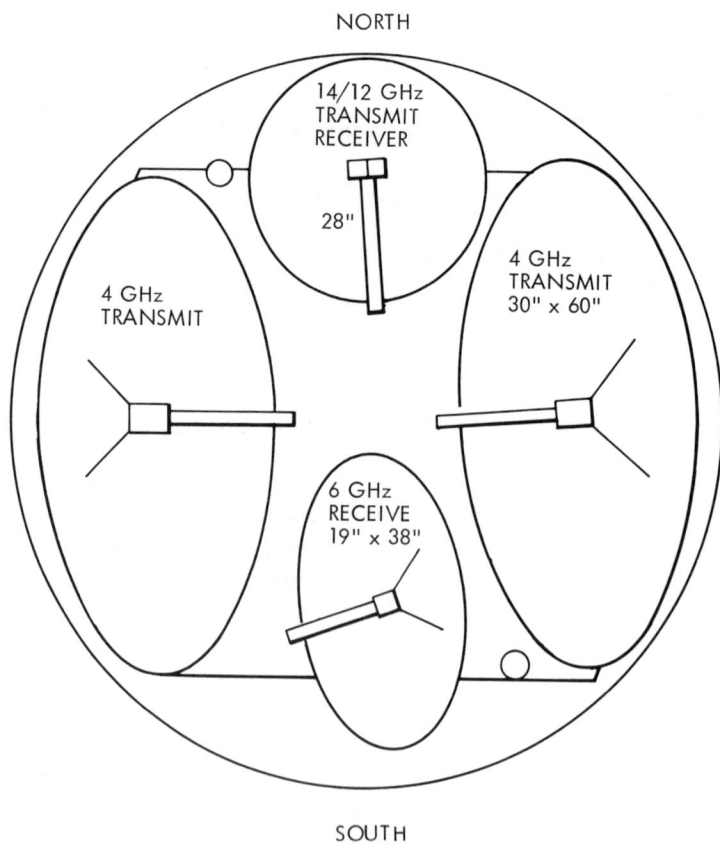

Fig. 6 Antenna arrangement for phase 1 satellite.

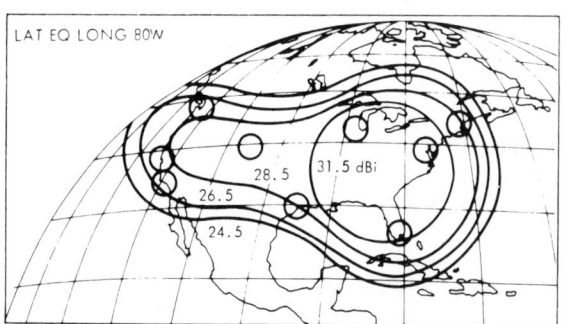

Fig. 7 14/12-GHz antenna coverage of contiguous United States.

Each transponder channel uses a 20-w TWTA and an 80-MHz bandwidth. Twenty-watt TWTA's for the 11-GHz band have been qualified for space use and will be flown on the Canadian Communications Technology Satellite in 1975. The higher rf power and the increased Earth station G/T combine to provide the additional rainfall attenuation margins required at 14/12 GHz as compared to 6/4 GHz. Earth station space diversity is assumed for the heavy rainfall areas of the Southeast.

Figure 8 shows a simplified block diagram of the transponder. Table 5 gives typical link parameters and capacity for single channel per carrier operation for this version of the satellite, as well as for two other variations to be discussed later. The total capacity of this version of the satellite is 23,800 50-kbit/sec channels.

The operating margin of Table 5 is intended to account for all transmission losses other than fixed space loss. The primary loss is caused by rainfall, with relatively minor contributions from oxygen and water vapor absorption. Variations in antenna gain caused by pointing errors and manufacturing tolerances are accounted for by the small circles around key cities in Figs. 5, 7, and 9. The shaped beam pattern of Fig. 7 provides a margin of at least 11 db in the Eastern United States.

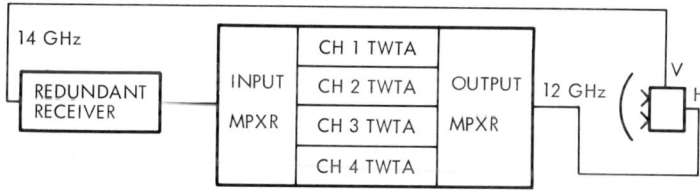

Fig. 8 14/12-GHz phase 1 transponder.

Table 5 14/12-GHz link parameters

	CONUS beam 32-ft Earth station	Spot beam 32-ft Earth station	Spot beam 16-ft Earth station
Satellite EIRP (beam edge), db-w	38.0	46.5	46.5
TWTA backoff, db	1.5	4.0	1.5
Operating margin, db	6.0	6.0	6.0
Space loss, db	205.9	205.9	205.9
Earth station G/T, db/°K (rain)	33.5	33.5	27.5
Received C/T, db-w/°K (rain)	-141.9	-135.9	-139.4
Capacity/transponder[a]			
4-phase PSK	...	3090	...
2-phase PSK	850	...	1380

[a]50-kbit/sec data channels, rate 3/4 coding, 85% activity factor.

Phase 1 Spacecraft

The phase 1 platform is very similar to the baseline platform. A subsystem-by-subsystem comparison follows:

1) Structure and thermal control: provide the space, thermal control, and strength for phase 1 payload, new apogee motor, and larger batteries.

2) Electrical power subsystem: modify solar panels, add battery capacity, and modify wire harness.

3) Attitude control subsystem: identical to baseline.

4) Command and telemetry: baseline (modify instrumentation).

5) Propulsion subsystem: baseline (load more hydrazine).

The payload-related mass is 310 lb, and the d.c. power is 464 w. The efficiency (M_p/M_s) is 33%.

V. Phase 2 Satellite

The phase 2 payloads and their spacecraft platform are described below.

Phase 2 Communications Payloads

The greater mass and power available to the communications payload in the phase 2 version of the satellite allows two different payloads to be accommodated. The 2A payload is a dedicated 14/12-GHz payload; the 2B payload is the baseline 6/4-GHz payload plus a pilot operation at 30/20 GHz.

The 2A payload comprises 12 14/12-GHz transponders. Six operate with 80-MHz bandwidths (68 usable) over the CONUS coverage beam (see Fig. 7). The other six operate with 160-MHz bandwidths over two 2° spot beams as shown in Fig. 9. The frequency band is reused two times in the spot beams, as indicated in the frequency plan of Fig. 10. Figure 11 shows the antenna arrangement for the 2A satellite. This arrangement provides general coverage with good connectivity and also high capacity between major metropolitan areas. The link parameters for the spot beams are shown in Table 5. The total capacity of this version of the satellite is 23,640 50-kbit/sec channels. Table 5 also shows the link calculations for operation of the spot-beam transponders with 16-ft Earth stations (miniterminals), such as might be used on rooftops. For this mode of operation, two-phase PSK modulation is used, and 1.5 db of TWT backoff is sufficient. The total capacity of the satellite if all spot-beam transponders were used with miniterminals would be 13,380 channels.

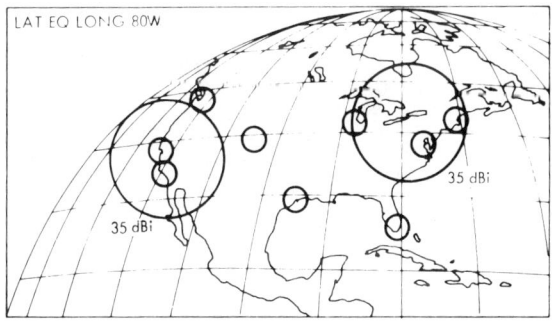

Fig. 9 14/12-GHz phase 2 spot beam antenna coverage.

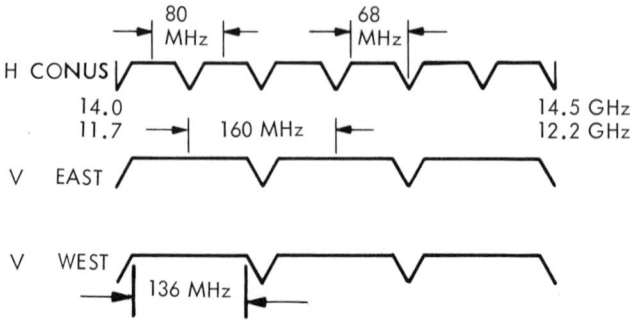

Fig. 10 14/12-GHz phase 2 frequency plan.

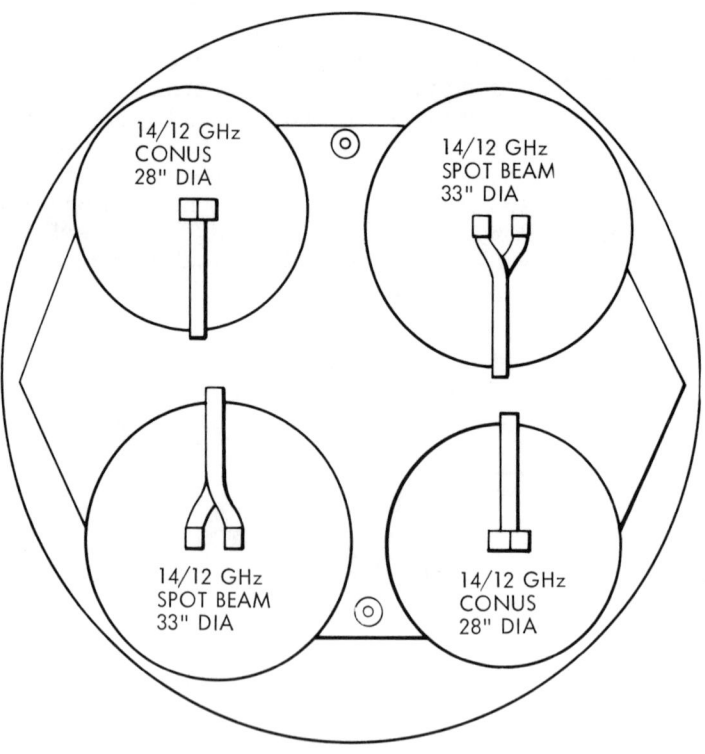

Fig. 11 Antenna arrangement for dedicated 14/12-GHz satellite.

HIGH CAPACITY COMMUNICATIONS SATELLITES

The 2B payload consists of the baseline 6/4-GHz transponders and antennas plus a pilot system operating at 30/20 GHz. This approach allows operating experience to be gained with the 30/20-GHz bands while the satellite pays its way using the 6/4-GHz bands. The 30/20-GHz part of the payload consists of four 320-MHz transponders using 20-w TWT's transmitting over two 1° spot beams. The 20-w TWTA's, operating at 20 GHz, will require development and qualification. A general block diagram and frequency plan are shown in Fig. 12. The antenna arrangement for this payload is similar to that shown in Fig. 6, but with a 44-in. 30/20-GHz antenna replacing the 14/12-GHz antenna shown. The 30/20-GHz link parameters are shown in Table 6; the capacity calculation assumes single-carrier TDMA operation. The link margin shown provides for general rain conditions but assumes that Earth station diversity is used in heavy rainfall areas. The capacity of the 2B satellite is 20,400 in the 6/4-GHz band and 21,200 in the 30/20-GHz band, for a total of 41,600 channels.

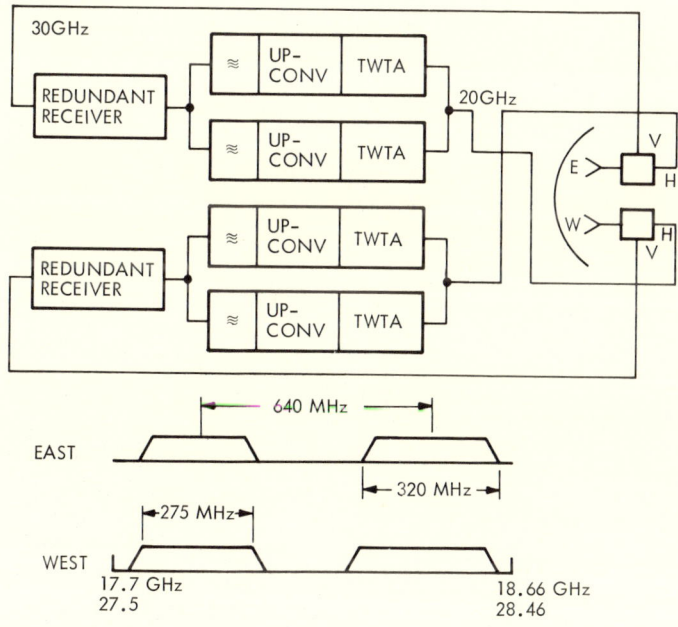

Fig. 12 30/20-GHz transponder arrangement and frequency plan.

Table 6 30/20-GHz link parameters

Satellite EIRP (beam edge), db-w	53.5
TWTA backoff, db	0.0
Rain margin, db	12.0
Space loss, db	210.3
Earth station G/T (32 ft, rain), db/°K	35.7
Received C/T, db-w/°K	-133.1
Capacity (50-kbit/sec data channel, 4-phase PSK, rate 3/4 coding), channels	5300

Phase 2 Spacecraft

The phase 2 spacecraft platform has the following similarities and differences compared to the phase 1 platform:

1) Structure and thermal control: modify to accommodate the increased size, mass, and power of the phase 2 payload. Provide for the increased volume requirement of the nickel-hydrogen batteries.

2) Electrical power subsystem: replace the rigid panel solar arrays with flexible substrate arrays and their stowage and extension mechanisms. Provide increased capacity with nickel-hydrogen batteries. Modify wire harness.

3) Attitude control subsystem: identical to phase 1.

4) Command and telemetry: phase 1 (modify instrumentation).

5) Propulsion subsystem: phase 1.

The payload-related mass for the phase 2 spacecraft is 305 lb, and the d.c. power is 611 w. The efficiency (M_p/M_s) is 32%. It is interesting to note that the payload-related mass of this spacecraft version is less than phase 1, even though the actual payload mass is increased by 50 lb and the power by 144 w. This results from the greatly reduced battery weight when nickel-hydrogen batteries are used.

HIGH CAPACITY COMMUNICATIONS SATELLITES

VI. Phase 3 Satellite

Phase 3 represents the last major step in the orderly evolution from the current technology baseline to a highly efficient domestic communications satellite. The advanced phase 3 payloads and the platform to support them are described below.

Phase 3 Payloads

Three different types of payloads are possible with the phase 3 satellite, resulting from the greater efficiency of this version. The 3A payload is a dedicated 14/12-GHz system, essentially identical in function to the 2A payload described earlier. The chief difference is the use of parallel 20-w TWT's in the six transponders covering the continental United States. This change permits the use of four-phase PSK and higher capacity with 32-ft Earth stations, or, alternatively, the CONUS coverage beams may be used with 16-ft Earth stations, providing a capacity of 620 channels/transponder. The total capacity with 32-ft terminals is 28,240 channels. Although the gain in capacity is not startling, the increased flexibility is significant. The frequency plan and antenna arrangement for this version of the satellite are the same as shown previously in Figs. 10 and 11.

The 3B payload consists of the baseline 6/4-GHz transponders augmented by a pilot satellite-switched TDMA system operating in the 30/20-GHz bands over three 1° spot beams. Figure 13 depicts the antenna beam coverages. A simplified block diagram of the transponder, as shown in Fig. 14, indicates how the satellite switching unit controls the routing of signals. The switching unit operates at the intermediate frequency, which would be in the 2- to 4-GHz range. The required bandwidth for the receivers and transmitters is 550 MHz, and, because this bandwidth is reused in each beam, the total bandwidth of the system is 1.65 GHz. The functional operation of SS/TDMA systems has been described in the literature.[2] The capacity of the SS/TDMA system is 31,800 channels, yielding a total capacity of 52,200 for this version of the satellite.

The 3C payload represents the final stage of the satellite evolution and is later in time than the on-orbit operation of the 3B payload. This version of the satellite is a fully operational satellite-switched TDMA system, operating with eight beams within the contiguous United States. Figure 15 shows the antenna coverage pattern. The selection of beamwidth for this satellite is constrained primarily by the launcher shroud diameter, which in this case is 86 in., allowing a 0.5° beam at 20 GHz. An optimum beamwidth from the standpoint of satellite stabilization accuracy would be 0.35°, but the loss in link margin by using the 0.5° beam is only 1 db. The satellite antenna is a rigid offset reflector, with eight feeds producing the eight beams, as depicted in Fig. 16.

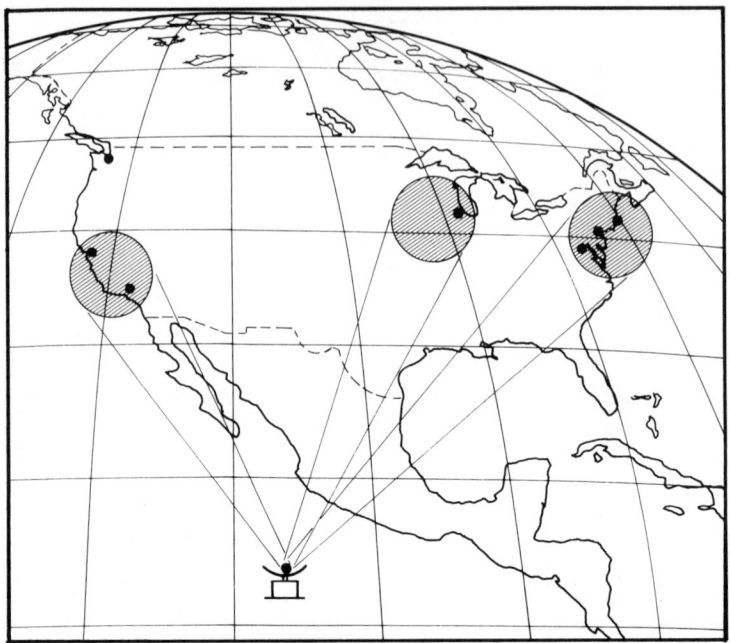

Fig. 13 30/20-GHz antenna coverage, phase 3 pilot system.

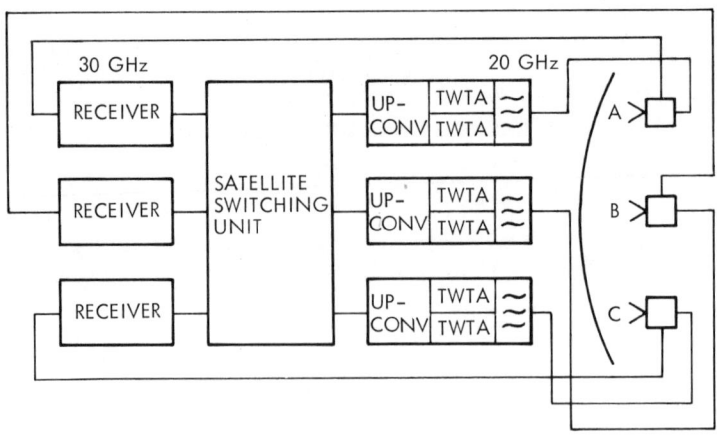

Fig. 14 Transponder and antenna arrangement for pilot satellite-switched TDMA system.

HIGH CAPACITY COMMUNICATIONS SATELLITES

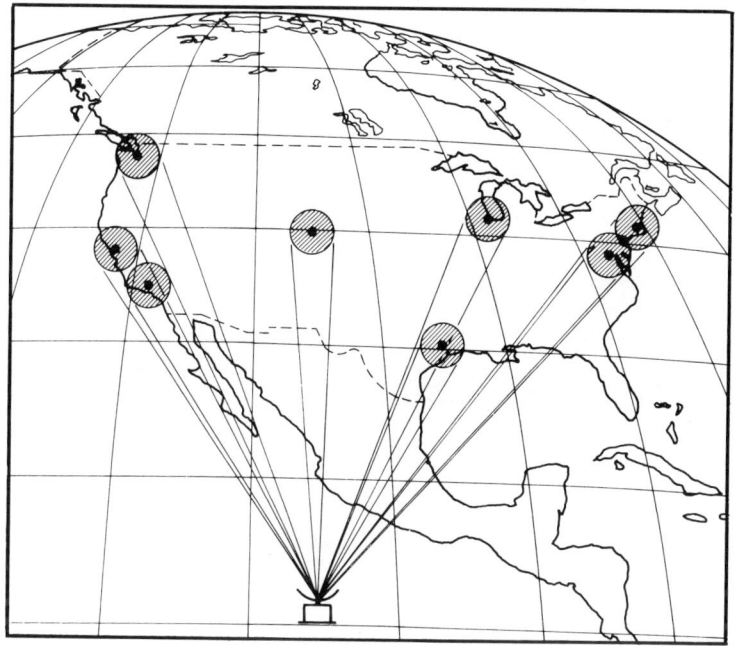

Fig. 15 30/20-GHz antenna coverage, phase 3 operational system.

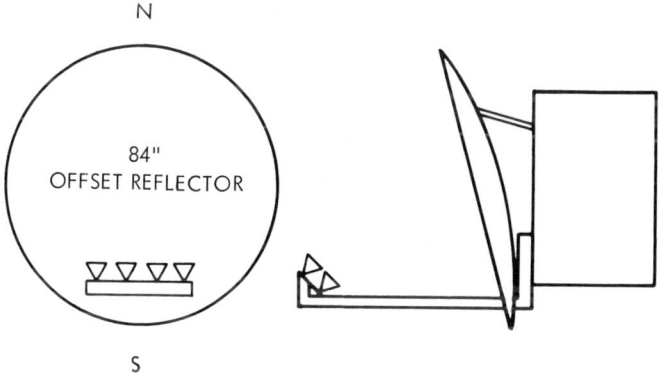

Fig. 16 Antenna arrangement for operational SS/TDMA system.

A simplified block diagram of the transponder-antenna system is shown in Fig. 17. As in the case of the pilot system, the cross-connection is accomplished at the intermediate frequency. The TWTA's provide 20 w of rf power. The pertinent link parameters are given in Table 7. It is apparent from Fig. 15 that several pairs of beams are spaced too closely to provide the necessary isolation required for frequency reuse. Two solutions are available to solve this problem. Two separate frequency bands may be used for adjacent beams, with the necessary frequency offsets provided in the frequency down-converters and up-converters. Obviously, the cross connection must be done in the same band. Alternatively, adjacent beams may be cross-polarized to provide isolation. If necessary, both solutions may be used simultaneously. The total capacity of the 3C payload is 97,200 channels.

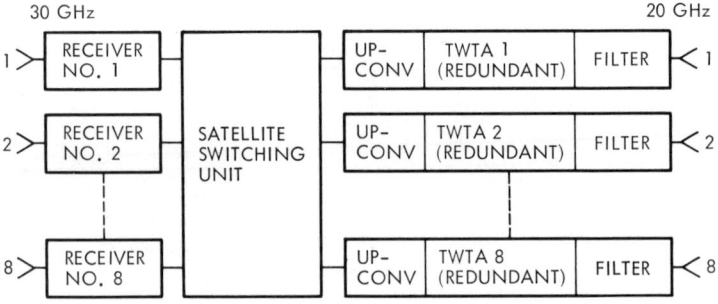

Fig. 17 Block diagram of operational SS/TDMA system.

Table 7 30/20-GHz link parameters (8 spot beams, frequency reuse)

Satellite EIRP (beam edge), db-w	59.5
TWTA backoff (single carrier), db	0.0
Operating margin, db	12.0
Space loss, db	210.3
Earth station G/T (32 ft, rain), db	35.7
Received C/T, db-w/°K	-127.1
Capacity (50-kbit/sec channel, 4-phase PSK, rate 7/8 coding), channels	12,150

HIGH CAPACITY COMMUNICATIONS SATELLITES 21

Phase 3 Spacecraft

Phase 3 platform changes are detailed as follows:

1) Structure and thermal control: modify to accommodate the ion engine subsystem, the phase 3 payload, and larger batteries.

2) Electric power subsystem: lengthen the flexible substrate solar arrays to supply the ion engines and the increased payload power. Provide increased battery capacity. Modify wire harness.

3) Attitude control subsystem: Incorporate interface with ion engines to provide inverse modulation and unloading of the momentum wheel gimbal actuators during north-south stationkeeping.

4) Command and telemetry: phase 2 (modify instrumentation).

5) Propulsion subsystem: Add four ion engines for redundant north-south stationkeeping, plus associated propellant tank, power control and conditioning unit, valves, lines, and fittings. Install hydrazine tanks with 28% of the capacity of the replaced phase 2 tanks.

The payload-related mass of the phase 3 spacecraft is 399 lb, and the d.c. power is 896 w. The efficiency (M_p/M_s) is 42%.

VII. Conclusions

The cautious evolution of communications satellites just described is a minimum-risk approach to the achievement of maximum communications capability. The payoff is summarized graphically in Fig. 18. The baseline satellite's communications subsystem is extremely efficient by today's standards; by employing 80-MHz bandwidth transponder channels and dual polarization with frequency reuse, it fully utilizes the 6/4-GHz frequency bands to achieve the respectable capacity, 20,400 50-kbit channels, shown in Fig. 18. However, a decade later, the phase 3C payload achieves the more remarkable total of 97,200 50-kbit channels (8 x 12,150, Table 7). It also is noteworthy that the phase 2 and 3 payloads provide the higher EIRP needed to exploit the 14/12- and 30/20-GHz frequency bands, an exploitation that the anticipated early saturation of the 6/4-GHz bands makes highly desirable.

The growth of spacecraft platform capability in step with the development of advanced payloads is illustrated by Fig. 19, which shows payload growth from 116 lb and 256 w for the baseline to 279 lb and 896 w for phase 3. As previously noted, the phase 3 platform achieves an efficiency (M_p/M_s) of 42%. Consideration of Fig. 2 in conjunction with Fig. 19 gives rise to the following observations:

1) Beginning in phase 2, the flexible substrate solar arrays and nickel-hydrogen batteries effectuate a dramatic reduction in the growth rate of power subsystem mass with increasing payload power.

2) The phenomenal specific impulse of the ion engine makes a substantial contribution to the mass and power available for the phase 3 payload.

3) The baseline satellite's selected attitude control subsystem, which features a double-gimballed wheel for stabilization in orbit, adapts to all three satellite development phases with virtually no change in mass or in function. The three-axis stabilized satellite readily accommodates payload flexibility and growth, and the double-gimballed wheel is inherently superior for long life and precise pointing accuracy.

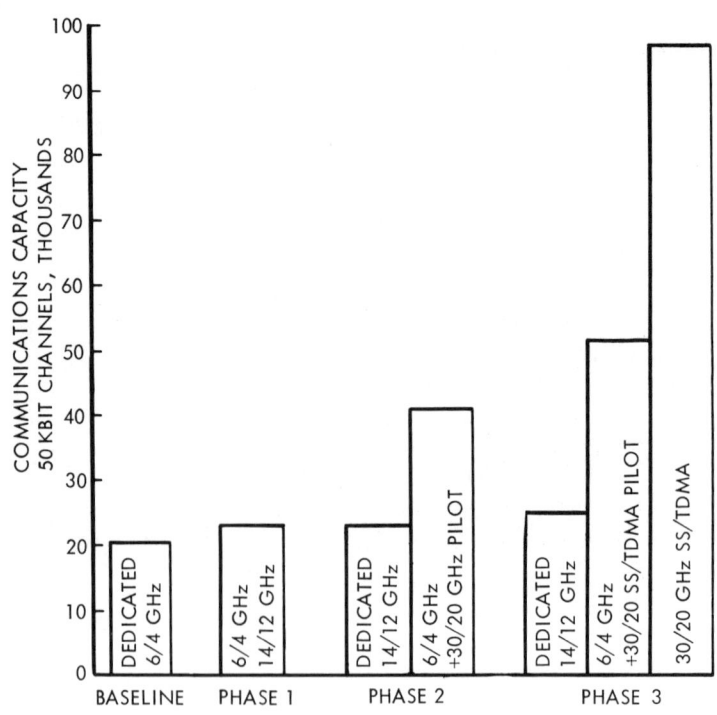

Fig. 18 Communications capacity improvement with technology evolution.

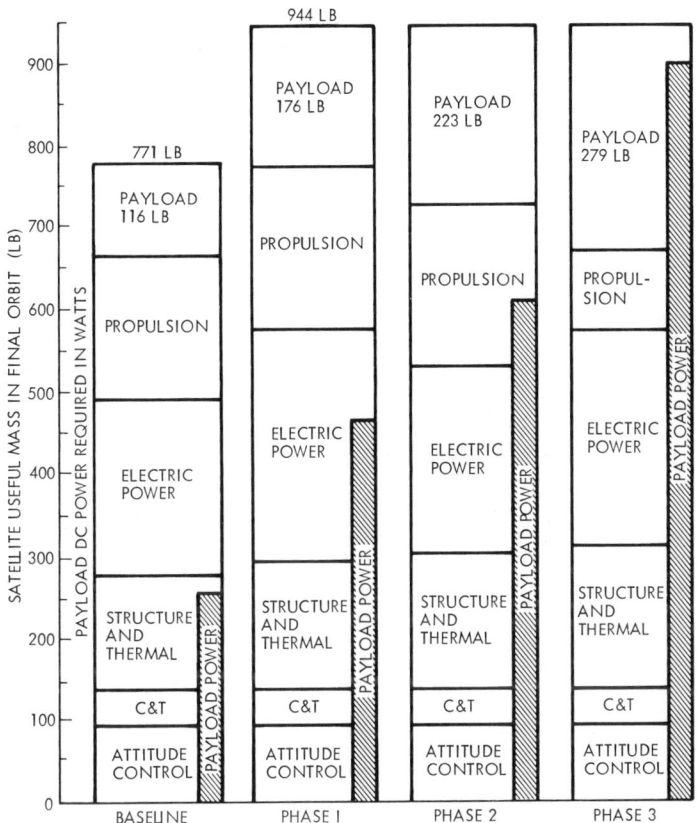

Fig. 19 Payload and platform mass summaries.

References

[1] Kiesling, J. D., Elbert, D. R., Garner, W. B., and Morgan, W. L., "A Technique for Modeling Communications Satellites," COMSAT Technical Review, Vol. 2, No. 1, Spring 1972, pp. 73-103.

[2] Schmidt, W. G., "An On-board Switched Multiple-access System for Millimeter-wave Satellites," Proceedings, INTELSAT/IEE International Conference on Digital Satellite Communication, Nov. 1969, pp. 399-407.

A CONSTRAINED LENS ANTENNA FOR
MULTIPLE-BEAM SATELLITES

H. S. Lu,[*] W. G. Scott,[+] T. Smith,[#] and A. Smoll[§]

Aeronutronic-Ford Corporation, Palo Alto, Calif.

Abstract

A wide-band microwave constrained lens and feed array uses printed circuit technology for simultaneous multiple co-frequency band pencil beams from a single aperture. This frequency reuse multiple-beam concept requires sidelobes below -34 db for at least 37 beam positions over an 18° conical field of view. With such sidelobes, any one of several sets of six beams can be used at one time while maintaining high interbeam isolation. The lens is a space-fed array consisting of a large number of thin, printed circuit cards arranged in a circular grid. Computed and measured radiation characteristics are presented. The construction techniques are described.

I. Introduction

Advanced communication satellites will obtain greater usage of the assigned transmit and receive frequency bands via the technique called "spatial frequency reuse" whereby multiple

---

Presented as Paper 74-465 at the AIAA 5th Communications Satellite Systems Conference, Los Angeles, Calif., April 22-24, 1974.
   [*]Senior Engineering Specialist.
   [+]Section Supervisor, Multibeam Antenna Section.
   [#]Engineer, R & D.
   [§]Principal Engineer.

separate cofrequency pencil beams will illuminate isolated spots on Earth. Avoidance of cochannel interference requires that radiation from all adjacent beams on a given spot be below a predetermined level relative to the radiation of the intended beam on that spot. This ratio is called interbeam isolation and is typically required to be -27 to -30 db. It is easy to see that, as the number of simultaneous cofrequency beams per satellite increases, the sidelobe levels per beam must decrease to meet a given isolation requirement. If only two beams are to be used, each covering its own 3-db beamwidth spot, then -30 db maximum sidelobes per beam will suffice to meet the aforementioned isolation specification. However, for six beams, the required sidelobe level per beam is -34 db. Achieving such sidelobes in a single-aperture multiple-beam antenna is the design problem of this paper.

II. Antenna Design Approach

A constrained transverse electromagnetic mode (TEM) lens with separate beam feeds on the quasi-focal surface is the approach chosen to provide multiple beams because of its prospect of low phase error with scan and thus potentially low sidelobes for off-axis beams. The lens is a figure of revolution so that three-dimensional, off-axis scan performance is obtained (rather than scan limited to one plane).

In this paper, we describe the electrical design of a 4-6-GHz TEM lens antenna, for synchronous satellite use, which has fixed cluster feeds, each cluster capable of generating a separately pointed pencil beam. Any one of several sets of six beams can be chosen to radiate simultaneously (but from incoherent sources) from the single lens aperture. The 37 total beam positions are arranged in a symmetrical triangular grid, with one beam position centered on the lens axis. The beam positions in a typical sector of Earth's field of view (FOV) (projected) are shown in Fig. 1. The dotted circles are edge-of-coverage gain contours that have points in the triangular grid as centers. Taken as a multiple-beam antenna, the feed system thus has 37 separate beam position ports. The maximum number of ports which can be excited simultaneously will be a function of interbeam isolation and gain system requirements, and antenna isolation and gain performance capability. For a given number and spacing of simultaneously excited beams, worst-case antenna isolation will depend upon the main beam slopes and sidelobe levels. An illustration of this condition is given in Fig. 1, where the first nulls of the beams marked M and N are seen to establish the minimum spacing between simultaneously active beams of the satellite antenna.

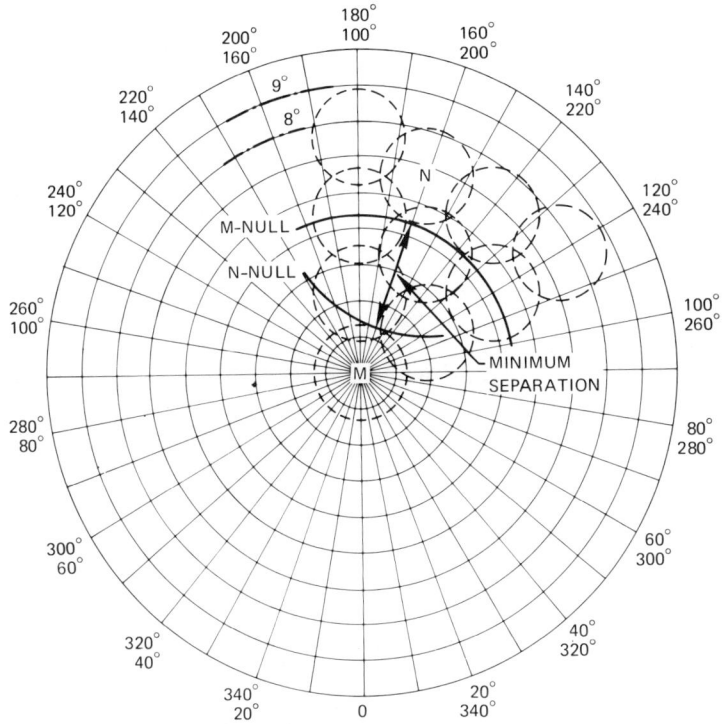

Fig. 1 Multibeam Earth coverage.

In the final analysis, actual isolation depends upon exact location of the network of Earth stations; beam positions are customized for a best fit thereto. In this paper, we discuss only worst-case isolation, since our intent is to develop a multiple-beam antenna having maximum flexibility to be customized easily to any fixed (or variable) Earth terminal network. The present goal is six simultaneous beams with 27-db worst-case interbeam isolation.

III. Constrained Lens Optics and Design

The lens itself consists of a large number of radiating elements arranged in a circular grid to form a primary receiving array of small pickup antenna elements, a secondary re-radiating array of elements, and a set of rf delay lines connecting, on a one-to-one basis, primary to secondary radiating elements. This microwave "lens" is a "space-fed" array, illuminated by a smaller primary feed array, which generates a group of multiple beams, each beam sharing the common lens aperture. The lens is designed so that the off-axis angular displacement of the phase center of each primary feed cluster causes a specific angular displacement off-axis of each corresponding secondary beam.

Considerable study of the optics of TEM lenses was performed by Aeronutronic-Ford prior to deriving a preferred multiple-beam design. The TEM lens takes its name from the characteristic transverse electromagnetic mode transmission lines connecting each primary surface reradiator. In contrast to the better-known waveguide (or "eggcrate") lens, the TEM lines, having no cutoff frequency/size characteristic, can be arbitrarily small in cross section and thus may be "snaked" or coiled, providing relative delay vs position between the two lens surfaces (S_1 and S_2) independently, and achieve focusing properties by proper choice of the snaked TEM "delay" lines. For example, S_1 and S_2 may be planar, as shown in Fig. 2a. This configuration, viewed in cross section, led to the name "bootlace lens." The off-axis scan properties of the planar-planar lens were shown to be inadequate for low-sidelobe operation. Another variable in the TEM lens is the relationship between the positions of interconnected primary and secondary radiators. Using these variables, it has been shown [1] that a two-dimensional "R-2R" TEM lens is feasible such that a point source feed may be scanned arbitrarily far off-axis on a circular focal line with no phase error in the secondary tilted linear phase front. However, our studies showed that a three-dimensional version of this lens does not exist.

Additional studies at Aeronutronic-Ford showed various TEM lens focusing and partial focusing possibilities for three-dimensional, small-angle scan requirements. A ring focus lying perpendicular to the lens axis was shown to be impossible. Various minimized phase error, three-dimensional lenses were derived and analyzed. For the beamwidth of the current lens, the planar-secondary/concave-primary shaped lens shown in Fig. 2b has been found to maintain very low phase error for up to 8.5° scan. For significantly narrower beams, other lenses may be optimum but suffer from more complex lens surface shapes.

IV. Computer Programs

Three computer programs were developed to enable radiation pattern calculations and evaluation of the scanning characteristics of the preferred three-dimensional TEM lenses. A brief description of these programs is given in Table 1. Some of

Table 1 Lens programs

Name and function	Method
Conlens: Models an array-fed constrained lens antenna. Input: Feed geometry excitations, patterns, positions, and orientations; lens line characteristics and lengths; zone boundaries within which a different element pattern may be assigned. Output: Directivities of elements and complete feed array; illuminations across lens faces; efficiencies; aperture, phase spillover, capture, lens transmission, polarizations; principal and cross-polarization secondary patterns.	Near-field array formulas are used to trace complex vector-polarized radiation fields from array feed to primary pickup elements through the lens to, and then from, secondary radiators to antenna's far field. Polarization property and effective area of each lens element are accounted for.
Symlens: Same as Conlens except that lens pickup and reradiation arrays are approximated by a continuous distribution. Input: Feed geometry, excitations, pattern; primary and secondary surface and line lengths as a function of secondary radius. Output: Voltage distribution across secondary aperture which serves input to Secpat.	Near-field array formula used to obtain incident voltage at primary surface, then transferred through lens and space to secondary aperture plane.
Synthesis: Automatic optimization of feed excitations that produce secondary aperture distribution best fitting specified distribution. Used with Minipat to obtain patterns. Input: Feed array and lens geometry; feed and lens element patterns. Output: Feed element excitations, degree of fit, aperture distribution.	Program models the lens antenna similar to Symlens. Least-squares method used for optimization search. Programed for time-sharing.

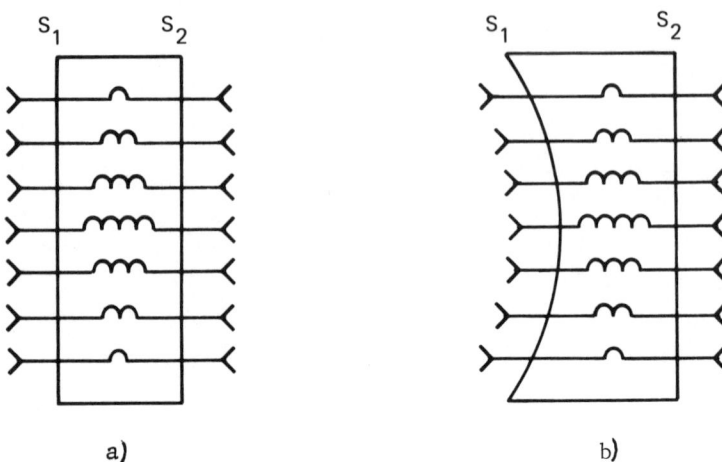

Fig. 2   TEM lens configurations.

the inputs to these programs consist of measured imbedded lens element and imbedded feed element patterns.

V.   Feed Cluster Design

In order to obtain very low sidelobes in the secondary patterns, it has been found necessary to utilize a coherently fed cluster of several point source feeds on the quasi-focal surface to form each beam. The computer program Synthesis solves for the optimum complex excitation coefficients of each specified feed cluster which provide the best rms fit to a specified Taylor illumination for the lens aperture. The feed excitation coefficients thus derived are used as inputs to Conlens, which calculates the secondary radiation characteristics of the lens. This process is repeated for all beam positions of interest. The feed cluster geometry and excitations for the 4-and 6-GHz bands are shown in Fig. 3. A seven-element cluster suffices at 6 GHz, but 13 elements are required at 4 GHz. Radiation patterns for various scan angles of the uplink or 6-GHz band are shown in Fig. 4. The cross-polarized component of these circularly polarized patterns also is given for the on-axis and maximum scan cases. The highest sidelobes in Earth's FOV are about 37 db down. Similar patterns for the 4-GHz downlink have higher FOV sidelobes, about 35 db down, and the cross-polarization level for maximum scan is -28 db. These patterns are shown in Fig. 5 and

CONSTRAINED LENS ANTENNA FOR SATELLITES

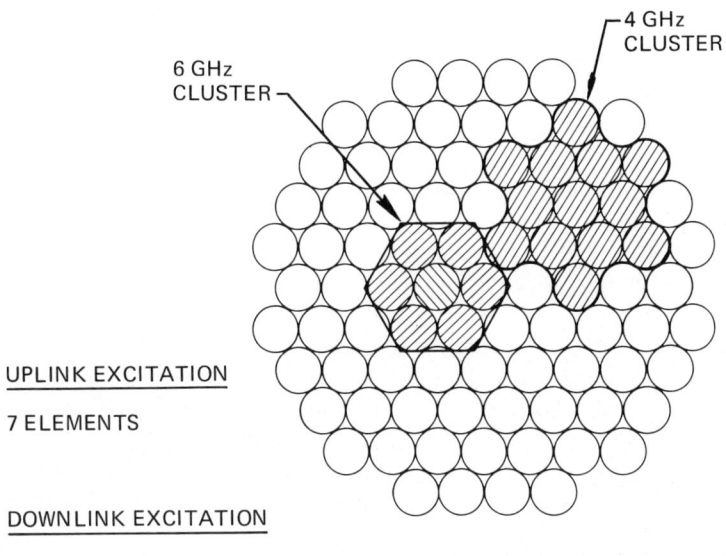

Fig. 3 Feed cluster geometry and typical excitations.

exemplify the low sidelobe performance capability of the TEM lens design.

Methods of reducing the cross-polarization level in the main-lobe regions have been studied. Computed patterns indicate that appropriate treatment of the feed elements, which are assumed to have 1.0-db axial ratio, can reduce significantly the cross-polarization level normally associated with this value of axial ratio. Special beam-shaping techniques also have been explored. A flat-topped beam with constant gain over a 6° cone of coverage is shown in Fig. 6. The highest sidelobe is -35 db. The 19-element feed cluster used to generate this pattern is indicated in the figure. Included for comparison is a standard pencil beam computed for the same frequency.

VI. Lens Elements

Each pair of primary and secondary lens elements and their interconnecting snaked TEM delay lines are all etched on the two flat surfaces of a thin, copperclad dielectric printed

circuit board. The lens is a circular collection of rings o:
such cards, resulting in a relatively lightweight and econom:
cal structure. A major component is the broadband printed ra
diator pair on each card. The imbedded voltage standing-wave
ratio (VSWR) of each radiator is less than 1.5 relative to tt
impedance of its delay line over an octave frequency band.
Orthogonal independent pairs of such PC radiator/delay line
cards are used in the lens to preserve the polarization of tt
primary feed wave. A sketch of an orthogonal card pair asser
bly is shown in Fig. 7.

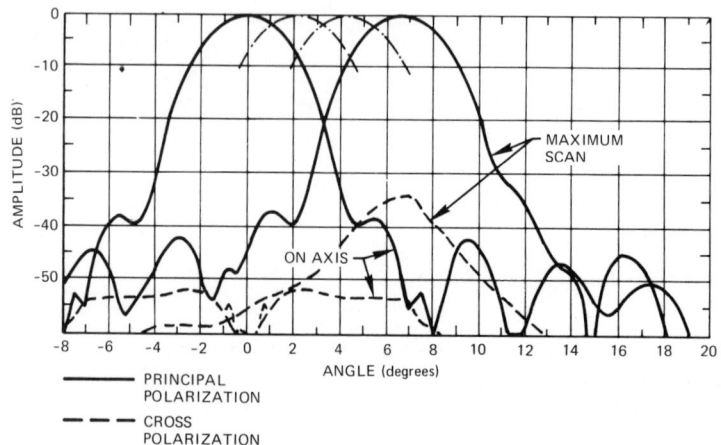

Fig. 4 Computed lens patterns at 6 GHz.

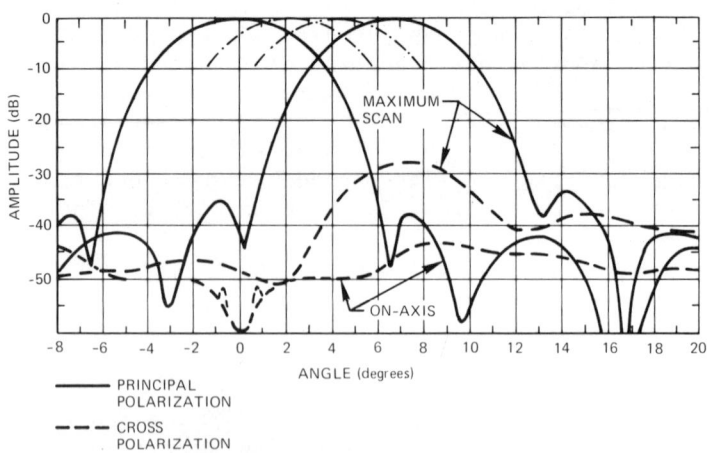

Fig. 5 Computed lens patterns at 4 GHz.

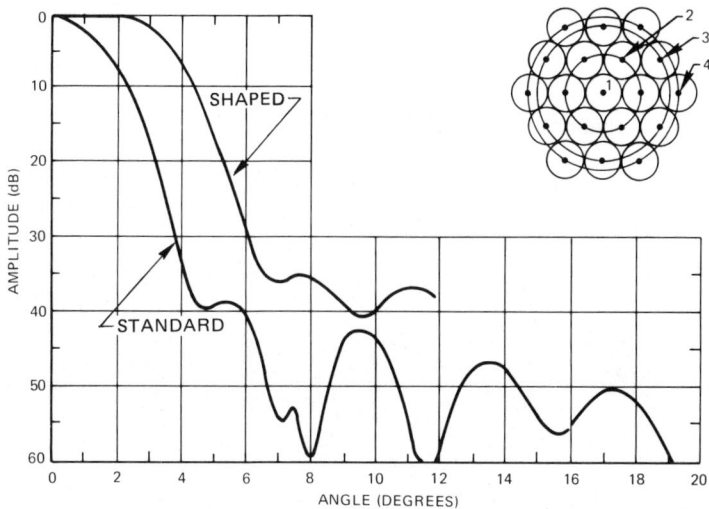

Fig. 6 Computed shaped-beam pattern at 6 GHz.

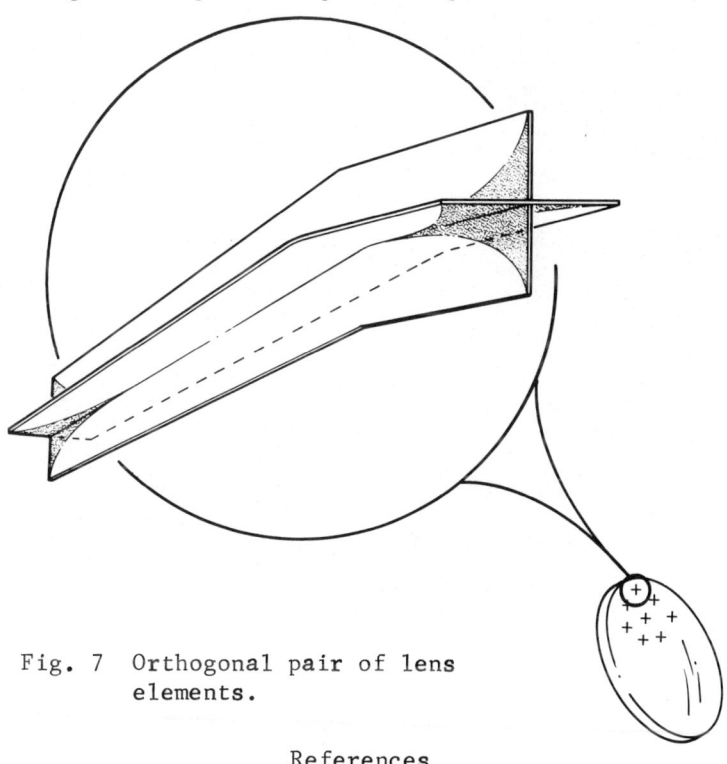

Fig. 7 Orthogonal pair of lens elements.

References

[1]Collin, R.E. and Zucker, F.J., <u>Antenna Theory</u>, Part II, Mc-Graw Hill, New York, 1969.

ATS-6 INTERFEROMETER:
A PRECISION, WIDE-FIELD-OF-VIEW
ATTITUDE SENSOR

M.A. Teichman,[*] F.L. Marek,[+]
J.J. Browning,[+] and A.K. Parr[≠]

IBM Federal Systems Division, Manassas, Va.

Abstract

A radio-frequency phase interferometer has been launched as part of the ATS-6 spacecraft attitude control system. Laboratory measurements indicated that the interferometer is capable of determining spacecraft attitude in pitch and roll to an accuracy better than 0.018° and to better than 0.1° in yaw over a field of view of ±12.5° about the spacecraft normal axis, with an angular resolution of 0.004°. Initial flight test results have demonstrated that the interferometer as a three-axis stabilized sensor is capable of holding the spacecraft attitude to better than 0.004° in pitch and roll and to 0.028° in yaw for extended periods of time. The system is completely solid state, weighs 17 lb, and consumes 12.5 w of d.c. power. This paper presents the system design, error analysis, laboratory test, and preliminary flight test results for the system.

Presented as Paper 74-486 at the AIAA 5th Communications Satellite Systems Conference, Los Angeles, Calif., April 22-24, 1974. This work was supported by Fairchild Space and Electronics Company under Subcontract SC71-3 and by NASA Goddard Space Flight Center, Greenbelt, Md., under Contract NAS5-21100. The authors would like to acknowledge the cooperation and assistance of A. Kampinsky, the technical officer representing Goddard Space Flight Center; and J. Mudano, the technical officer representing Fairchild Industries, Germantown, Md., the applications technology satellite prime contractor. The helpful assistance of G. Chick and R. Lashenka of IBM during the critical testing phase also is acknowledged.
*Staff Engineer.
+Advisory Engineer.
≠Senior Associate Engineer.

I. Introduction

With the increased use of high-gain antennas on spacecraft, precise stabilization of the spacecraft becomes crucial. It is imperative that the attitude of the spacecraft be sensed with a high degree of accuracy and reliability. These requirements can be met presently by radiometric means using a phase interferometer. This paper describes the Applications Technology Satellite (ATS)-6 interferometer developed for NASA Goddard Space Flight Center, Greenbelt, Md., which was launched in May 1974. The interferometer is a precision, wide-field-of-view attitude sensor designed to support the achievement of two of the ATS preliminary missions: 1) providing spacecraft fine pointing and slewing, and 2) providing an oriented stable spacecraft at synchronous altitude for advanced technology experiments.

The system has demonstrated the capability for determining the spacecraft attitude to better than ±0.018° (3σ) over an unambiguous field of view of ±12.5°. Over a field of view of ±30°, the measured accuracy of the system was better than 0.025° (3σ). The interferometer offers the following advantages over presently available attitude sensors: 1) all solid-state instrument (no moving parts), 2) high resolution and accuracy for complete three-axis sensing, 3) wide field of view (off-Earth pointing and acquisition), 4) fast response (permitting the tracking of high spacecraft jitter rates), 5) accuracy and resolution not limited by Earth atmosphere variables, and 6) in-flight calibration of receiver phase biases.

II. Theory of Operation

The ATS-6 interferometer measures the phase difference of C-band ground station signals received at paired antenna elements and converts this information into digital data. In very fundamental terms, the interferometer phase measurement may be expressed as the dot product between a vector baseline L, whose magnitude is determined by the distance between antenna pairs, and the vector line of sight (LOS). Figure 1 illustrates the vector equation

$$\cos \theta = \text{LOS}_j \ (L_j/|L_j|) = (\lambda/2\pi) \ (\gamma_j/|L_j|) \quad (1)$$

where

$\lambda/|L|$ = ratio of rf wavelength to baseline length

γ_j = electrical phase measured across jth baseline

ATS-6 INTERFEROMETER

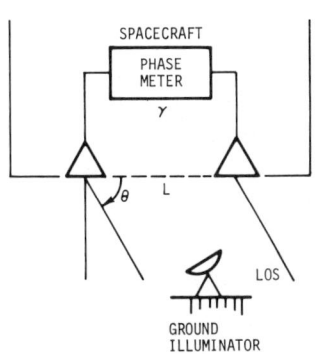

Fig. 1 Basic interferometer relationship; $\cos\theta = (\lambda/2\pi)(\gamma/L)$.

Since the separation between antennas is several wavelengths, the transfer characteristic of space angle vs measured phase is ambiguous. To resolve these ambiguities, a second baseline (coarse) is established using a third antenna. The coarse baseline exhibits an unambiguous transfer characteristic over a ±17.5° field of view.

Given $|\overline{L}|$, determined by optical alignment, and determining γ by interferometer measurement, the object is to solve for the spacecraft attitude. Given the $\overline{LOS}$ vector to the ground station and the satellite ephemeris data, the pitch and roll attitude of the spacecraft can be determined. By utilizing two ground stations simultaneously (using frequency multiplexing), complete pitch, roll, and yaw can be determined for the spacecraft from the four-phase measurements.

The interferometer can be used either as an onboard closed-loop attitude sensor or for spacecraft attitude determination using telemetry data and ground computations. Closed-loop onboard control of pitch and roll utilizes one ground transmitter and an onboard digital processor. Yaw control is accomplished with the Polaris sensor. Complete spacecraft pitch, roll, and yaw attitude can be determined utilizing two ground transmitters. Telemetry data are used to compute the spacecraft attitude on the ground. In summary, three-axis attitude determination using the interferometer can be accomplished by the following:

1) Two-station interferometer: provide satellite ephemeris, ground station locations, baseline lengths; measure four electrical phases; compute pitch, roll, yaw.

2) Interferometer and Earth sensor: provide satellite ephemeris, ground station locations, baseline lengths; measure two electrical phases; compute pitch, roll, yaw.

3) Interferometer and star tracker: provide satellite ephemeris, ground station locations, baseline lengths; measure two electrical phases, one star tracker angle; compute pitch, roll, yaw.

III. System Description

A block diagram of the interferometer system is shown in Fig. 2. The antenna array assembly, which receives the rf signals at 6.150 and 6.155 GHz, consists of six antennas, transmission lines, and the switch module. The six linearly polarized antenna elements are arranged to form two orthogonal baselines, parallel to the spacecraft pitch and roll axes, each containing three antenna elements (reference, coarse, and vernier). The spacing between the vernier and reference antennas is nominally 38.292 in. (20 λ). In order to resolve the ambiguities inherent in the long baseline, a short baseline is established between the reference and coarse antennas at a spacing of nominally 3.191 in. (1.66 λ).

The signals from the antennas are routed to the switch module by means of six transmission lines. These are semirigid coaxial cables that exhibit excellent phase tracking characteristics with temperature. The switch module provides signal selection from pairs of antennas in accordance with the interferometer operating modes.

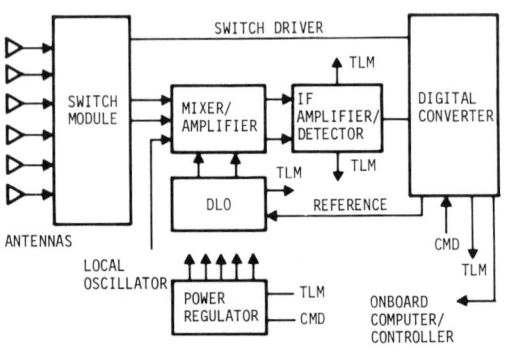

Fig. 2 Interferometer block diagram (TLM means to telemetry subsystem, CMD from command subsystem).

Signals from the switch module enter the dual channel mixer/amplifier module in the receiver converter assembly. One channel contains the reference signal, and the other channel contains the signal to be compared against the reference. (These signals are reversed by the switch module during calibration.) The interferometer employs dual conversion with two mixer/amplifiers in series in each channel. Signals at the input are down-converted in the first mixing action to an intermediate frequency (nominally 150 MHz). The resultant signals are amplified and down-converted in the second mixer/amplifier to a lower intermediate frequency (nominally 30 MHz), where further amplification takes place.

ATS-6 INTERFEROMETER

To obtain the first IF signal at 150 MHz, a local oscillator at 6.0 GHz is utilized. This local oscillator signal is provided by the spacecraft transponder frequency synthesizer. A dual local oscillator (DLO) module is used for the second mixer local oscillator operating at a nominal frequency of 120 MHz. The DLO provides two outputs, differing in frequency by a low-frequency reference (2 kHz) received from the digital converter. The phase of the low-frequency offset which results from the DLO's two outputs is related coherently to its reference input, by means of a phase-locked loop, and forms a firm basis for data extraction. The DLO also transmits lock condition to the telemetry system to indicate when data may be regarded as valid. The outputs of the two channels are combined and fed to the IF amplifier/detector module.

The requirement that signals be received simultaneously from the two ground stations is implemented by providing two separate, tuned IF amplifiers in the IF amplifier/detector module. The two IF amplifiers are tuned to 27.5 and 32.5 MHz. By spacing the transmitted frequencies by 5 MHz, the requisite simultaneous reception of two signals is accomplished.

Signals from the mixer/amplifier module are detected in the IF amplifier/detector module after filtering and amplification. The two carrier frequencies employed for dual frequency operation are f_1 (6.150 GHz) and f_2 (6.155 GHz), and so the outputs of the first mixer/amplifiers in both comparison and reference channels are 150 and 155 MHz, respectively (LO frequency is 6.000 GHZ). The DLO frequency to the reference channel is 122.500 MHz, and the corresponding comparison channel DLO frequency is 122.502 MHz. The "reference" mixer outputs are 27.500 and 32.500 MHz, respectively, for the 6.150- and 6.155-GHz carriers, and the "comparison" frequency outputs are 27.498 and 32.498 MHz, respectively. The four signal frequencies are combined and filtered in the IF amplifier/detector module.

Each channel in the IF amplifier/detector module contains a main IF amplifier, an IF detector, a bandpass filter, and a low-frequency amplifier. Automatic gain control (AGC) affords the system a sufficient dynamic range to control the expected levels of the received signal. The AGC voltage is sent by means of telemetry to ground stations as an indicator of received signal strength. The IF envelope detector extracts the 2-kHz difference frequency from reference and comparison signals. This 2-kHz data signal is filtered and amplified to adequate levels by an operational amplifier and then shaped by a zero crossing detector in the digital converter. These signal operations apply equally to the f_1 and f_2 channels.

The digital converter contains an input data switch, a zero crossing detector (ZCD), a digital phase comparator, an averaging counter, discontinuity detector logic, a receiver clock, and a format generator. The receiver clock acts as a stable time reference for the converter's circuits, whereas the format generator acts as the master control for the converter's operations. The format generator contains logic circuits that generate timing waveforms that are used to achieve converter and switch module synchronous operation.

The 4.096-MHz output of the receiver clock in the converter is transformed by the clock divider (a 2048 divider) to 2 kHz. This signal serves as the reference for the format generator, digital phase comparator, and DLO module phase-locked loop. The f_1 and f_2 data signals from the detector module are sampled alternately in sequence and are timed by the format generator. The 2-kHz digital data signals are compared in the digital phase comparator with the 2-kHz reference (from the clock divider). The resultant 2-kHz signal containing the phase information is used in the phase count gate to stop counter action previously initiated by the 2-kHz reference signal from the format generator. The number of pulses supplied by the receiver clock and allowed to exist in the phase count gate is in proportion to the phase difference between the data signal and the reference and therefore is a function of spacecraft attitude angle (360° is equivalent to 2048 counts). To reduce the variance of the phase jitter caused by random noise, 64 vernier data counts are averaged.

An interferometer high-speed data link (IHSDL) output is provided by a direct connection between the digital converter phase count gate and the spacecraft communication subsystem prime focus feed. These data are provided to the ground at a high data rate on a sample-by-sample basis at a 1-msec sampling

Table 1 Frequency parameters

Frequency parameter	Receiver/converter function
F1 & F2	Alternately measure inputs f_1 and f_2
F1 only	Measure input f_1 only
F2 only	Measure input f_2 only

rate. This mode of operation bypasses the standard spacecraft telemetry link, which samples the interferometer data every 3 sec. The IHSDL will be utilized for experiments concerned with high spacecraft jitter rates. In addition to the telemetry and IHSDL data links, an onboard link is provided to the attitude control system. The data consist of 11 bits of digital data representing pitch and roll information. Power for all components is supplied by the power regulator, which contains the necessary regulating and signal conditioning circuits to interface with the spacecraft power system.

IV. Operational Modes

The interferometer performs either a coarse-type or a vernier-type measurement of the input signal phase. These types of measurements are defined as follows:

1) Coarse-type measurement: One sample of the input phase is digitally encoded to a 5-bit quantity (resolution: 1 part in 32). The coarse-type measurement is associated only with input signals originating from a reference-coarse antenna pair. The coarse-type measurement is accomplished in 9 msec.

2) Vernier-type measurement: 64 samples of the input phase are averaged and digitally encoded to an 11-bit quantity (resolution: 1 part in 2048). The vernier-type measurement normally is associated with input signals originating from a reference-vernier antenna pair, except in a coarse-only mode of operation, where a reference-coarse antenna pair is the source of the input signals. The vernier-type measurement is accomplished in 75 msec.

The interferometer has 27 modes of operation, as determined by the combination of frequency, axis, and resolution parameters. These parameters, controlled by commands from the ground, establish a predefined input selection and measurement sequence and thereby control the operation of the receiver converter. Any of the 27 modes can be selected in either the normal or calibrate position of the switch module. This flexibility has been designed into the system in order to support spacecraft experiments and also to assist in fully determining the interferometer's capabilities. The interferometer operating modes are shown in Tables 1-3.

The frequency parameter controls the gating of input signal f_1 and f_2 to the phase-measuring circuit in accordance with the information given in Table 1. The axis parameter

Table 2 Axis parameters

Axis parameter	Receiver/converter function
P & R	Alternately select pitch/roll axis
P only	Select pitch axis only
R only	Select roll axis only

controls the selection of the source signal for f_1 and f_2 inputs such that antenna elements along either the pitch or the roll axis are selected for phase measurement. The relationship is shown in Table 2. The resolution parameter controls the selection of coarse or vernier antenna elements in accordance with the information given in Table 3.

The C & V command alternately selects a coarse antenna element, performing a coarse-type measurement, and a vernier element, performing a vernier-type measurement. The V-only command selects only vernier antenna elements and performs the vernier-type measurement. The C-only command selects only coarse antenna elements and performs the vernier-type measurement (64 samples averaged) of coarse data. To perform a complete cycle of F1 coarse and vernier, pitch and roll, followed by F2 coarse and vernier, pitch and roll takes 336 msec.

V. Interferometer Performance

The error budget for the interferometer is divided into random, boresight, and off-boresight errors. The random errors consist of the noise-dependent errors associated with

Table 3 Resolution parameters

Resolution parameter	Receiver/converter function
C & V	Alternately measure coarse & vernier
V only	Measure vernier only
C only	Measure coarse only

the phase measuring receiver. Also included are the switch module phase jitter, which consists of the switch-phase jitter introduced by d.c. bias fluctuations and the DLO phase noise. The random errors are reduced by averaging 64 samples of data. As the field of view is increased beyond ±12.5° these errors tend to increase, because the antenna gain decreases, thus reducing the received signal strength.

The boresight errors consist of errors that are independent of attitude angle. These errors, which are largely a function of temperature, introduce differential phase shifts in the antenna array assembly. The differential phase shifts introduced in the receiver converter portion of the system are removed by inflight calibration. The calibration of the receiver converter phase shifts is accomplished by comparing the interferometer phase measurements at a given attitude in the normal configuration with the phase measurements taken in the calibration configuration. In the latter configuration, the switch module reverses the inputs to the receiver converter.

In the normal switch configuration, the measured phase is

$$\phi_{M_1} = \phi_o + \phi_A + \phi_R$$

where

ϕ_M = measured phase
ϕ_A = differential phase bias introduced by antenna array
ϕ_R = differential phase bias introduced by receiver converter
ϕ_o = desired phase

With the switch in the calibration position,

$$\phi_{M2} = -(\phi_o + \phi_A) + \phi_R$$

so that the receiver converter differential phase bias can be determined by

$$(\phi_{1M} + \phi_{2M})/2 = \phi_R$$

The off-boresight errors consist of errors that are a function of spacecraft attitude angle relative to the ground transmitter. These errors are composed of the following:

1) Antenna phase tracking errors: relative symmetry of antenna phase center position as a function of angle. The

relative phase center position is relatively constant over narrow fields of view and becomes progressively worse with wide angles.

2) Isolation from 30-ft reflector: represents the interference caused by coupled signals diffracted from the 30-ft parabolic reflector of the ATS-6. Experiments have indicated a worst-case error of 0.9°.

3) Switch leakage: the finite isolation between switch ports causes cross-coupling of adjacent channels. These cross-coupling errors of 0.8° represent absolute worst-case for the measured isolation.

4) Distortion: distortion of the baseline geometries will cause interferometer errors. The most severe errors are caused by temperature expansion of the antenna mounting platform. Changing the antenna separation results in a different proportionality constant for Eq. (1).

The random error total represents the resolution of the interferometer. The relative accuracy of the interferometer is determined from the root sum square (rss) total of the random and off-boresight errors. The total attitude error is determined by the rss total of all three error sources. Table 4 presents a summary of the error sources for several fields of view.

VI. Laboratory Test Results

The evaluation of the system performance as an attitude sensor was performed in an anechoic chamber with the system mounted and optically aligned in a mockup of the ATS-6 spacecraft (Fig. 3). The mockup was mounted on an elevation over azimuth pedestal which contained drive motors and position sensors that positioned the mockup in azimuth and elevation with an angular resolution and repeatability of 0.0025°. The tests consisted of measuring the interferometer digital output as a function of position relative to the transmitting antenna in the anechoic chamber. These data, along with the pedestal position data, were recorded automatically on magnetic tape as the pedestal was positioned automatically in increments of 0.5° over a field of view of ±30° in azimuth and elevation. The data were compared with a mathematical model developed for the chamber and the resultant 3σ error determined.

Table 4 Interferometer error budget

Error sources, electrical deg	Field of view, deg		
	±12.5	±20	±30
Random			
1) System noise (ERP + 73 dbw)	2.64	3.54	6.3
2) Switch jitter & DLO stability	0.76	0.76	0.76
RSS subtotal	2.75	3.68	6.35
Average 64 samples	0.45	0.60	1.03
3) Quantization	0.09	0.09	0.09
RSS subtotal	0.46	0.60	1.03
Space angle resolution	0.004	0.005	0.009
Boresight			
1) rf component mismatch	0.40	0.40	0.40
2) Transmission line phase tracking	0.50	0.50	0.50
3) Switch phase tracking	0.70	0.70	0.70
Total in-flight drifts	1.6	1.6	1.6
Off boresight			
1) Antenna phase tracking	0.7	0.9	1.1
2) Reflector interaction	0.9	1.0	1.1
3) Switch leakage	0.8	0.8	0.8
4) Mechanical & thermal baseline distortions	0.6	0.9	1.3
RSS total	1.5	1.8	2.2
Attitude accuracy (all 3 error groups)	0.0187	0.021	0.027
Relative accuracy (random and off-boresight errors)	0.0127	0.016	0.023

The results were
compared with the pre-
dicted performance for
the interferometer in
the anechoic chamber.
The error sources
(Table 5) included
only those inter-
ferometer errors
that pertained to
anechoic chamber
tests and any errors
introduced by the
test environment.
The results indicated
that the measured
performance of the
interferometer was
well within the pre-
dicted error.

Fig. 3 Interferometer in mockup of spacecraft.

VII. Flight Test Results

Under the direction of Endres and Isley,[1] interferometer co-investigators at Goddard Space Flight Center,
Greenbelt, Md., the performance of the interferometer is being evaluated by analysis of telemetered interferometer data.
Mathematical modeling has included the formulation of a hardware representation of the interferometer that converts raw
phase data into corrected readings which account for system
biases and nonlinearities. Several different calibration
models were specified to permit 1) self-check for a
frequency configuration, 2) check against orbit-derived line-
of-sight reference, 3) check against other sensor references,
and 4) check against IHSDL data. The algorithms were developed with support from IBM Corporation under Contract NAS5-
20026. Preliminary test results indicate that the interferometer
during a 22-min test cycle period maintained spacecraft attitude
to better than 0.004° in pitch and roll and to 0.028° in yaw.[1]

VIII. Conclusions

An attitude sensor using radiometric interferometer techniques has been described. This system has been developed for
NASA's ATS-6 spacecraft and represents an initial attempt at
spacecraft stabilization using an interferometer. The test

Table 5 Interferometer error sources: anechoic chamber tests [a]

Interferometer errors (3σ)	Field of view, deg	
	+12.5	+30
Random (noise induced)	0.16	0.16
Antenna phase center tracking	0.70	1.1
Switch interchannel coupling	0.80	0.80
Anechoic chamber errors		
Angular quantization	0.16	0.16
Repeatability	0.17	0.17
RSS subtotal	1.10	1.39
Chamber wall reflections	0.33	0.33
Mockup distortions	0.67	1.42
Total error, electrical deg	2.10	3.14
Space angle error	0.017	0.029
Measured results	0.014	0.023

[a] All error sources were determined by direct measurements.

program results indicate an attainable accuracy of better than 0.025° over a ±30° field of view. The interferometer method of attitude sensing offers the highest degree of presently available accuracy coupled with flexibility for experimental analysis.

References

[1] Endres, D.L. and Isley, W.C., "ATS-6 Interferometer," Aerospace and Electronic Systems, Nov. 1975 (special issue on ATS-6 Spacecraft).

DEVELOPMENT AND AIR BEARING TEST
OF A DOUBLE-GIMBALLED MOMENTUM WHEEL
ATTITUDE-CONTROL SYSTEM

J. J. Kalley Jr.[*] and H. L. Mork[+]

TRW Systems, Redondo Beach, Calif.

Abstract

A three-axis attitude-control system, suitable for present and future generation communication satellites, is described with emphasis upon physical test activities. The system employs a double-gimballed momentum wheel, roll/pitch attitude sensing, a digital processor, and thrusters. Scaled physical tests provide capture and multiorbit performance, including investigations into momentum unloading, capture from large yaw error, and gimbal servo control. The test results, demonstrating accurate roll/pitch pointing (0.05°) and analytically predictable yaw error, affirm the compatibility of the processor-wheel-gimbal implementation concept.

Nomenclature

$a_{ix}, b_{ix}, a_{iz}, b_{iz}$ = generalized torque compensator gains
B_i, B_o = viscous friction of inner and outer gimbal
B_{i1}, B_{i2} = inner gimbal structural damping
B_{o1}, B_{o2} = outer gimbal structural damping

Presented as Paper 74-488 at the AIAA 5th Communications Satellite Systems Conference, Los Angeles, Calif., April 22-24, 1974. This paper is based upon work under the sponsorship of the International Telecommunication Satellite Organization. Any views expressed are not necessarily those of Intelsat. Appreciation is extended to J. Dohogne of Sperry Flight Systems Division, Sperry Rand, Phoenix, Ariz., for the timely suggestion of the gimbal precession control concept.
[*]Member Technical Staff, Design Analysis & Simulation Department.
[+]Section Head, Design Analysis & Simulation Department.

H_c, H_B	= wheel momentum and momentum bias
H_w	= wheel momentum resolved to body coordinates
I_o, I_i	= total outer and inner gimbal inertia
I_{i1}	= inner gimbal ring inertia about inner axis
I_{i2}	= inner gimbal trunnion and drive shaft inertia about inner axis
I_{01}	= outer gimbal ring and inner gimbal ring inertia about outer axis
I_{02}	= outer gimbal trunnion and drive shaft inertia about outer axis
I_{xx}, I_{yy}, I_{zz}	= body principal axes inertia
J_w, I_w	= transverse and polar wheel inertia
K_{i1}	= transverse rotor web, rotor bearing, and rotor shaft stiffness about inner axis
K_{i2}	= inner gimbal trunnion, drive shaft, and gimbal ring stiffness about inner axis
K_{01}	= inner gimbal shaft, inner ring and transverse rotor web, rotor bearing, and rotor shaft stiffness about outer axis
K_{02}	= outer gimbal trunnion, drive shaft, and gimbal ring stiffness about outer axis
K_{p1}, a_{p1}, b_{p1}	= pitch rate filter gains
K_{p2}, a_{p2}, b_{p2}	= pitch position filter gains
K_s, a_s, b_s	= pitch position sensor filter gains
K_r	= effective pitch rate gain
$K_x, K_z, \overline{K}$	= gimbal precessional control gains
$K_1, K_2, \overline{K}_1, \overline{K}_2$	= body roll and yaw rate and position torque gains
K_3, K_3'	= body no-yaw sensor gains
$K_\phi, K_{\dot\phi}, K_\psi, K_{\dot\psi}$	= body roll and yaw rate and position control gains
T_f, T_{wd}	= wheel friction and windage torques
T_w	= wheel motor torque
T_{dx}, T_{dy}, T_{dz}	= body disturbance torques
γ_{01}, γ_{02}	= motion of I_{01} and I_{02} with respect to the outer gimbal
γ_{i1}, γ_{i2}	= motion of I_{i1} and I_{i2} with respect to the inner gimbal
$\hat\gamma_{ox}, \hat\gamma_{iz}$	= outer and inner gimbal angle
γ_{ox}, γ_{iz}	= outer and inner gimbal resolver angles
$\gamma_{oxc}, \gamma_{izc}$	= outer and inner gimbal angle digital processor commands
$\gamma_{oxc}(0), \gamma_{izc}(0)$	= initial condition of outer and inner gimbal commands
ϕ, θ, ψ	= Euler angles
$\phi_\ell, \theta_\ell, \psi_\ell$	= body control position error limit
$\dot\phi_\ell, \dot\theta_\ell, \dot\psi_\ell$	= body control slew rate limit
ϕ_d, θ_d, ψ_d	= body control position deadzone

AIR BEARING TEST OF ATTITUDE-CONTROL SYSTEM

μ	= orbit angle
$\bar{\omega}_e, \omega_{ey}, \omega_{ez}$	= Earth rate and Earth rate pitch and yaw components in body coordinates
ω_0	= orbit rate
Ω, Ω_B	= wheel speed and wheel speed bias
$\Omega_\phi, \Omega_\psi, \Omega_\gamma$	= bandwidth of roll body, yaw body, and gimbal control system
τ	= gimbal precessional control nutation filter time constant
τ_{mo}, τ_{mi}	= outer and inner gimbal motor torque

Introduction

As the requirements of future communication satellites become more stringent with regard to antenna pointing, solar array-derived power, and reliability [pointing accuracy requirements of narrow-beam antennas will be in the $0.1°$ to $0.2°$ range, power requirements will be a minimum of 1 kw (end of life), with system life of from 5 to 10 yr], a fully stabilized spacecraft configuration becomes an attractive alternative to the more conventional dual-spin configuration. A fully stabilized configuration employs a three-axis controlled main body having one axis (yaw) pointed to nadir, a second axis (pitch) normal to the geosynchronous orbit plane, and the third axis (roll) in the direction of orbit velocity. Solar power is derived from symmetric solar arrays, which are counter-rotated at orbital rate to remain sun-pointing.

On-orbit attitude sensing of roll and pitch motion is achieved via Earth sensors. Yaw motion, however, cannot be sensed by observing Earth but could be sensed during portions of each orbit with sun sensors, continuously sensed with a star sensor, or with gyros. An alternate, superior method does not rely upon yaw sensing per se but provides accurate yaw control by employing a relatively small, speed-biased momentum wheel suspended in double gimbals which produces a yaw stiffness similar to that of a spinning spacecraft.

Figure 1 depicts a three-axis stabilized communication satellite and a conceptual representation of the double-gimballed momentum wheel (DGMW). The on-orbit fine control system is comprised of an Earth sensor assembly, double-gimballed wheel, momentum unloading and attitude-control thrusters, digital controls processor, and wheel/gimbal/thruster electronics. The Earth sensor provides roll and pitch spacecraft attitude errors by measuring the deviation of the yaw body axis from nadir. The gimballed momentum wheel assembly is comprised of the speed biased wheel, suspension gimbals, motors

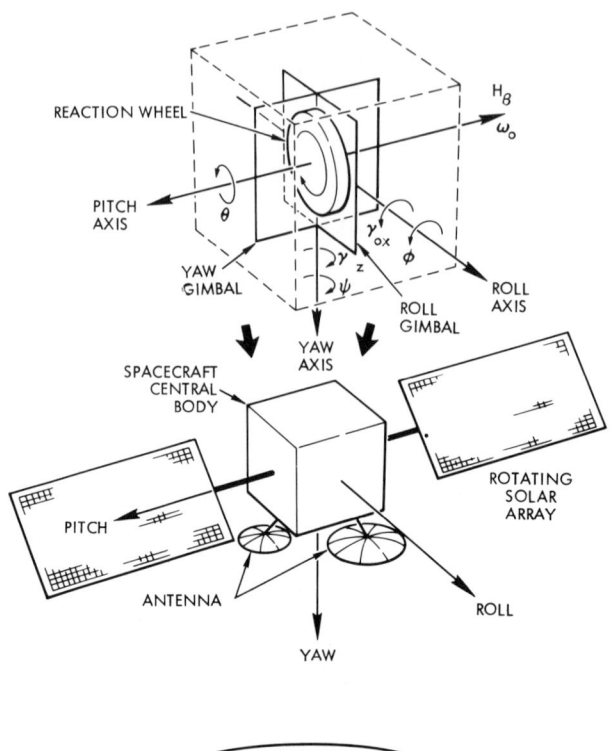

Fig. 1 Representative three-axis communication satellite showing conceptual representation of a double-gimballed momentum wheel.

(momentum wheel and two gimbal), wheel tachometer, and gimbal angle resolvers. This assembly is mounted in the spacecraft with the wheel spin axis nominally aligned with the pitch axis and gimbal axes nominally coincident with the spacecraft roll and yaw axes. The momentum unloading thrusters torque the spacecraft whenever the momentum stored by the gimballed wheel becomes excessive.

All control laws for gimbal and wheel control, sensor conditioning, telemetry and command conditioning, thruster logic, control mode switching logic as well as transfer orbit, acquisition, and velocity correction control reside in the controls processor. Control torques are generated by varying the wheel speed and gimbal angles. Pitch control torque results from changing the speed of the reaction wheel while roll and yaw control torques are generated by torquing the gimbals to rotate the direction of the wheel momentum. This active roll/yaw control is augmented by an inherent long-term gyroscopic torque due to the momentum bias and spacecraft orbital kinematics. Spacecraft maneuvers are accomplished by using the gimballed wheel as a momentum source, whereas long-term steady-state control is accomplished by absorbing disturbance momentum into the gimballed wheel.

In addition to this fine, on-orbit control system, additional sensors (sun and Earth) may be required during the transfer orbit phase prior to insertion into the geosynchronous orbit. The solar array drive mechanism and electronics (utilized in the on-orbit phase) as well as stationkeeping correction thrusters and associated electronics are required. The controls processor is sized to handle this processing (plus command and telemetry) and interfaces with this equipment.

Table 1 summarizes a typical flight application for such a system. Of note is the relatively small bias momentum required to achieve the yaw accuracy of $0.4°$. Wheel momentum sizing is a function of the cyclic environmental torques, number of momentum unloading cycles, and the no-yaw sensor error as a function of uncompensated torques, which is $\psi_e = \Delta T_{dz}/\omega_0 H_B$, where ΔT_{dz} is the uncompensated environmental disturbance torque (primarily solar pressure). Also of note is the possible requirement for spacecraft offset pointing from nadir which can be accommodated by the DGMW system.

The bulk of this paper is devoted to the description of the air bearing test (ABT) system, the test facility, and the results of the test. The system design considerations for the flight system and the ABT are fundamentally the same, with slight variations due to the physical parameters involved (e.g., the ABT disturbance torques are out of proportion to those experienced in flight). The testing of the DGMW concept passed through a number of stages, including 1) full digital computer dynamics simulation, 2) hybrid testing with the DPA

Table 1 Summary of geosynchronous communications satellite DGMW application

Spacecraft pointing accuracy (3σ), deg	
Pitch, roll	0.05
Yaw: with yaw sensor	0.20
without yaw sensor	0.40
Spacecraft offset point capability, deg	
Roll	±5
Pitch	±5
Solar array power (end-of-life), kw	≥1
Operating life, yr	7
Disturbance torque range, ft-lb	$(1-5) \times 10^{-6}$
Wheel bias momentum range, ft-lb-sec	7-30
High-level stationkeeping disturbance, ft-lb	
EW pitch or yaw	0.1
NS roll or yaw	0.4

computing the control signals and analog computing equipment simulating dynamics, and 3) motion testing. The motion test was performed on a three-axis air bearing, which was allowed to rotate continuously about the pitch axis as a means of reproducing orbit conditions. The motion was executed by rotating an "Earth" stimulus about the table and allowing the table pitch control system to sync the table to the stimulus motion.

Air Bearing Test System Description

Time Scaling Considerations

The motion test was scaled in the sense that the table moment of inertia was less than that of a satellite. The test was also time scaled to allow performance observation over several orbits. Real-time air bearing tests,[1] if the inertias are scaled down, have the disadvantage of placing table balance and disturbance torques out of proportion to those required by the scaling. Time scaling has the reverse effect and, for reasonable scalings, results in very large thruster and gimbal rate requirements that are impractical. Problems also exist with scaling the DPA processor time. Table pendulousity, balance, and Earth rate disturbances combined with a practical limit on gimbal control bandwidth and limited battery life are additional factors to be considered in test scaling.

With these conflicts in mind, a test philosophy was adopted in which it was felt practical to perform the thruster control, DGMW unloading, and the DPA processing in real time and to scale the yaw, roll, and gimbal control response. Because of the prohibitive battery capacity necessary to sustain long tests, full-up system operation time was limited to 3 hr. This yielded a lower bound on the orbit rate (hence the time scaling) of at least 1 rev/hr. The lower bound is also a function of the yaw stiffness with respect to disturbance torques. Specifically, air bearing disturbance torques in the yaw axis are roughly 3×10^{-4} ft-lb; this, with a momentum of 30 ft-lb, requires the orbit rate to be greater than 3×10^{-3} rad/sec to hold yaw errors to less than $0.2°$. This is roughly 30 min/orbit.

An upper orbit rate bound results from a limited gimbal control bandwidth and slew rate. Both are directly proportional to motor torque and inversely proportional to wheel momentum. To preserve real-time operation of the gimbals and to minimize heating of the gimbal torquers, gimbal bandwidth should be less than 10 rad/sec. The flight design places the gimbal bandwidth a factor of 100 above the yaw bandwidth; this,

AIR BEARING TEST OF ATTITUDE-CONTROL SYSTEM

coupled with the air bearing inertia scaling, indicated a maximum ABT orbit rate of 15 min/rev. The lower bound of 30 min/orbit was selected, with provisions available to increase the speed if disturbances became greater than expected.

Equipment Complement

The DGMW test system (see Fig. 2) consists of the DGMW, a digital processor (DPA), auxiliary electronics for the wheel, gimbal and thrusters (AEA), interface test electronics for power and telemetry and command functions (TSE), a cold-gas thruster system, and a battery power system. Specific features of the DPA and the DGMW appear in Table 2. The DGMW contains a limited-rotation d.c. torquer in each axis and an eight-speed resolver with 3-min (electrical) accuracy. The inner gimbal is aligned along the body z axis and the outer along the x axis. This orientation is selected to minimize the torquing requirements of the inner axis due to the poor heat dissipation in that axis. Heavy current requirements exist for the roll torquer (at maximum wheel speeds) during yaw gimbal unloading. Wheel rotor torques are generated by a two-phase a.c. motor driven in a full on or off configuration. Direction of applied torque is a function of the phasing between the two windings.

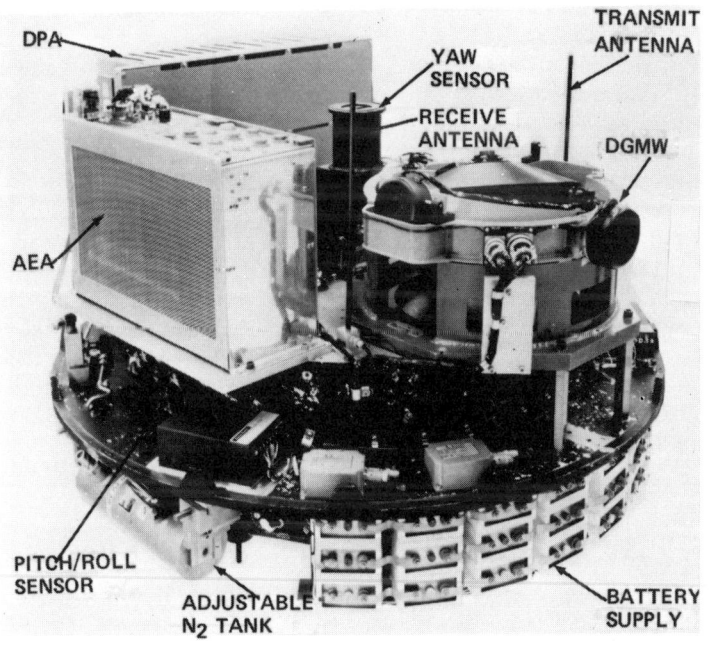

Fig. 2 DGMW air bearing system equipment.

Table 2 DGMW-DPA specifications

DGMW specifications		
Angular momentum, nominal, ft-lb-sec		30
Speed, nominal, Hz		80
Speed range, %		±10
Spin motor		Two-phase a.c.
Spin motor torque, in.-oz		10
Rotor stiffness, ft-lb/rad		5.4×10^4
Gimbal torquer	Limited rotation d.c. (60 oz-in.)	
Gimbal freedom, deg		±20
Gimbal caging		Electronic
Gimbal stiffness, ft-lb/rad		$>10^4$
DPA specifications		
Arithmetic	Fixed point, two's complement, fractional	
Work length	16 bits with double-precision capability	
Instructions		38
Speeds	Add time = 7.2 sec, 120 kbit/sec	
IOU inputs	32 analog (12-bit A/D conversion)	
	40 bilevel	
	32 serial data source	
	2 command source	
IOU outputs	36 bilevel	
	16 serial data destinations	
	2 telemetry destinations	
Ram memory	512 16-bit words (expandable)	
Rom memory	2048 16-bit words (expandable)	

The DPA was constructed by TRW specifically for flight control systems applications.[2] A 512-word Ram is addressable by ground command, allowing minor program or parameter changes to be performed during the processor cycle. The DPA telemetry interface consists of two Rams capable of delivering 16 words of 16 bits, one of which is being loaded while the other is read to the downlink data stream. The DPA is supported by a test set, teletype, and paper-tape reader. A complete acceptance test and interpretive computer simulation (ICS) exists for the DPA.

Body Control System

The roll/yaw body control consists of the elements shown in Fig. 3. The reference system for the gimbals, body axes, momentum, and orbit rate appears in Fig. 1. Control error signals are developed in the digital processor and then transmitted as serial data to the gimbal electronics, where a

AIR BEARING TEST OF ATTITUDE-CONTROL SYSTEM

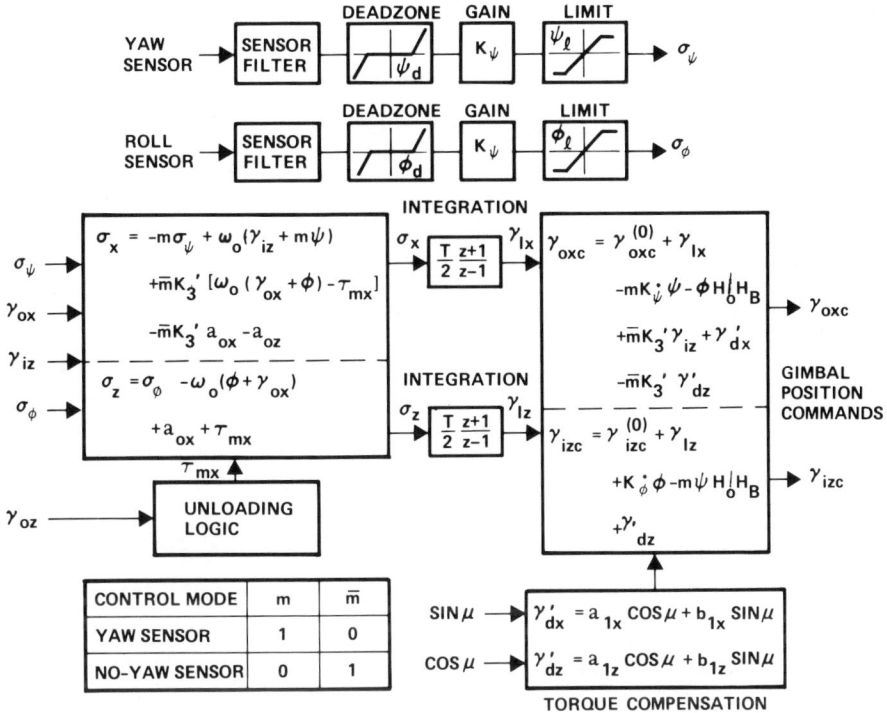

Fig. 3 Body roll/yaw control mechanization.

digital difference with the gimbal is performed. This signal, in turn, is converted to an analog voltage, which drives the gimbal torquers. The gimbals are driven as position loops rather than rate loops to take advantage of the properties that 1) computations of gimbal and body rates are not required, 2) position control accuracy is much greater than rate control accuracy, and 3) rate command resolution at the Digital-to-Analog Converter (DAC) interface to the gimbal control is inadequate for proper orbit rate and disturbance torque decoupling.

Roll/Yaw Body Control Equations

The body dynamic equations necessary in the development of the roll/yaw and pitch control are presented in the Appendix. The small angle and rate roll/yaw equations decouple from pitch and appear as

$$
\begin{aligned}
I_{xx}\ddot{\phi} + \omega_o H_B \phi + H_0 \psi + \omega_o H_B \gamma_{ox} + H_B \dot{\gamma}_{iz} &= T_{dx} \\
I_{zz}\ddot{\psi} + \omega_o H_B \psi - H_0 \dot{\phi} - H_B \dot{\gamma}_{ox} + \omega_o H_B \gamma_{iz} &= T_{dz}
\end{aligned} \quad (1)
$$

where

$$H_o = H_B + \omega_o(I_{yy} - I_{xx} - I_{zz})$$

When roll and yaw attitude error signals are available, an obvious control law decouples the dynamics so that the control dynamics equations become

$$I_{xx}\ddot{\phi} + K_1\dot{\phi} + K_2\phi = 0 \qquad (2)$$

$$I_{zz}\ddot{\psi} + \overline{K}_1\dot{\psi} + \overline{K}_2\psi = 0$$

This form is obtained, assuming that gimbal control dynamics are negligible, if the gimbals are driven with the commands

$$\gamma_{oxc} = \gamma_{oxc}(0) - K_{\dot{\psi}}\psi - \int_o^t [K_\psi \psi - \omega_o(\psi + \gamma_{iz})]d\sigma - H_0\phi/H_B - \gamma_{dx}$$

$$\gamma_{izc} = \gamma_{izc}(0) + K_{\dot{\phi}}\phi + \int_o^t [K_\phi \phi - \omega_o(\gamma_{ox} + \phi)]d\sigma - H_0\psi/H_B + \gamma_{dz}$$
(3)

<u>Normal Mode Control (No-Yaw Sensor)</u>

When a yaw sensor signal is not available (the normal mode), the control laws[3] are somewhat more complex and yield the dynamic equations

$$I_{xx}\ddot{\phi} + K_1\dot{\phi} + K_2\phi = -H_0\dot{\psi}$$

$$I_{zz}\ddot{\psi} + K_3\dot{\psi} + \omega_o H_B \psi = 0$$
(4)

This form is developed by driving the gimbals such that

$$\gamma_{oxc} = \gamma_{oxc}(0) + K_3'(\gamma_{iz} - \gamma_{dz}) + \int_o^t [K_3'\omega_o(\gamma_{ox}+\phi) + \omega_o \gamma_{iz}]d\sigma - H_0\phi/H_B - \gamma_{dx}$$

$$\gamma_{izc} = \gamma_{izc}(0) + K_{\dot{\phi}}\phi + \int_o^t [K_\phi \phi - \omega_o(\gamma_{ox}+\phi)]d\sigma + \gamma_{dz}$$
(5)

where

$$K_3' = \frac{K_3}{H_B}, \quad K_\phi = \frac{K_2}{H_B}, \quad K_\psi = \frac{\overline{K}_2}{H_B}, \quad K_{\dot{\phi}} = \frac{K_1}{H_B}, \quad K_{\dot{\psi}} = \frac{\overline{K}_1}{H_B}$$

To avoid saturating the gimbal rate during large body error transient conditions, the term $K_\phi\phi$ is limited at a value ϕ_ℓ such that, if $\dot{\phi}_\ell$ is the rate limit desired, then $\phi_\ell = K_{\dot\phi}\dot\phi_\ell$.

AIR BEARING TEST OF ATTITUDE-CONTROL SYSTEM

When the limit is exceeded by large pointing errors, the limit acts as a rate command, causing the body to slew at a constant rate until the error is less than the limit. This prevents saturation of the gimbal rates which, in turn, yields a more controlled performance. A similar limit is used for yaw control when the yaw sensor is available.

Disturbance Compensation

A generalized torque compensator for disturbances that are functions of orbit position or are constant in the body reference frame is

$$\gamma_{dx} = \int_0^t \frac{T_{dz}}{H_B} ds = \gamma'_{dx} + \int_0^t a_{ox} ds \tag{6a}$$

$$\gamma'_{dx} = \sum_{i=1}^M (a_{ix} \cos i\mu + b_{ix} \sin i\mu)$$

$$\gamma_{dz} = \int_0^t \frac{T_{dx}}{H_B} ds = \gamma'_{dz} + \int_0^t a_{oz} ds \tag{6b}$$

$$\gamma'_{dz} = \sum_{i=1}^M (a_{iz} \cos i\mu + b_{iz} \sin i\mu)$$

The coefficients M, a_{ij}, and b_{ij} are determined by detailed analysis of external disturbances.

Laboratory Test Earth Rate Compensation

Earth rotation produces a change in the relative alignment of the table inertial reference axes and the laboratory floor. The obvious effect is that, assuming no friction or control action, the table would rotate as the day progressed. The introduction of a table control loop stops this relative motion, with the effect of causing the table to have an inertial rate opposing Earth rate. The gimbal angles are measured relative to the table; thus, for fixed gimbal angles, the wheel momentum (which is nominally along the -y table axis) will have an inertial rate equal to minus the Earth rate. To hold the wheel momentum vector fixed relative to the table, the gimbal motors must exert a torque on the table equivalent to $\omega_e \times H_B$, where ω_e is Earth rate. The table control system opposes this torque by generating a table attitude error, which, because of the body control loops, will cause the gimbals to precess at a rate that holds the wheel momentum vector fixed in inertial space.

As a means of preventing the table attitude error offset, the gimbals can be driven in an open-loop fashion given sufficient information with respect to the gimbal axes relative to the Earth rate vector and the orbit angle μ. The inertial rate of the momentum vector for an orbit rate ω_o with small roll and yaw gimbal and body angles is

$$\omega_{Hx} = \dot{\phi} - \omega_o \psi + \omega_{ez} \sin\mu + \dot{\gamma}_{ox}$$
$$\omega_{Hy} = \dot{\theta} - \omega_o - \omega_{ey} \qquad (7)$$
$$\omega_{Hz} = \dot{\psi} + \omega_o \phi + \omega_{ez} \cos\mu + \dot{\gamma}_{iz}$$

Clearly, to compensate for Earth rate, the following decoupling terms must be added to the gimbal commands for $\mu = \omega_o t$:

$$\gamma_{xd} = - (\omega_{ez}/\omega_o)(\cos\mu - 1)$$

$$\gamma_{zd} = - (\omega_{ez}/\omega_o)\sin\mu$$

where $\omega_o = 3.5 \times 10^{-3}$ rad/sec. The generalized torque disturbance compensator is used to implement these correction terms. If compensation is not applied, a cyclic yaw error with a maximum value of 1° will occur.

Pitch Dynamics and Control

For the purpose of control system design, the pitch dynamics are expressed as

$$I_y \ddot{\theta} = -T_w + T_{dy} + T_f + T_{wd} \qquad (8)$$

The wheel dynamic equation is

$$I_w \dot{\Omega} = -T_f + T_w - T_{wd} \qquad (9)$$

The pitch control is a relatively straightforward application of a lead filter with an on-off modulator. Modifications to the filter allow a rate-limiting feature, which improves large signal error performance. A block diagram of the wheel-pitch control system is shown in Fig. 4. The effects of sensor or electronic pickup noise are minimized by a low pass filter and

AIR BEARING TEST OF ATTITUDE-CONTROL SYSTEM

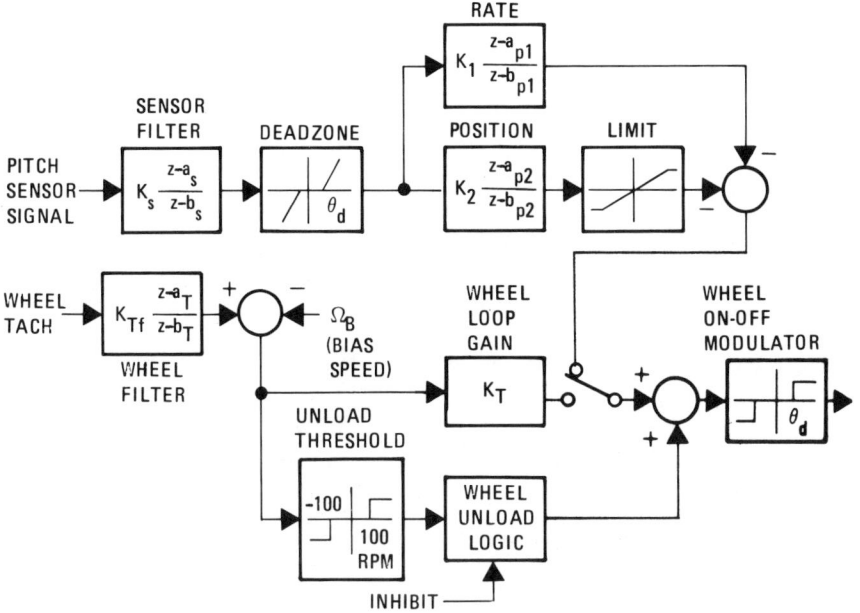

Fig. 4 Wheel/pitch control mechanization.

a deadzone. The modulator deadzone is selected equal to this deadzone. A continuous system version of this modulator would have hysteresis to avoid chatter at low signal levels; however, the computation and sampling delay are sufficient to avoid this problem.

Wheel speed control is utilized only to bring the wheel up to bias speed or during thruster control modes. Speed control is not used during normal DGMW control. The modulator presents a very large stiffness with respect to disturbance torques; therefore the control does not require torque compensation or other corrective action.

Gimbal Control

The hybrid gimbal control system developed for the DGMW (Fig. 5) is an automatic-switching dual-mode control in which the specific operational mode is a function of wheel speed. Zero and low-wheel-speed (<17 Hz) gimbal control is a conventional direct servo with lead filter compensation. High-wheel-speed (>17 Hz) control is performed with a precession control[4] similar in concept to that used for spinning spacecraft. The dual-mode approach allows well-behaved gimbal control over the entire wheel speed range and eliminates the need for a conven-

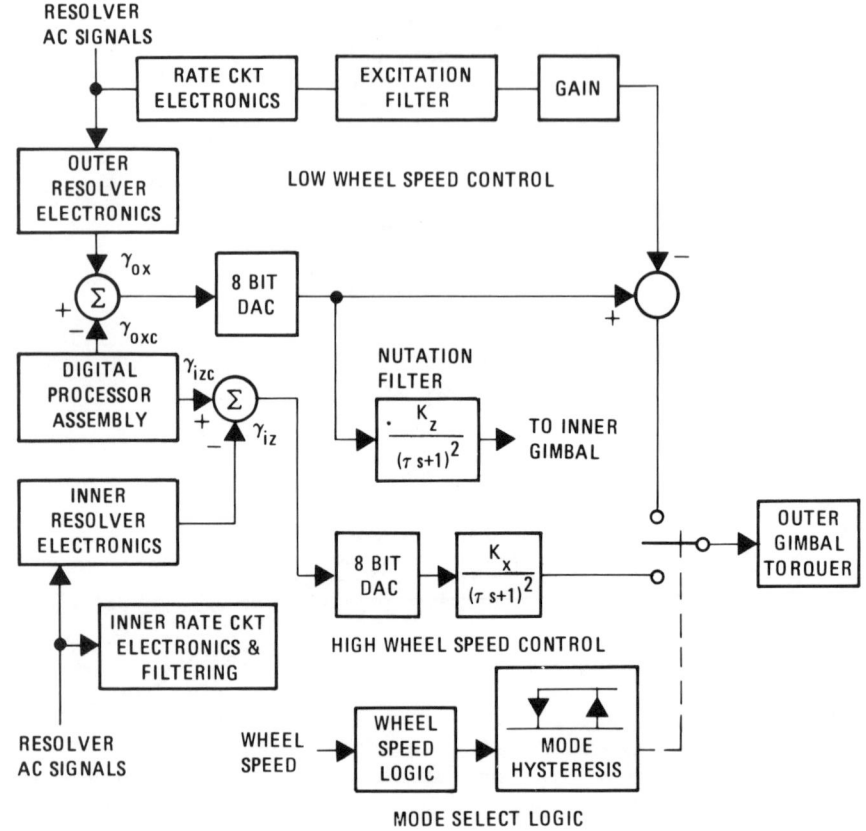

Fig. 5 Roll (outer) gimbal control mechanization.

tional mechanical gimbal tachometer. The removal of the tachometer allows greater freedom in mechanical assembly, a smaller yoke configuration, and reduced cost.

Gimbal pointing commands originate in the processor and, after being summed digitally with the resolver signals, are transmitted through a DAC to the analog electronics. The scaling of the error signal is performed before entering the DAC by left-shifting the data. The scale factor selected results in a least significant bit (LSB) weighting that is large enough so that errors generated by the drifts and biases of the operational and power amplifiers are minimal.

Limiters following the DAC act as rate commands during large position error signals and limit the slew rate to acceptable levels. The slew limiting, in turn, prevents control torque saturation limit cycles, which could occur as a result of the large error signals. This is particularly critical in the low-speed mode because of the relatively large cross-coupling torques between gimbal axes.

A primary difference between the direct servo (low-speed mode) and the precession[7] (high-speed mode) control is the cross-strapping of the gimbal position error signals in the precession control. In addition, with appropriate lag filtering to suppress the DGMW nutation dynamics, the precession

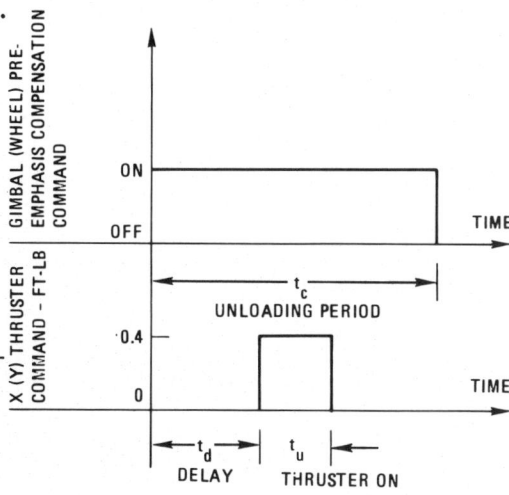

Fig. 6 Wheel or gimbal unloading sequence.

control does not require a rate signal or lead filters. This feature allows a dramatic reduction in control error noise and gimbal torquer heating.

Gimbal and Wheel Momentum Unloading Technique

The unloading of the wheel or the gimbals is performed using the pulse sequence given in Fig. 6. Accelerating the wheel or gimbal before the thruster pulse is applied allows a body rate to be built up which will oppose that produced by the thruster firing. This reduces the maximum pointing error for a given momentum unloading, thereby allowing a longer thruster pulse to be used, and hence an increase in the momentum dumped. Through the use of the pre-emphasis technique, high-level thrusters, necessary to control the ΔV stationkeeping disturbances, can be used to unload the gimbals and the wheels, removing a possible requirement for a special set of thrusters to perform only unloading. Clearly, this technique need be used only when the thruster torque exceeds the maximum wheel or the gimbal reaction torque.

To suppress dynamic effects produced by the body control laws, the unloading sequence should be completed well within the time constant of the pitch or roll control. Specific unloading parameters are obtained with a knowledge of the body inertia, disturbance torques, and the largest acceptable position error during unloading. The thruster pulse is roughly centered with respect to the pre-emphasis pulse.

Gimbal unloading is performed whenever the yaw gimbal exceeds $2°$ by firing the table roll axis thruster. Air bearing test maximum body torques are estimated at 3×10^{-4} ft-lb which produce a peak cyclic momentum of 0.8 ft-lb-sec, equivalent to $1.6°$ gimbal angle. The major secular torque is due to Earth rate and produces a momentum build up per 1800 sec orbit of $\Delta H_s = \omega_{ez} H_B t = 3.3$ ft-lb-sec, which is equivalent to a gimbal angle of $\Delta \gamma = \omega_{ez} t = 6.3°$. Simulations have shown that, by setting the unloading at $2°$, slightly greater than the cyclic component, there is sufficient time to dump the secular component completely during each orbit.

Wheel unloading for the ABT is executed for wheel speed variation of ±100 rpm off the bias. The maximum pitch secular torques due to air ball turbine and room air currents are estimated to be less than 2×10^{-4} ft-lb. If 15 rpm is removed every unloading cycle, for the air bearing tests the time between cycles is estimated at 7500 sec, or roughly once every four orbits. To insure pitch momentum unloading during a data run, an intentional disturbance was imparted to the table.

Test Control Design

Air Bearing Test Design Goals and Constraints

The gimbal pointing design goals are $0.022°$ in each axis during normal pointing and $0.10°$ in roll during yaw gimbal unloading. Additional design conditions are 1) maximum uncompensated disturbance torque in each axis 3×10^{-4} ft-lb, 2) gimbal command resolution of $0.011°$, 3) maximum gimbal bandwidth less than 10 rad/sec at operating wheel speed, and 4) maximum gimbal rate less than 0.25 deg/sec at operating wheel speed. The command resolution is the result of processing the data at 12 bits and restricting the maximum angle at ±22.5°. The disturbance torques are primarily due to the resolution in balancing the table. The gimbal bandwidth and rate limitation result from torquer heating considerations, since the gyroscopic cross-coupling torques (the major load) on the gimbals are relatively large at operating wheel speeds. Gimbal angle errors must be held to a minimum in the no-yaw sensor mode, because if $\Delta \gamma_x$ and $\Delta \gamma_z$ are the gimbal position errors, the yaw error is

$$\psi_e \leq \Delta \gamma_z + K_3' \Delta \gamma_x \qquad (10)$$

Design accuracy goals for the air bearing test roll/yaw control correspond to those of the satellite application given in Table 1.

Given a specific momentum bias, the selection of gains is a direct function of the yaw axis no-yaw sensor bandwidth and the gimbal unloading parameters.[3] Fundamentally, for the no-yaw sensor mode the roll and yaw bandwidths must be widely separated to assure reasonable yaw damping. A factor of 10 is acceptable in most cases. The no-yaw sensor bandwidth is

$$\Omega_\psi = (\omega_o H_B / I_{zz})^{\frac{1}{2}}$$

Roll/Yaw Control

The gain K_3 is set to yield reasonable yaw damping, with an additional constraint set by the LSB of the gimbal command. A review of Eq. (4) shows that $\dot{\gamma}_{iz} \simeq \dot{\psi}$ given ideal compensation of the orbit rate coupling terms and disturbance torques. Therefore, the term $K_3 \dot{\gamma}_{iz}$ in Eq. (5) represents the means by which the yaw damping is transmitted to the body. If $q_{\gamma x}$ represents the LSB weight of the γ_{ox} command, then the effective quantization of the yaw damping signal is $\Delta\psi_e \leq H_B q_{\gamma x}/H_0 K_3'$, which implies no yaw damping within the incremental bounds of $\Delta\psi_e$. Since some disturbance torques will exist, the yaw error can be expected to wander within this bound about some nominal level. For typical ABT physical parameters, $K_3' \simeq 0.1$. (It is desirable to hold K_3' to the smallest value possible to minimize the effects of disturbances and gimbal errors and to insure an adequate roll/yaw transient response.)

The yaw bandwidth is Ω_ψ = 0.055 rad/sec. Roll bandwidth is selected to be Ω_ϕ = 0.55 rad/sec. Furthermore, a gimbal bandwidth Ω_γ of Ω_γ = $10\Omega_\phi$ = 5.5 rad/sec allows sufficient separation between the gimbal and body controls without leading to excessive gimbal drive power. The rate gains are selected to yield critical damping. Deadzones are set at $0.02°$.

Pitch Control Design

The pitch control design is relatively simple, as a result of using a bang-bang modulator. The control has large stiffness with respect to disturbances due to modulator gain; hence the gain selection is primarily a matter of satisfactory transient performance.

Pitch Wheel Lead Filter Design

System stability is obtained through the use of a digital lead filter. Since the bandwidth of the pitch performance is much lower than the sampling rate, a useful estimate for transient performance is

$$\theta(t) = \theta(t_o)e^{-(1/K_r)(t-t_o)} \qquad (11)$$

A time constant in the range 1-10 sec is acceptable for the air bearing system test. The accuracy of the estimate is reasonable until K_r forces the bandwidth to be larger than that compatible with the maximum wheel torque available. This condition is satisfied if

$$K_r \leq \sqrt{I_y \theta(t_o)/(T_w + T_d)} \qquad (12)$$

DGMW Dynamics Model

A number of tests on the DGMW hardware were made to gather sufficient inertia and compliance data to construct a mathematical model[5] that could be used in the control system design and analysis. A model was selected assuming small angle and rate motion of the form shown in Fig. 7. The system equations representing the gimbal compliances and gyroscopic coupling are given in Ref. 7.

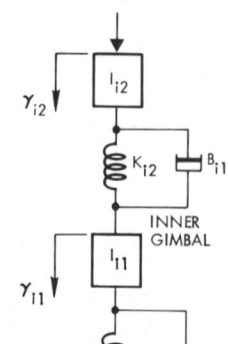
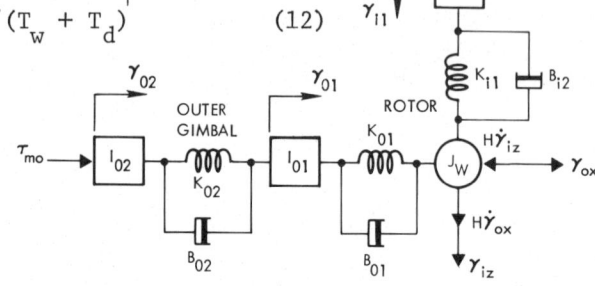

Fig. 7 DGMW gimbal compliance model.

Gimbal viscous damping is difficult to measure directly in that it is a function of the bearing properties, static friction, and the wheel-induced vibrations. The damping information is necessary before an accurate means of computing control stability at low frequencies can be gained. Viscous damping was obtained by exciting the gimbal nutation at several wheel speeds with no gimbal control and computing the frequency and the decay of the gimbal rate perturbation. The real part of the nutation root α in most cases varied between 0.1 and 0.3 rad/sec. Comparisons to the mathematical model at these speeds (see Fig. 8) indicate a gimbal damping term of $B_o = B_i = B = 0.015$ ft-lb-sec. With the selection of gimbal damping, the comparison of the test data with the model showed very good correlation, indicating that a satisfactory model had been selected. The precession control gains were sized using the model

$$\begin{aligned} I_o \ddot{\gamma}_{ox} + H_B \dot{\gamma}_{iz} + B \dot{\gamma}_{ox} &= \tau_{mo} \\ I_i \ddot{\gamma}_{iz} - H_B \dot{\gamma}_{ox} + B \dot{\gamma}_{iz} &= \tau_{mi} \end{aligned} \qquad (13)$$

Gimbal Control Design

The DGMW design considerations that led to the precession control concept are 1) wheel-induced gimbal vibrations, 2) gimbal resolver excitation noise, 3) gimbal nutation stability, 4) open-loop gain dynamic range requirements, and 5) gimbal pointing error goal of $0.022°$. Wheel-induced vibrations are due to spin-bearing runout[6] and produce a gimbal mechanical rate motion at spin frequency of roughly 0.5 deg/sec. If the rate signal is extracted by analog techniques from the resolver sine-cosine signals, considerable filtering must be wedged in between the resolver excitation frequency of 1000 Hz and the DGMW nutation pole to remove the

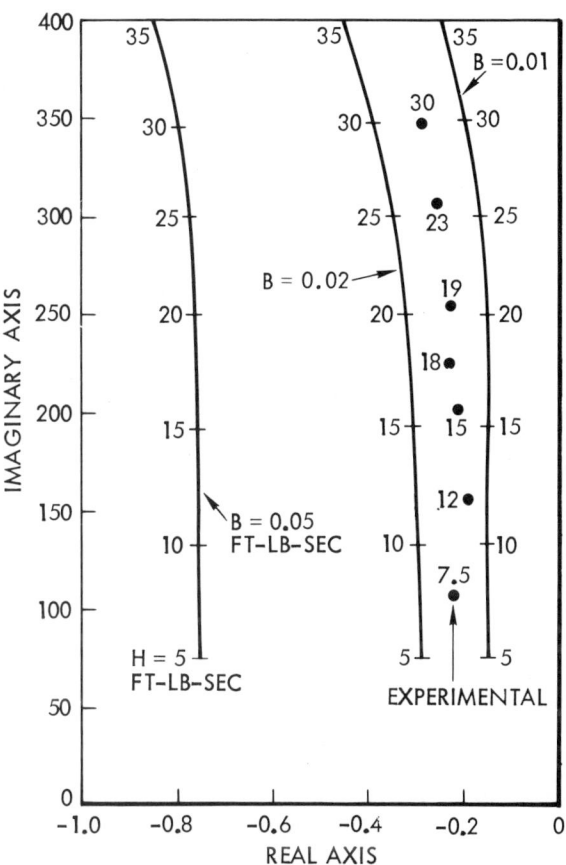

Fig. 8 DGMW no-control nutation pole as function of gimbal damping B and wheel speed. Dark circles indicate experimental data.

excitation noise on the lead filter output. Similar problems exist with a lead filter after the error DAC in that the lead gain and the quantization steps produce large torquer voltages at a relatively high frequency. The high-frequency content, in either case, without proper filtering causes excessive torquer heating. Clearly, since the gimbal precession control has no rate signal feedback, the preceding problems disappear. The control laws can be approximated as

$$\tau_{mi} = -K_x(\gamma_{iz} - \gamma_{izc})$$
$$\tau_{mo} = +K_z(\gamma_{ox} - \gamma_{oxc})$$
(14)

Combining these and solving for γ_{ox} and γ_{iz} yields

$$\begin{bmatrix} \gamma_{ox} \\ \gamma_{iz} \end{bmatrix} = \frac{1}{D(s)} \begin{bmatrix} (I_i s^2 + Bs) & -(H_B s + K_x) \\ (H_B s + K_z) & -(I_o s^2 + Bs) \end{bmatrix} \begin{bmatrix} K_x \gamma_{izc} \\ -K_z \gamma_{oxc} \end{bmatrix} \quad (15)$$

$$D(s) = \text{DET} \begin{bmatrix} s^2 I_o + Bs & (H_B s + K_z) \\ -(H_B \dot{s} + K_x) & I_i s^2 + Bs \end{bmatrix}$$

For large H_B, the denominator term can be approximated as

$$D(s) = [I_o I_i s^2 + B(I_o + I_i)s + H_B^2] \cdot \left[s^2 + \frac{(K_x + K_z)}{H_B} s + \frac{K_x K_z}{H_B^2} \right] \quad (16)$$

Although this often is not accurate at small H_B, the lower frequency root estimate proves to be good for determining the magnitude of the required control gains at the nominal high operating speed.

Control gains are selected with considerations for unloading disturbance, bandwidth, and stability. When the unloading commands are executed, the outer gimbal motor must produce the load torque of $H_B \dot{\gamma}_{izL}$, where $\dot{\gamma}_{izL}$ is the unloading rate. Since the gimbal torque is a function of the gimbal error $\Delta \gamma_{ox}$, the position gain must be sized such that $K_z \geq H_B \dot{\gamma}_{izL}/\Delta \gamma_{ox}$. To have a suitable response during unloading and to have sufficient separation from the roll and yaw controls (less than 0.6 rad/sec bandwidth), a bandwidth of $\Omega_\gamma = 5.5$ rad/sec was considered acceptable. The control gains necessary to satisfy this were determined from the control root approximation such that $K_x K_z \geq H_B^2 \Omega_\gamma^2$.

A simple cross-position feedback is not satisfactory for loop stability over the operating wheel speed range of the control. This is true for even the simple dynamic model of Eq. (15). For example, the nominal parameters yield the roots

$$D = (s - 4.7 \pm j\, 424.)(s + 4.99 \pm j\, 0.0056) \quad (17)$$

which indicate that the control roots are roughly those estimated, but the nutation pole is unstable. Stability is provided by phase-stabilizing the nutation pole with two first-order filters in each feedback path. The stability effect of three values of filter pole locations are developed in Fig. 9 from the model of Fig. 7. The time constant selected for the design was $\tau = 0.02$ sec. This value insures an acceptable stable wheel speed range, and, since it is a decade above the gimbal control bandwidth at a wheel speed equivalent to H_B = 30 ft-lb/sec, it also insures predictable, well-behaved system performance. Removal of one filter in each loop can be seen to greatly reduce the stability margin.

To insure minimal effects due to gimbal and wheel compliances, the gimbals were constructed to have stiffness greater than 10^4 ft-lb/rad. This placed the wheel/gimbal compliance frequencies greater than 1200 rad/sec for all wheel speeds, allowing ample separation from the nutation and spin frequencies.

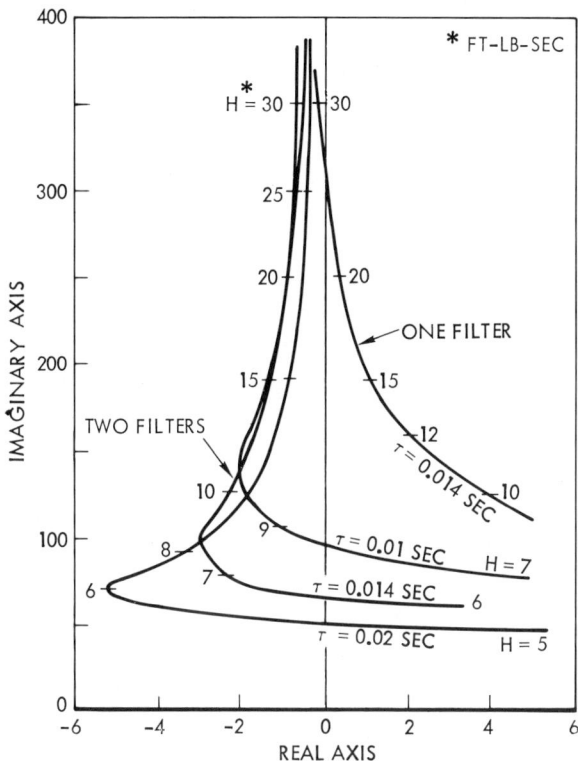

Fig. 9 DGMW nutation pole loci as function of control filter τ and wheel momentum. Loop gain $K_x = K_z = K = 150$ ft-lb/rad.

No-Yaw Sensor Error Prediction

Air bearing table disturbance torques derived by preliminary tests are shown in Table 3. Air currents and turbine torques are mostly about the pitch axis. Table and gimbal imbalance torques fall predominately in the roll-yaw plane. Table balance drift was due to the thermal loading of the table batteries, power converters, and the major electronics packages. ABT runs were not initiated until a relatively steady-state thermal condition was reached; however, in most cases, a balance drift was noted for most runs independent of the length of preheating the table.

Simulation models have shown that, on the average, the total yaw disturbance torques were 3×10^{-4} ft-lb; thus, if a stiffness of $\omega_o H_B \psi$ is assumed, the expected no-yaw sensor con-

Table 3 Disturbance torque summary

Disturbance source	1-σ torque
Room air currents on table, ft-lb	10^{-4}
Ball turbine torques, ft-lb	10^{-4}
Table static balance, ft-lb	2×10^{-4}
Table balance drift (one orbit), ft-lb	2×10^{-4}
Table pendulosity, ft-lb/deg	-4×10^{-4}
Gimbal balance with gimbal angle, ft-lb/deg	4×10^{-5}
Gas supply unbalance torque, ft-lb/lb	5×10^{-4}
Wheel friction torque + windage, ft-lb	5×10^{-3}

trol error is $0.163°$. Roll and pitch control accuracies dependent upon wheel friction and disturbances will be minimal because of the control gains selected. The total root sum squared (rss) error due to control inaccuracies ($0.1°$) and disturbance torques is then $\psi_{eT} = 0.191°$.

Test Facility Description

Test Objectives

System level motion tests were performed to verify 1) computer software/body dynamic interface, 2) subsystem unit level electrical compatibility, 3) operational mode sequencing, 4) DGMW performance margins, 5) telemetry and command interfaces, and 6) gimbal control performance. For performance evaluation, it was the purpose of the test to establish 1) roll, pitch, and yaw minimal pointing errors with compensation for known disturbances; 2) effects of realistic thruster misalignment upon unloading performance; 3) effects of realistic DGMW and sensor misalignments upon normal mode and large yaw angle performance; and 4) effects of hardware implementation anomalies such as electrical interference, computer quantization, computation delay, power supply regulation, and thruster delays.

Facility Description

The air bearing table (Fig. 10) consisted of a rotating pedestal supporting a 10-in. spherical air bearing. The pedestal and ball are capable of levitating a 1200-lb load. The pedestal and an Earth stimulus support boom were mounted to a common bearing and structure driven by a geared d.c. motor with 1% speed regulation capability.

AIR BEARING TEST OF ATTITUDE-CONTROL SYSTEM

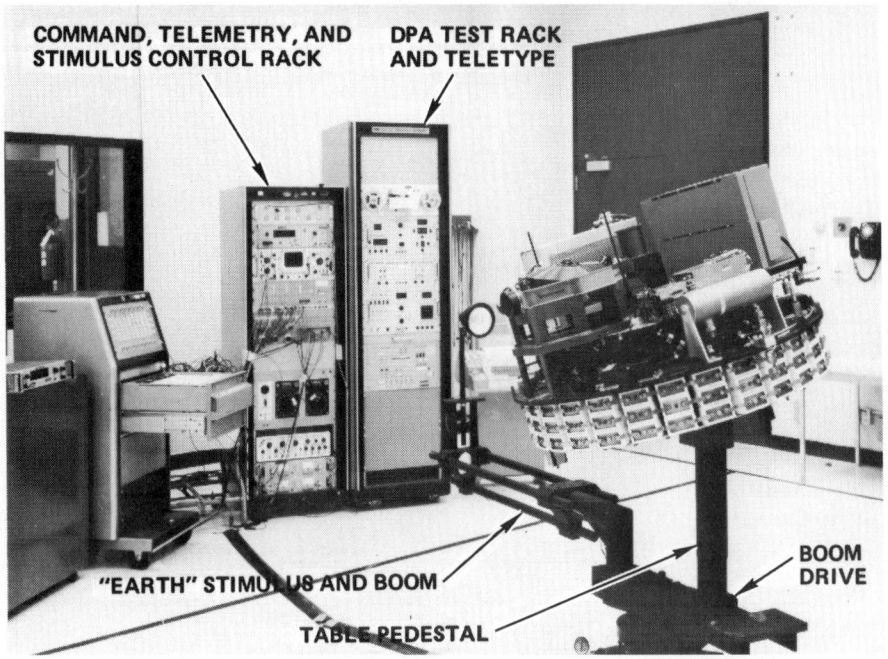

Fig. 10 DGMW air bearing system test facility.

The roll, pitch, and yaw sensors are differentially operated solar cells with linear ranges roughly $\pm 1°$, with proportional ranges out to $\pm 3°$. The stimulus for the roll/pitch sensor was a floodlamp mounted on the Earth stimulus boom. The yaw sensor was fixed to the ceiling directly over the yaw sensor.

The DPA, DGMW, and necessary support electronics were mounted (see Fig. 2) on the table and powered by a 28-v battery system comprised of three parallel strings of 23 Gould P23 nicad batteries for total capacity of 69 amp-hr. The useable voltage range was 24-28 v. The table average electrical load was 200 w, with peaks of 670 w. The table gas system was dry nitrogen held in two symmetrically placed bottles with a total capacity of 1.6 lb at 3000 psig.

An rf command system interfaced directly to the DPA with the capacity of 16 words of 16 bits each at a rate of 100 kbit/sec. Telemetry originating in the DPA consisted of 16 words of 16 bits and was transmitted serially at a rate of 2 - 100 kbit/sec. Ground-processing equipment commutated the serial telemetry stream into serial words with parallel bits, which were D/A converted and displayed on recorders. The overall air

Table 4 Air bearing table and DGMW physical properties

Table	
I_{xx}, slug-ft^2	28
I_{yy}, slug-ft^2	47
I_{zz}, slug-ft^2	34
I_{xy}, slug-ft^2	0.483
I_{xz}, slug-ft^2	0.127
I_{yz}, slug-ft^2	0.305
Table weight, lb	900
DGMW	
I_o, slug-ft^2 (total)	0.1022
I_i, slug-ft^2 (total)	0.0440
J_w, slug-ft^2	0.0314
I_w, slug-ft^2	0.0605
Ω_B, Hz	80
Thruster	
Torque, ft-lb	0.4

bearing power and signal flow appear in Fig. 11. A summary of the air bearing table physical characteristics appears in Table 4.

Air Table Balance

Table balance was obtained in two stages, starting with a coarse balance with the table electronics inoperative. The second stage was a fine balance performed with the three-sensor DGMW system operative. The slope of the gimbal angle signals gave a direct indication of the table balance to accuracies better than 10^{-4} ft-lb. Pendulosity measurements were made by tilting the table 10° and again monitoring the gimbal history. Manual adjustments for balance were made by a micrometer with resolution of 0.1×10^{-4} ft-lb, with a three-axis balance taking 10-15 min. Although balances less than 10^{-4} ft-lb could be obtained, the balance drift (a result of a complex thermal mechanism) was 2×10^{-4} ft-lb in 30 min.

Table and DGMW Mechanical Alignment

The table reference axes were based on a 3-arc-sec cube placed at the center of rotation of the table. The table position sensor, the thrusters, and the DGMW were mounted with respect to the cube and local vertical. Alignment of the table consisted of adjusting the table mounting screws to insure that the plane of the "Earth" stimulus and the table would be perpendicular to the local gravity vector. The table pedestal and

AIR BEARING TEST OF ATTITUDE-CONTROL SYSTEM

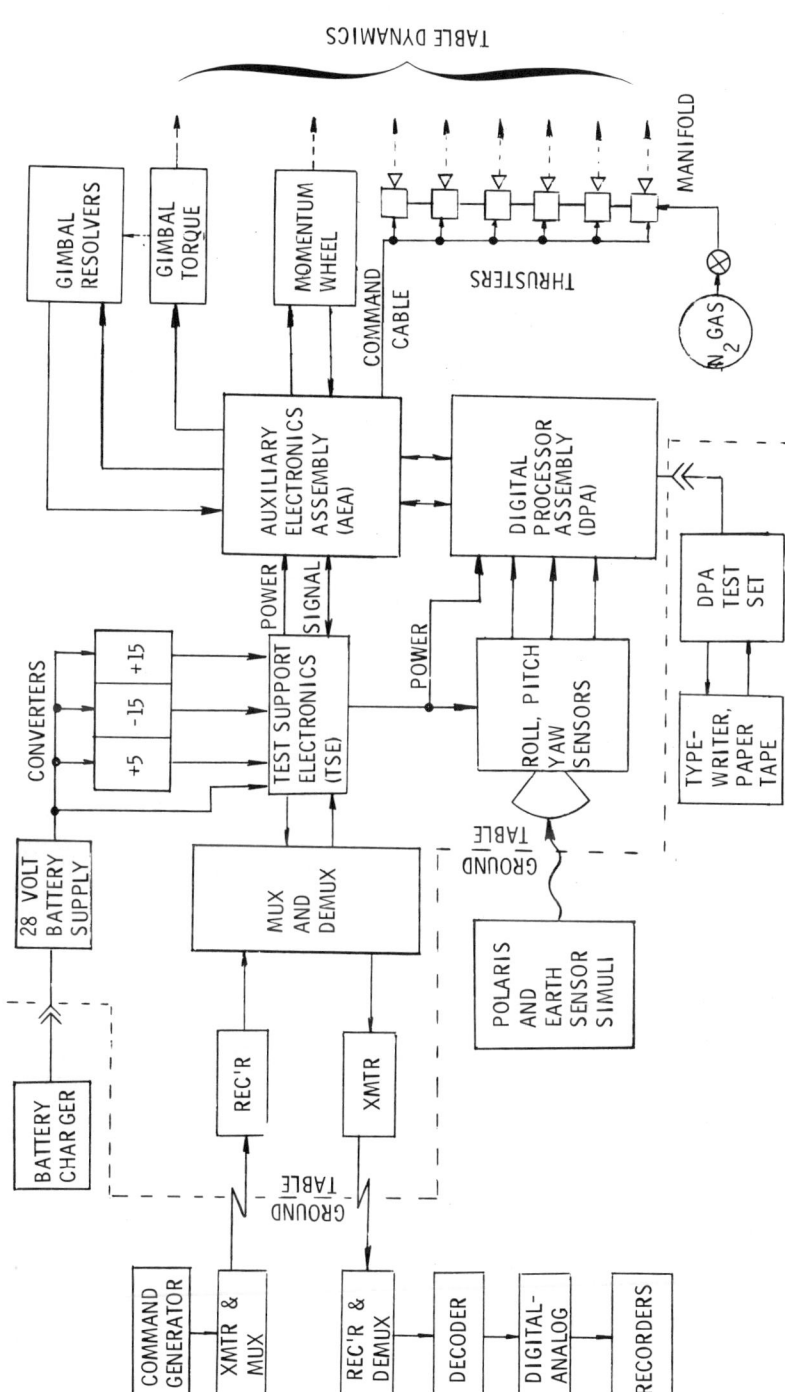

Fig. 11 Air bearing test signal flow diagram.

Table 5 ABT alignment and sensor errors

Sensor	Maximum alignment error	Signal error[a]
Roll/pitch sensors	180 arc-sec at yaw angle of 45°	Less than 0.01°
Yaw sensor	± 180 arc-sec	Less than 0.01°
DGMW	± 30 arc-sec	Both gimbal angles less than 0.05°
Thrusters	Less than ± 2°	N/A

[a]Telemetry readout error less than signal errors.

the Earth stimulus were mounted on large conventional ball bearings, which had a runout of 2 arc-min. It was possible to align the table and the yaw stimulus until this was the dominant error. The absolute mounting accuracies of the sensors, DGMW, and the table are shown in Table 5, together with the signal accuracies.

Test Results

DGMW bench tests were performed initially to determine control performance at both low and high wheel speeds. Of interest was the gimbal nutation stability at the high wheel speeds. Body control tests followed, with emphasis on relative pointing accuracy and unloading performance without a yaw sensor. The functional modes for the test were 1) roll, pitch, and yaw sensing with wheel and gimbal unloading; 2) no-yaw sensor control with wheel and gimbal unloading; and 3) large yaw angle transient response in the no-yaw sensor control mode.

Gimbal Precession Control

Figure 12 demonstrates the gimbal control response with the wheel at 80 Hz for a 5.6° gimbal command. The entrance into and exit from the constant rate mode occur with a first-order (overdamped) characteristic. Coupling between channels is virtually nonexistent. Relative electrical error accuracy is ±1 LSB or 0.011°.

Gimbal Nutation Stability

Figure 13 represents closed-loop precession control with the wheel speeds at 40 and 80 Hz. The gimbal yoke was struck to excite the nutation dynamics as a means of checking the control stability at the higher frequencies and to gather model data. Decay time for each case is roughly 4-5 sec, indicating (as predicted by the model) that the real part of the nutation root is virtually unaffected by the control loop or the wheel speed. This indicates that little gimbal power goes into controlling the nutation frequencies.

Three-Axis Body Control with Yaw Sensor

Typical transient performance shown in Fig. 14 for $1°$ commands illustrates the success in decoupling roll and yaw through the control mechanization. In either roll or yaw, the error commands are sufficient to saturate the gimbal rates; however, the slew-limiting produces a well-controlled response over the complete range of error signals. Response is overdamped, with time constants equivalent to those set by the original design. The Earth stimulus remained fixed for this test.

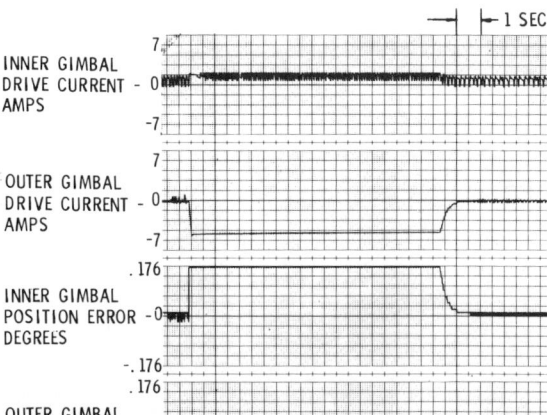

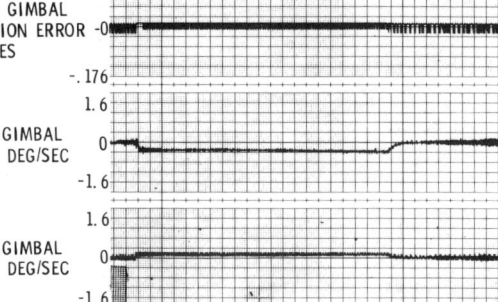

Fig. 12 DGMW response to a command = $5.6°$ at wheel speed = 80 Hz.

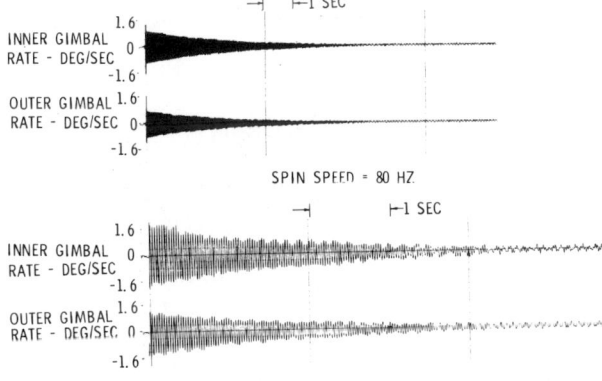

Fig. 13 Gimbal control response to an excited nutation oscillation for wheel speeds of 40 and 80 Hz.

A two-orbit Earth stimulus tracking run is shown in Fig. 15 with a demonstration of wheel and gimbal unloading. Earth rate compensation is used, as well as compensation for known table unbalance torques. The errors in all axes during periods of no unloading are

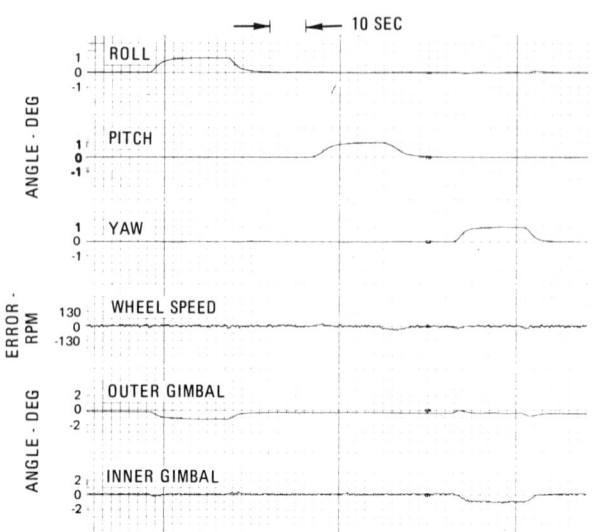

Fig. 14 Three-sensor control response to step commands.

well within $0.03°$. The unloading gains are sized to yield $0.05°$ errors, which are the levels seen in the figure.

No-Yaw Sensor Control

The normal Earth-pointing mode is shown in Fig. 16 for approximately 1.5 orbits. Roll and pitch performance between unloading intervals seldom reached levels greater than $0.03°$. Roll pointing errors are slightly greater than those with the yaw sensor, because of the increased coupling from the yaw axis. Pointing errors, nevertheless, remain less than $0.05°$.

As predicted by analysis, yaw error drifts between $\pm 0.1°$, when large torques are present such as during gimbal unloading. The peak yaw errors for this run are within $\pm 0.2°$. The momentum transfer between roll (outer) and yaw (inner) gimbals is that expected, with the outer gimbal trailing the inner by 1/4 orbit. Pitch/wheel performance is adequate, with pitch errors less than $0.03°$. Pitch disturbance torques, which are roughly 10^{-4} ft-lb, are insufficient to cause unloading during the run.

Capture from Large Initial Yaw Errors

The test yaw sensor field of view is roughly $\pm 5°$, the proportional region is $\pm 3°$, and the sensor is linear for angles less than $2°$. These characteristics make a complete demonstration of the large-angle yaw performance difficult, but basic features can be shown. Figure 17 is a case in which the initial yaw error is $5°$. The table starts from rest, with 10 sec necessary to bring it up to speed. The decay to $0.5°$ is roughly 150 sec. Ideally, this should be accomplished in $t_d = 5/\Omega_\psi$, where yaw bandwidth $\Omega_\psi \simeq 0.055$ rad/sec. The discrepancy is due to the negative pendulosity.

The yaw (inner) gimbal response compares well with the theoretical results, by increasing with the yaw angle until $2°$

AIR BEARING TEST OF ATTITUDE-CONTROL SYSTEM

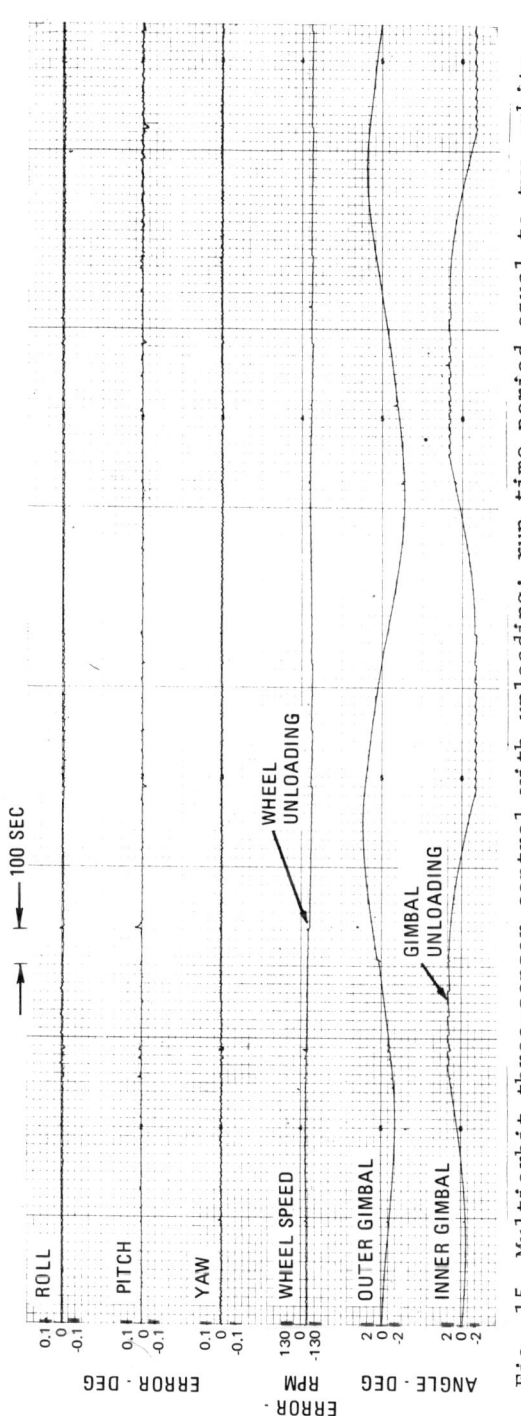

Fig. 15 Multiorbit three-sensor control with unloading; run time period equal to two orbits.

is reached and then initiating an unloading cycle until the yaw angle is less than $0.2°$. The roll error as a function of yaw rate can be expressed by the relationship $\phi_\varepsilon = \dot\psi/K_\phi \cong \gamma_{oz}/K_\phi$. From the yaw (inner) gimbal data, the gimbal rate before unloading is roughly 0.06 deg/sec, $K_\phi = 0.288$; thus the estimated roll error would be $0.208°$. This agrees closely with the observed results.

Conclusions

The air bearing test has shown that, with appropriate considerations for table imbalance and Earth rate coupling, the DGMW hardware and software mechanization concept is a viable three-axis control technique. This conclusion is based upon a test with a full complement of possible performance degrading hardware elements.

Pointing accuracy with near-perfect three-axis attitude sensors was better than $0.03°$ between periods of wheel or gimbal unloading, demonstrating that, given a sufficiently tight body control, the effect of the relatively large dis-

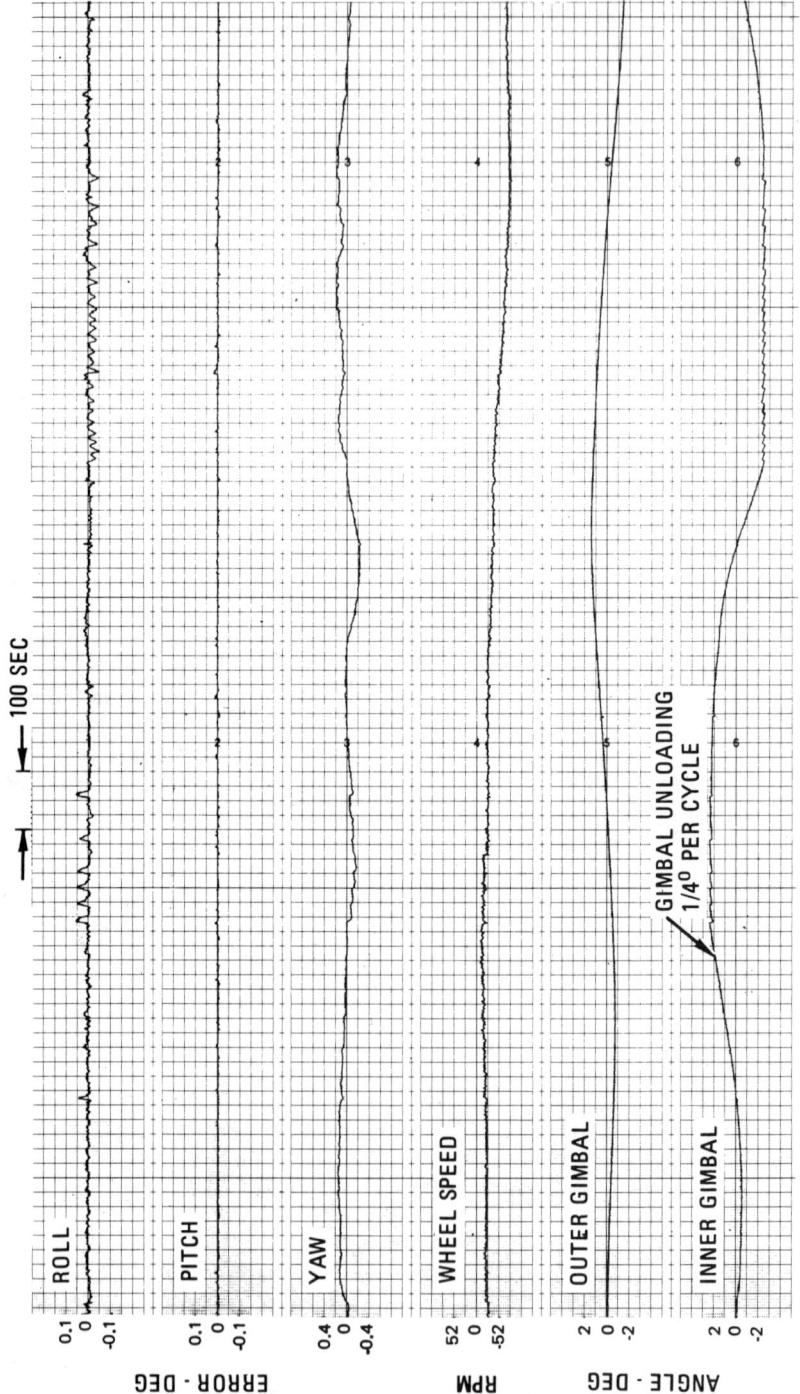

Fig. 16 Nominal no-yaw sensor control with unloading.

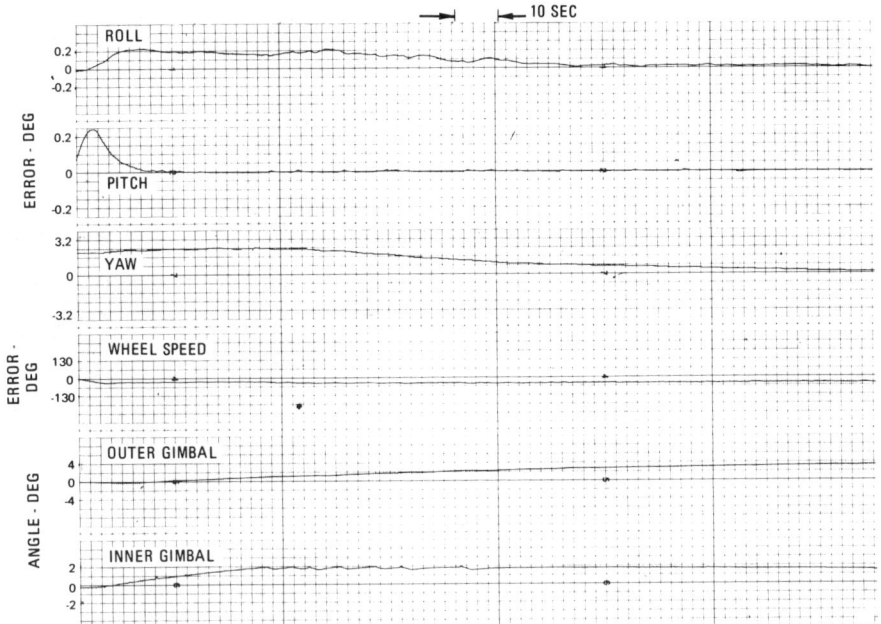

Fig. 17 No-yaw sensor control response from initial yaw angle of 5°.

turbance torques could be minimized. This demonstrated, among other things, that the processor did not restrict the level of accuracy obtainable by the system. Unloading performance, executed in real time, for the wheel or gimbal was very close to that predicted by analysis, with roll and pitch pointing errors held to less than 0.05° while significant increments of momentum were dumped.

The tests demonstrated that active electronic gimbal caging could be performed at any wheel speed. In addition, gimbal rate tachometers were not required, but they could be replaced, when required, by lead filters operating on the gimbal angle. The gimbal precession control performance, with appropriate stabilization, was quite predictable, producing a minimum of torquer heating and no noticeable coupling between gimbal axes or with the body control systems.

Normal mode no-yaw sensor performance was close to that predicted, even though the table was under substantial disturbance loads due to imbalances and Earth rate coupling. Yaw pointing errors were held to less than 0.2°. This is roughly the accuracy lower bound for the present mechanization concept because of the resolution of the gimbal data and Earth rate decoupling accuracy. The present gimbal angle processing is at 12 bits. By expanding to 13 or 14 (the present state of the art for resolver processing), the yaw error could be reduced significantly.

Appendix: Derivation of Spacecraft and Gimballed Wheel Equations of Motion

Denote the momentum of the rigid spacecraft plus the two-degree-of-freedom gimballed reaction wheel by $\bar{H}$. The motion of the spacecraft is determined by the equation

$$d\bar{H}/dt = \dot{H} + \bar{\omega} \times \bar{H} = \bar{T}_d \tag{A1}$$

where T_d is the external torque, $d/dt(\,)$ denotes a time derivative with respect to inertial space, $(\dot{\,})$ denotes a time derivative with respect to the spacecraft body coordinates, and ω is the spacecraft angular velocity with respect to inertial space.

By inspection of Fig. 1, the momentum of the gimballed reaction wheel resolved in body coordinates is

$$\begin{aligned}
H_{wx} &= (\sin\gamma_{iz})H_c \\
H_{wy} &= -(\cos\gamma_{ox}\cos\gamma_{iz})H_c \\
H_{wz} &= -(\sin\gamma_{ox}\cos\gamma_{iz})H_c
\end{aligned} \tag{A2}$$

where the wheel momentum H_c is defined positive along the negative pitch axis for $\gamma_{ox} = \gamma_{iz} = 0$. The system momentum (rigid spacecraft and gimballed wheel) is then

$$\begin{aligned}
\bar{H} = &[I_{xx}\omega_x + (\sin\gamma_{iz})H_c]\hat{x} \\
+ &[I_{yy}\omega_y - (\cos\gamma_{ox}\cos\gamma_{iz})H_c]\hat{y} \\
+ &[I_{zz}\omega_z - (\sin\gamma_{ox}\cos\gamma_{iz})H_c]\hat{z}
\end{aligned} \tag{A3}$$

Substitution of (A2) and (A3) into (A1) yields the three equations of motion:

$$\begin{aligned}
I_{xx}\dot{\omega}_x + \dot{\gamma}_{iz}(\cos\gamma_{iz})H_c + (\sin\gamma_{iz})\dot{H}_c \\
+ \omega_y[I_{zz}\omega_z - (\sin\gamma_{ox}\cos\gamma_{iz})H_c] \\
- \omega_z[I_{yy}\omega_y - (\cos\gamma_{ox}\cos\gamma_{iz})H_c] = T_{dx}
\end{aligned} \tag{A4a}$$

$$I_{yy}\dot{\omega}_y + \dot{\gamma}_{ox}(\sin\gamma_{ox}\cos\gamma_{iz})H_c + \dot{\gamma}_{iz}(\cos\gamma_{ox}\sin\gamma_{iz})H_c$$
$$- (\cos\gamma_{ox}\cos\gamma_{iz})\dot{H}_c + \omega_z[I_{xx}\omega_x + (\sin\gamma_{iz})H_c]$$
$$- \omega_x[I_{zz}\omega_z - (\sin\gamma_{ox}\cos\gamma_{iz})H_c] = T_{dy} \qquad (A4b)$$

$$I_{zz}\dot{\omega}_z - \dot{\gamma}_{ox}(\cos\gamma_{ox}\cos\gamma_{iz})H_c + \dot{\gamma}_{iz}(\sin\gamma_{ox}\sin\gamma_{iz})H_c$$
$$- (\sin\gamma_{ox}\cos\gamma_{iz})\dot{H}_c + \omega_x[I_{yy}\omega_y - (\cos\gamma_{ox}\cos\gamma_{iz})H_c]$$
$$- \omega_y[I_{xx}\omega_x + (\sin\gamma_{iz})H_c] = T_{dz} \qquad (A4c)$$

These equations are linearized about a nominal orientation having axis $\hat{x}$ pointed along the spacecraft velocity vector, axis $\hat{y}$ pointed normal to the orbit plane, and axis $\hat{z}$ pointed at the geocenter. The small-angle kinematic relationships are

$$\omega_x = \dot{\phi} - \omega_o\psi, \quad \omega_y = \dot{\theta} - \omega_o, \quad \omega_z = \dot{\psi} + \omega_o\phi \qquad (A5)$$

With the additional approximation that $H_c \simeq H_B$, $\sin\gamma \simeq \gamma$, and $\cos\gamma \simeq 1$, dropping the products of variables, the substitution of (A5) into (A4) yields the linearized dynamical equations

$$I_{xx}\ddot{\phi} + [\omega_o H_B]\phi + [\omega_o(I_{yy} - I_{xx} - I_{zz}) + H_B]\dot{\psi}$$
$$+ \omega_o H_B \gamma_{ox} + H_B \dot{\gamma}_{iz} = T_{dx}$$

$$I_{yy}\ddot{\phi} - \dot{H}_c = T_{dy} \qquad (A6)$$

$$I_{zz}\ddot{\psi} + \omega_o H_B \psi - [\omega_o(I_{yy} - I_{xx} - I_{zz}) + H_B]\dot{\phi}$$
$$- H_B \dot{\gamma}_{ox} + \omega_o \gamma_{iz} H_B = T_{dz}$$

References

[1] Lebsock, K. L., Flanders, H. R., and Rodden, J. J., "Simulation and Testing of a Double-Gimballed Momentum Wheel Stabilization System for Communication Satellites," *Journal of Spacecraft and Rockets*, Vol. 11, No. 6, June 1974, pp. 395-400.

[2] Sorensen, A. A., Elbert, J. D., and Wheeler, P. C., "Spacecraft Digital Attitude Control," *1972 WESCON*, Sept. 19-22, 1972.

[3] Mork, H. L., "Synthesis and Design of a Gimballed Reaction Wheel Attitude Stabilization Package (GRASP)," AIAA Paper 71-950, Aug. 16-18, 1971, Hempstead, N. Y.

[4]Briggs, R. W., "Stability of a Two-Degree-of-Freedom Gyro with External Feedback," *IEEE Transactions on Automatic Control*, Vol. AC-10, No. 3, July 1965, pp. 244-249.

[5]Maunder, L., "Natural Frequencies of a Free Gyroscope Supported in Gimbals on an Elastic Shaft," *Journal of Mechanical and Engineering Science*, Vol. 3, No. 4, March 1961, pp. 318-323.

[6]Phillips, J. P., "Control Moment Gyro Characteristics and their Effects on Control System Performance," AIAA Paper 68-875, Aug. 12-14, 1968, Pasadena, Calif.

[7]Kalley, J. J., Jr., "Control System Design For Double Gimballed Wheels," AIAA Paper 74-895, Aug. 5-9, 1974, Anaheim, Calif.

SOLAR VS NUCLEAR POWER: IS THERE A CHOICE?

Bernard Raab[*] and Jay J. Karlin[+]

Fairchild Space and Electronics Company, Germantown, Md.

Abstract

The anticipated development of advanced high-performance radioisotope power systems has motivated a new evaluation of their possible utility in Earth-orbit missions. These have been compared, in detailed spacecraft design studies, with current and improved oriented-solar-array/battery systems in commercial synchronous-orbit application. It is concluded that the nuclear systems would be superior economically to current solar systems, including violet cells, and potentially competitive with systems employing nickel-hydrogen batteries. This results in part from significant reductions in housekeeping power and in the weights of other subsystems, which can be achieved with nuclear power.

I. Introduction

All communications satellites to date, and indeed the vast majority of Earth-orbiting satellites of any kind, have been based on the use of solar-array/battery systems for prime electric power. Although nuclear-isotopic power systems, developed primarily for deep-space (or lunar and planetary surface) missions, have been considered periodically for Earth-orbit applications, these have generally been rejected on first

---

Presented as Paper 74-489 at the AIAA 5th Communications Satellite Systems Conference, Los Angeles, Calif., April 22-24, 1974. This work was supported under U.S. Atomic Energy Commission Contract AT(49-15)-3063.

*Program Manager, Nuclear Applications.
+Principal Systems Design Engineer.

principles, mainly for excessive cost and weight as compared to solar-array/battery systems. In general, this rejection has been well justified. With the possible improvement in solar arrays and batteries represented by the violet cells and nickel-hydrogen systems in particular, the case for nuclear power would appear all but hopeless, in particular for commercial satellites where weight and cost are dominant considerations.

However, various improvements in nuclear-isotopic systems also now appear practical. In addition, most previous evaluations of nuclear vs solar power have been based on comparisons of specific weight and cost of the power systems alone, rather than on comparative evaluations of complete spacecraft designs and overall system factors. Therefore, this study was initiated to re-evaluate the possible utility of the advanced nuclear systems in Earth-orbit missions from an overall systems standpoint.

Commercial communications satellites were selected as a primary objective of the study, because this is a basically new class of mission and one likely to generate a continuing requirement for satellites over an extended period of time. Also, relative value could be established more readily on a quantitative basis because of the commercial nature of the mission, permitting comparisons to be based on clear economic factors.

However, in choosing this mission, a built-in handicap to nuclear power was recognized. The geosynchronous orbit is among the most favorable for solar power primarily because of three factors:

1) In the equatorial orbit, the sun angle is confined within $\pm 23.5°$ of the spacecraft axis at all times.

2) Eclipse occurs relatively rarely during the year at this altitude. During maximum eclipse periods, which occur only near solar equinox, solar-array output is maximum for an array oriented along the north-south axis, and about 23 hr are available for battery charging between successive occultations. This tends to minimize both array and battery size. In general, an array oriented along the north-south axis and sized for full mission power at solar solstice automatically will produce enough additional power at solar equinox to charge the batteries when, fortuitously, these are most needed.

SOLAR VS NUCLEAR POWER: IS THERE A CHOICE?

3) The beam-pointing requirements of this mission require a three-axis control system for the antennas. Therefore, the use of an oriented solar array introduces little additional attitude-control requirements to the mission.

The study was based on a detailed design for a 24-transponder solar-powered satellite, designed by a Fairchild-TRW team for launch on the uprated Delta 3914. This design, described briefly in Sec. III, was used as the basis for competing nuclear-powered designs, described in Sec. V. Although various uncertainties can be expected to exist in any purely paper design, however detailed, the comparative evaluation of the two approaches based on the same assumptions is believed to be more significant than the absolute values of either of the designs themselves.

Based on the most optimistic future projections of both solar and nuclear technology, a comparison of power subsystems alone would continue to favor solar power. However, it is shown that reduction in the power consumption and weight of other spacecraft subsystems as a direct result of using nuclear power can result in a totally competitive or superior nuclear spacecraft. The key to economic superiority is the achievement of a net weight advantage for the complete spacecraft. Such a weight saving can be transformed into a system revenue increment of $ 1/2 to 1×10^6/lb. in present value.[1] This revenue potential is seen to be capable of producing a net economic advantage for the nuclear systems, even when compared to the most advanced solar-battery combinations now foreseen.

In summary, and in answer to the question posed in the title, successful development of one or more of the advanced nuclear power systems described in Sec. II would provide the system designer with a commercially viable alternative to solar power in synchronous-orbit communications satellites.

II. Review of Power Systems

Nuclear

Nuclear-isotopic power systems have rendered reliable and predictable service in a number of missions over recent years. For example, SNAP-27 on the lunar surface,[2] SNAP-19 on Pioneers 10 and 11 in interstellar orbits,[3] and Transit/RTG on the Triad Earth-orbit spacecraft[4] all have performed up to or exceeding expectation. Since November 1969, a total of 14 radioisotope thermoelectric generators (RTG's) have been launched and operated in space without a single failure.

The state-of-the-art generators all employ thermoelectric solid-state converters made of doped lead-telluride or silicon-germanium and use plutonium-238 radioisotope fuel in blast-, fire-, and re-entry-proof containers. Despite the rigid fuel containment requirements, current versions of these generators achieve up to 2 w(e)/lb in the range of 150 to 200 w(e)/generator.[5] Converter efficiencies fall in the range of 5 to 7%. Because the plutonium fuel, at $650/thermal watt, has been the most expensive part of these RTG's, converter efficiency has a strong influence on the unit cost of these devices. Therefore, a strong incentive has existed to increase conversion efficiency. Increased conversion efficiency now is believed to be achievable in either of two ways: 1) new high-efficiency thermoelectric materials, or 2) turbine-generator conversion.

The first approach has resulted in the development, under Atomic Energy Commission (AEC) sponsorship, of a new class of thermoelectric materials based on selenide alloys.[6] Current "standard" selenide materials are believed to be capable of achieving 10 to 11% conversion efficiencies reproducibly, with an ultimate efficiency of 13 to 15% expected from this material. Life-testing indicates nondegrading performance with time, promising a long life potential for these selenide materials. Two parallel studies on generator designs utilizing selenide converters recently have been completed, in which overall generator efficiencies of 10 to 11% were predicted based on thermoelectric efficiencies of approximately 12%.[7,8] One of these designs is shown in Fig. 1.

The other approach now believed to be feasible in the several-hundred-watt range uses miniature turbine-generator units. The turbines are driven by gas or super-heated vapor, which has been heated by the nuclear fuel, and which then is returned to the heater in a closed cycle. Low- and fractional-horsepower turbines and a new form of noncontacting hydrodynamic gas-film bearing for the rotating unit have been developed for aircraft auxiliary power units and other applications. Successful long-term operation under multiple start-stop conditions has imparted confidence in the readiness of this approach.

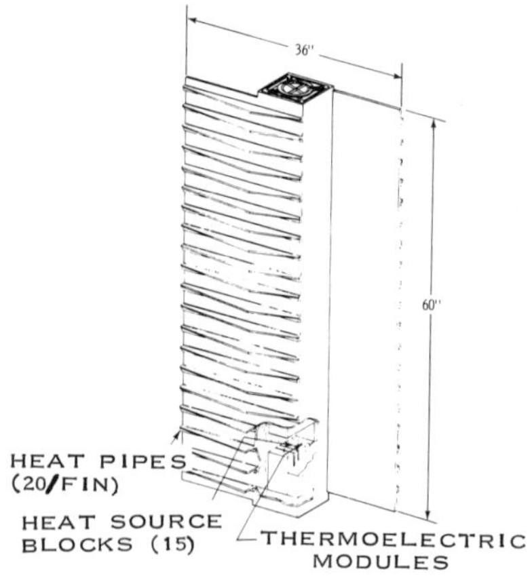

Fig.1 Selenide RTG (courtesy of Teledyne/Isotopes[7]).

SOLAR VS NUCLEAR POWER: IS THERE A CHOICE?

Design studies on each of two different turbine-generator systems were completed recently. One uses a Brayton conversion cycle, in which the working fluid is in inert gas[9]; the other uses a Rankine cycle, in which an organic hydrocarbon working fluid is used.[10] A number of these organic fluids have thermodynamic properties that result in high cycle efficiencies at lower peak cycle temperatures than, for example, steam or metal-vapor working fluids, which were proposed for space applications in former years. A satisfactory body of experience with the applicable organic fluids has been acquired in recent years in experimental automotive, artificial heart, and other engines in a variety of sizes.

Both Brayton and organic-Rankine systems promise overall system conversion efficiencies of approximately 20%. Each has its advantages and particular virtues. However, in the size range of interest to the reference mission, the organic-Rankine system is significantly lighter and was, therefore, selected as the dynamic-system representative in this study.

Figure 2 shows the basic components of the organic-Rankine system. In the nuclear spacecraft, the components are rearranged to fit within the available volume. In particular, the largest single component, the condenser-radiator, is specially designed for the spacecraft, as is discussed in Sec. IV.

Two different radioisotope heat sources were considered in these studies: plutonium-238 and curium-244. The former has been the standard space nuclear fuel; the latter is potentially available in larger quantities and at a lower cost from the growing supply of nuclear-reactor waste products.

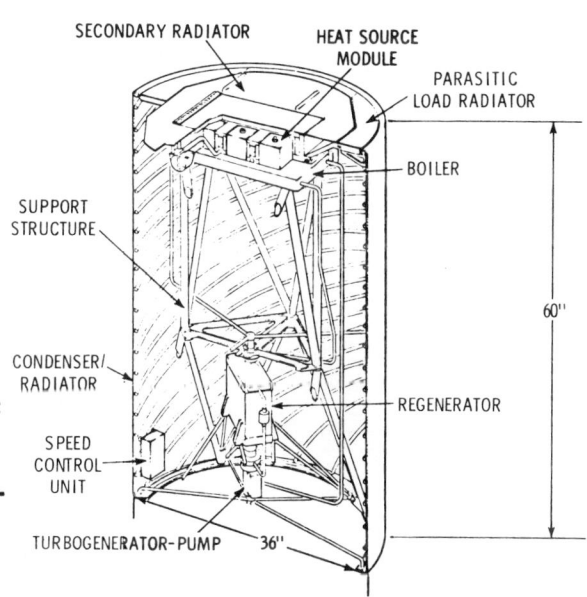

Fig. 2 Organic-Rankine system (courtesy of TRW Systems[10]).

The heat source secondary radiator shown in Fig. 2 is required with a curium heat source and serves two functions: it power-flattens the heat source by radiating excess isotope heat to compensate for natural decay of the curium-244 (18.1-hr half-life); and it provides for standby cooling of the heat source prior to system start-up or in case of system failure. The longer half-life of plutonium-238 (86.4 yr) and the use of a vented heat source obviate the need for this secondary radiator with plutonium fuel.

The major features of each of these power systems are summarized in Table 1 for a nominal power output of 400 w(e) after 5 yr in orbit. In general, the weight numbers shown are useful for scoping purposes; in calculating overall satellite weight (Sec. VI), the manner in which the power system was integrated into the spacecraft ultimately determined the power-system weight.

Concerning costs, it is obvious that the availability of curium-244 fuel at \$100/w(t) would result in sharply reduced unit recurring costs. However, facilities for the separation of this material from reactor wastes are not now available, and no commitment for the construction of such facilities has been made to date. As an alternative, the possibility of reducing plutonium cost while increasing its production rate is being studied under AEC sponsorship. However, the conclusions presented in Sec. VII do not rely on such possible cost reductions.

Solar

The reference-design solar spacecraft was based on conventional current technology. In order to permit longer-ranged comparisons, the possible improvements in solar-array and battery technology have been estimated, as shown in Table 2.

Direct Comparisons

Based on the above information, direct comparisons of solar and nuclear power system weights can be made on a watts-per-pound basis (Fig. 3). Some historical information also is shown for perspective. From Fig. 3, it can be seen that the advanced nuclear systems can compete on a weight basis with state-of-the-art solar-battery systems, even including violet cells.

Table 1 Nuclear power systems major parameters [400 w(e) EOM[a]]

Parameter	Selenide RTG		Org.-Rankine	
Efficiency, EOM	10.5%		20.9%	
Max. cycle temp.,°F	1472		600	
Radiator temp.,°F(BOL/EOM)	300/260		108/108	
Envelope, in.	Box; 60x36x8		Cylinder; D36xL60[b]	
Working material	N&P-type selenides		Hexaflourobenzene	
Fuel	Pu-238	Cm-244	Pu-238	Cm-244
BOL[c] power, w(e)	421	519	400	400
Nominal weight, lb	134	104	161[d]	153[d]
Low-weight version, lb	110[e](80)[f]	(N/A)	90-100[e]	120[d]
Unit recur. cost,[g] K$	3182	1288	2000	665
Excluding fuel, K$	822	822	245	245

[a] End of mission (5 yr).
[b] Determined by radiator surface area requirement.
[c] Beginning of life.
[d] Based on unvented heat sources.
[e] The nominal designs of both power systems were based on minimum cost rather than minimum weight. These were then investigated for variations that could reduce weight. The low-weight version of the organic-Rankine system is based on the use of a beryllium radiator and several other changes of lesser significance. For the plutonium heat source, the fuel must be vented, allowing a sizeable reduction in the weight of this heat source. This reduction possibly could be achieved in curium as well. At present, however, only plutonium heat sources may be said to have flight-proven vents. (All recent RTG's have contained vented plutonium heat sources.) The selenide RTG design was based on a vented heat source in both curium and plutonium versions. The low-weight version is based mainly on the achievement of a smaller envelope at some slight sacrifice in efficiency (~1/2%) by operating at a higher radiator temperature.

[f] The 80-lb selenide RTG is based on a Fairchild study[11] that was initiated with the specific intention of designing a minimum-weight RTG. However, this design was not far enough along at the time of the present study to permit its inclusion in detail.

[g] Based on Pu at $600/w(t) and Cm at $100/w(t), nominal generator version.

Table 2 Solar power system parameters

Solar array: power, 7-7R; solstice/equinox: 580/620 w(e)			
Technology	Conventional hard-mount Si	Flexible roll-up Si	Violet cells
Area, ft^2	86.2	98.2	71.8
Weight, lb	55.9	55.6	44.5

Battery: capacity, 24 transponders plus housekeeping: 618 w-hr		
Technology	Conventional nickel-cadmium	Potential nickel-hydrogen
No. cells per battery	22	6
Depth of discharge, %	65	85
Specific capacity, w-hr/lb	6.5	18.3
Weight, lb	95.6	33.7

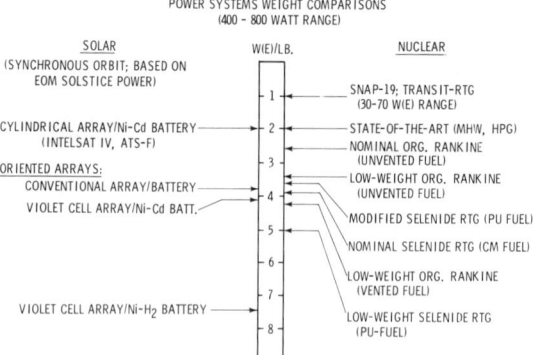

Fig. 3 Specific power of solar and nuclear systems.

SOLAR VS NUCLEAR POWER: IS THERE A CHOICE?

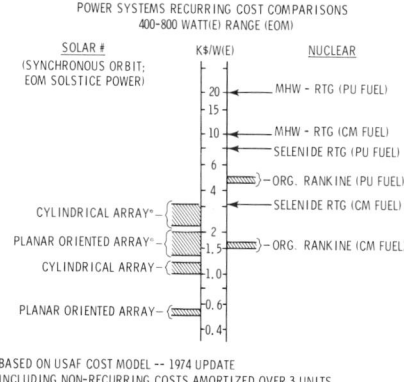

Fig. 4 Specific cost of solar and nuclear systems.

However, the appearance of nickel-hydrogen batteries in the same time frame would re-establish the weight advantage of the solar-battery systems.

The situation with regard to cost is hardly more encouraging (Fig.4). However, there is some justification for including non-recurring costs for solar systems. In general, a new solar-battery system development-test program accompanies each new spacecraft design, whereas the same nuclear system may be adapted to a variety of different spacecraft without redevelopment. For example, the MHW-RTG will be used in Earth orbit (Lincoln Experimental Satellites 8 and 9) as well as in interplanetary missions (Mariner/Jupiter-Saturn).

Although the advanced nuclear systems make a better showing in both cost and weight than the current systems, one clearly would not be justified in considering nuclear power for synchronous-orbit missions based on these direct comparisons alone. However, compared on a systems basis (Secs. VI and VII), quite the opposite conclusions can be drawn.

III. Solar-Powered Spacecraft Characteristics

The basis for this study is a 24-transponder C-band communications satellite designed for domestic service by a Fairchild-TRW design team. It is typical of the new generation of three-axis-stabilized, oriented-solar-array craft and utilizes the upgraded Delta (3914) launch vehicle as the synchronous orbit delivery system. Table 3 lists the major performance parameters of the satellite.

As shown in Fig. 5, the spacecraft is hexagonal in shape. The outer structure consists of honeycomb sandwich panels and framing, connected by graphite-fiber-reinforced-plastic (GFRP) trusses to a beryllium thrust cone. This, in turn, attaches to the Delta launch adapter to form the primary structural load path. The internal faces of the honeycomb panels serve as both structural and heat-transfer surfaces for equipment

Table 3 Major spacecraft parameters

Parameter	Characteristic
Size (main body)	50-in. height; 86 in. across corners
Weight, at launch in orbit	2000 lb 1015 lb
Electric power, solstice	833 w(e) BOL, 580 w(e) EOM, 86.2 ft^2 array, 28 amp-hr
Attitude control	± 0.15°, 3-axis with bias-momentum and solar-array drive assembly
Reaction control	Hydrazine monopropellant: 0.1- and 0.5-lb thrusters
Telemetry and control	C-band
Power distribution	Redundant 20-65-v unregulated buses: 20-33-v regulated buses
Thermal control	Selective coatings plus heater/thermostat units for cold extremes

mounting. Therefore, aluminum sheet with thicknesses that vary between 0.010 and 0.080 in. is used for these internal faces, the latter thickness where heat dissipation is concentrated. The outer facing material of the honeycomb sandwich is 0.040-in. beryllium.

The two large deployable solar array panels are wrapped around the satellite body while the satellite is in a spin-stabilized mode, prior to synchronous orbit injection. The deployed arrays project on booms from the north and south faces. The array panels are clock-driven by a solar-array drive assembly to maintain a near-normal sun-line for maximum solar energy input, whereas the antenna mounting surface is continuously Earth-directed through the gyroscopic action of the momentum wheel.

The hydrazine-fueled reaction control system (RCS) is used periodically to unload the momentum wheel, realign attitude in roll, and correct and maintain orbit station. The tracking, telemetry, and command (TT&C) subsystem operates on C-band with a nearly omnidirectional antenna shown extended from the

SOLAR VS NUCLEAR POWER: IS THERE A CHOICE?

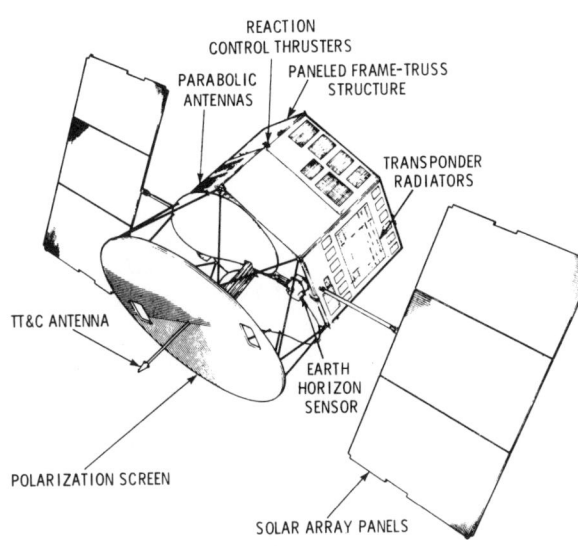

Fig. 5 Solar-powered domestic communications satellite.

center of the polarization grid (Fig.5). It functions throughout the transfer orbit and apogee injection maneuver, as well as at all times on station.

The attitude control system (ACS) operates in two modes: 1) spin-stabilized passively damped with thruster control during transfer, and 2) as a three-axis control system after apogee injection and despin. The satellite has a favorable moment of inertia for spin stability, in addition to being statically and dynamically balanced for the orbit transfer and injection sequence. The 960-lb apogee kick motor (AKM) that protrudes from the aft end (not visible in Fig. 5) provides a velocity increment sufficient to perform plane change and orbit circularization. The burnout weight of the satellite including the spent AKM case is approximately 1015 lb. The electric power system (EPS) and the thermal control system (TCS) are of particular interest to this study, since these are subsystems undergoing major change in converting to nuclear power. Therefore, these are discussed separately below.

Electric Power System

This system provides adequate electric power for 24-transponder operation for 7 yr, including eclipse periods, and employs both nonregulated and regulated redundant buses (Table 3). The regulated bus supplies all "housekeeping" functions, whereas the communications payload is supplied with unregulated power, which then is conditioned within the individual transponder power supplies. A separate 28-v regulator is provided for the thermal-control heaters. Also, most housekeeping modules include additional voltage regulation for their particular requirements.

Table 4 Electrical power subsystem equipment summary

Equipment	Quantity	Total weight, lb
Solar panels	6	55.9
Battery assembly	2	95.6
Power distribution	1	25.4
28-v regulator	1	7.0
Ordnance control	1	2.0
Bus tie	1	1.0
Other components	...	15.4
Interconnection harness	...	48.0

The solar array uses standard n-on-p 2x4-cm silicon cells with 0.010-in. fused-silica cover glass. These are mounted to a fiberglass-faced 1/2-in.-thick aluminum honeycomb substrate. There are two batteries, each with 14-amp-hr nickel-cadmium cells.

The electric power budget is discussed in Sec. VI. Table 4 gives the detailed EPS weight summary. Figure 6 shows the location of the EPS elements relative to the major system elements.

Thermal Control System

The basic approach to equipment thermal control is the use of a variety of surface coatings and heater/thermostat sets. An insulation blanket of alternating layers of Mylar and Kapton covers most external surfaces. Cutouts in the blanket expose second-surface fused-silica mirrors wherever dissipative equipment is mounted, including those portions of the north and south walls on which are mounted the traveling-wave-tube amplifiers (TWTA's) and their power supplies. Heat-load variations on the TWTA radiators result from two major sources: amplifier operation either at saturation (full signal transfer) or zero drive (no signal transfer), resulting in a 30% variation in local heat load; and variations of solar incidence angle of $\pm 23\text{-}1/2°$. These produce gross temperature extremes of $20°$ to $120°F$ on these radiators. Surface heaters are provided to maintain the lower limit, if necessary.

SOLAR VS NUCLEAR POWER: IS THERE A CHOICE?

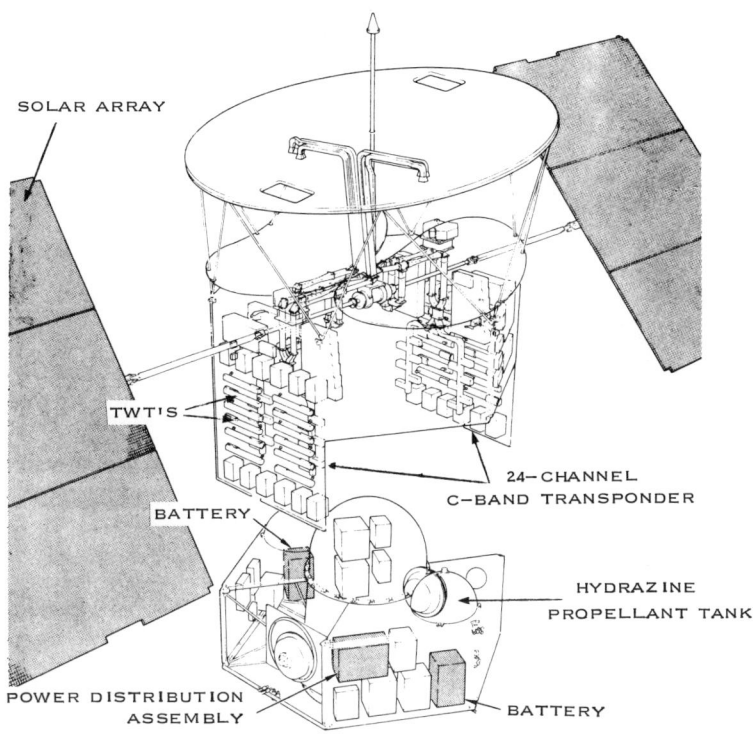

Fig. 6 Equipment locations: solar-powered spacecraft.

Batteries have tighter temperature restrictions than the other components, 10° to 40° F. They are insulated from the interior and given an isolated heat-sink radiator on two of the east-west walls. A charge-rate-regulating thermal switch and heater combination provide active thermal control for the batteries. The east-west panels undergo maximum sun-angle variation, but these support the relatively low-dissipation housekeeping modules and are provided with heaters to maintain the lower temperature limit. Table 5 shows the heater allocations for various regions of the satellite.

Thermal considerations also have influenced the design of satellite appendages. To prevent thermal distortion of the polarizing screen, the antenna array and screen are covered with a 0.003-in. Kapton film coated internally with thermatrol white paint. In addition, the solar array panels must be extended from the north and south faces with booms at least 4 ft in length to avoid blockage of the critical north- and south-face radiators.

Table 5 Heater locations

Location	Number	Watts, each
TWT[a]	24	9.3
TWT power supply[a]	8	7.5
Transponder receiver	2	3.0
Tunnel diode preamp	2	1.5
East/west panels	8	2.5, 3.5, 6.0, 6.5 (2 of each)
Batteries	2	15
AKM	1	12
RCS tanks	2	2
RCS lines	2	1.5
RCS valves	18	0.25
RCS thrusters	18	1.5

[a]Used only in the event of transponder failure.

IV. Nuclear Spacecraft Design Approaches

In designing nuclear-powered spacecraft, the nuclear systems traditionally have been appended to or cantilevered from one face or one end of the vehicle. This is the so-called "separated" approach in which little or no benefit from the waste nuclear heat usually is realized. (An exception is the Mars/Viking-75 lander vehicle, which employs active thermal switches to bleed waste heat from the RTG's into the lander during the Martian night.)

The nuclear Brayton and organic-Rankine systems, however, offer a combination of characteristics which makes a so-called "integrated" design approach possible, wherein the power system radiator is used as the external skin of the spacecraft. The combination of high efficiency ($\geq 20\%$) and low radiator temperature ($< 120°F$) must be achieved simultaneously in order for the integrated approach to be feasible.

SOLAR VS NUCLEAR POWER: IS THERE A CHOICE?

The low radiator temperature makes it practicable to mount the spacecraft equipment within the radiator, thereby achieving a controlled and nearly constant thermal environment for the equipment. This eliminates almost all thermal control problems normally generated by varying sun angles and equipment-dissipation variations. The high power conversion efficiency limits the total thermal power that must be radiated for a given electrical power and hence limits the spacecraft in size and weight to competitively acceptable values.

The integrated design approach first was investigated for the Brayton system in an earlier study for space-shuttle-launched spacecraft.[12] The organic-Rankine system is even more readily adapted to an integrated design, because the condenser-radiator is nearly isothermal, varying only a few degrees between the inlet and outlet ends. (The Brayton-system radiator, not normally isothermal, can be isothermalized deliberately, as described in Ref. 12.) The helically wound radiator tubes (Fig. 2) serve to distribute solar heat input effectively over the entire radiator, hence spacecraft, surface, making all interior surfaces equally available for equipment mounting.

Alternative Approaches

Integrated. The basic fully integrated approach would assume the configuration shown in Fig. 7. The external radiator shell would be configured into a number of flat panels so that dissipative equipment can be mounted to the interior walls with good heat conduction. Both top and bottom surfaces are thermally insulated, so that all electrically generated heat is radiated through the power system radiator, an approach that is seen to result in the least equipment temperature variation. The only part of the total heat load which is not radiated by the power system radiator is that which is radiated as rf energy by the antennas.

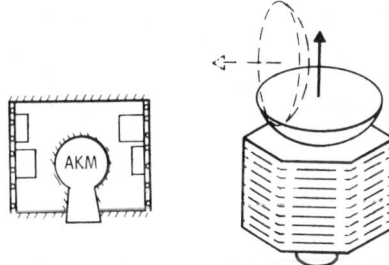

Fig. 7 Integrated approach.

In this approach, the main communications-system antennas can be oriented in just about any direction permitted by packaging and viewing considerations, since the spacecraft itself has no preferred orientation with respect to the ecliptic.

Semi-Integrated. The so-called "semi-integrated" approach

is to use the sidewalls of the
spacecraft as the power system
radiator, while placing equipment on the "top" surface only
(Fig. 8). This surface, if
north- or south-facing, contains
adequate surface radiating area
for the full equipment complement.

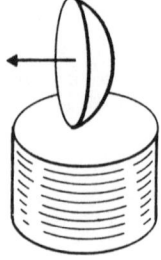

Fig. 8 Semi-integrated approach.

The major advantage to this approach is the ease with
which the equipment modules can be separated from the power system and subsequently installed as a unit. Although equipment
temperature can vary through wider limits than in the fully
integrated approach, this variation typically will be less
than that experienced with the solar-powered or separated designs. The radiative coupling between the power-system radiator and the equipment shelf generally is adequate to smooth out
the major low-temperature excursions without requiring heaters
or other active thermal-control devices in most cases.

The major problem with this approach results from the use
of a polarizing screen above the antenna reflectors in the
reference design. This screen, and its associated thermal blanket, results in almost complete blockage of the view to space
of the equipment panel radiator. Therefore, this approach did
not receive major consideration in this study. However, it
deserves more detailed consideration when other polarizing devices are used or where this blockage is otherwise reduced.

Separated. The separated design approach is the more traditional, except that in this case the power-system radiator
is custom-shaped and sized to fit best within the shroud diameter (Fig. 9). Essentially, this approach uses the solar-powered spacecraft with a minimum of change, replacing the
solar-battery system with the nuclear system.

An area of concern with this approach is the possible upsetting effect of the lengthened configuration on the spin stability of the spacecraft, needed for injection. However, this
was seen ultimately not to be a problem. An apparent penalty
for this approach is the interior heat shield required to protect the turbo-generator unit and radiator from the plume of
the AKM.

The separated design approach also was viewed as representative of the RTG-powered spacecraft, since no integrated or

SOLAR VS NUCLEAR POWER: IS THERE A CHOICE?

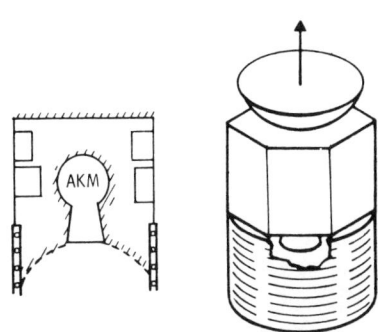

Fig. 9 Separated approach.

semi-integrated versions of an RTG-powered spacecraft appear feasible for this mission. In this case, the cylindrical radiator and other organic-Rankine system components would be replaced by two selenide RTG's, each approximately 280 w(e), on opposite sides of the AKM, with their extending fins shaped to conform to the shroud envelope, if necessary. Both integrated and separated designs were selected for more detailed study.

V. Nuclear-Powered Satellite Reference Designs

The basic arrangement of the separated configuration is shown in Fig. 10. Except for the heat source, the power system is contained within the cylindrical radiator, which is cantilevered from the bottom of the basic hexagonal satellite structure.

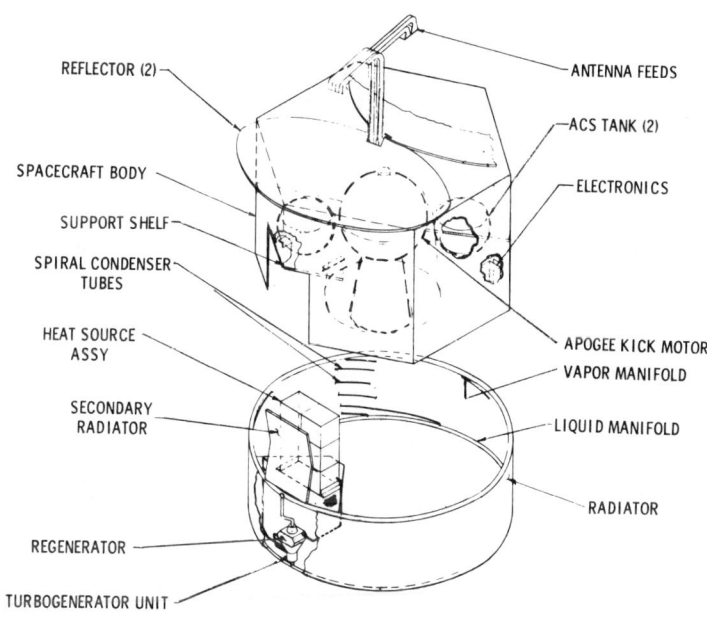

Fig. 10 Nuclear separated reference design.

The heat source itself protrudes upward into the satellite equipment bay. The local cutout in the face panels provides a direct thermal view to space for the secondary radiator. Dynamic balancing of the spacecraft is maintained by a rearrangement of several of the housekeeping equipment modules onto panels opposite the heat source.

This location for the heat source provides a number of benefits: 1) it does not upset the spin-stability of the spacecraft; 2) it permits ready access to the heat source mounting structure both before and after integration of the power-system module; and 3) it removes some of the more radiation-sensitive components from the direct vicinity of the heat source.

The use of curium heat source would produce somewhat higher levels of external neutron and gamma radiation than would a plutonium heat source of the same thermal power. This leads naturally to a concern about long-term radiation damage to electronic equipment and the possible need for shielding. Location of the more sensitive components on the sides of the spacecraft opposite the heat source is helpful in this regard. Of course, the use of a plutonium heat source completely obviates the problem.

For this design, the organic-Rankine system radiator is reduced slightly in size by using a different working fluid (Flourozene-M), which produces peak efficiency at a slightly higher temperature (140°F). All other external power system parameters remain unchanged.

The 140°F radiator is a cylindrical shell, 85 in. in diameter and 25 in. high, for an area of 46 ft^2. The shell is a 0.047-in. beryllium sheet bonded to 0.035-in. wall aluminum condensor tubes, as required by meteorite penetration criteria. The outer surface is covered by a silver-on-teflon thermal control coating. The entire assembly displays satisfactory margins of safety for Delta launch loads.

The interface with the Delta third stage and payload envelope is shown in Fig. 11. The radiator extension protrudes somewhat below the strictly defined limits of the payload envelope. However, this is not expected to present a problem in assembly of the vehicle to the third-stage adapter.

In this configuration, the satellite thermal control is identical to that of the solar-powered design, with the exception of a thermal-insulation barrier between the 140°F power-system radiator and the 100°F equipment radiators. Overall system weight and power comparisons will follow discussions of the integrated design.

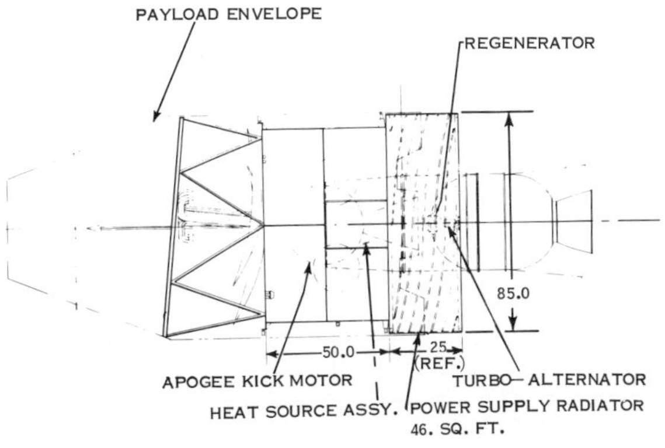

Integrated Design Configuration

In this version, the organic-Rankine system is used with hexaflourobenzene as the working fluid, at a nominal radiator temperature of 104°F. Because of the lower radiating temperature and the inclusion of the full equipment heat load, the total radiator area requirement is 76 ft². However, this is less than the total external area of the hexagonal solar-powered satellite, approximately 83 ft². In this design, all of the internal surface is available for equipment mounting, whereas in the solar-powered version only the two north-south panels feasibly could be used for mounting the TWTA's.

Fig. 11 Delta interface: separated design.

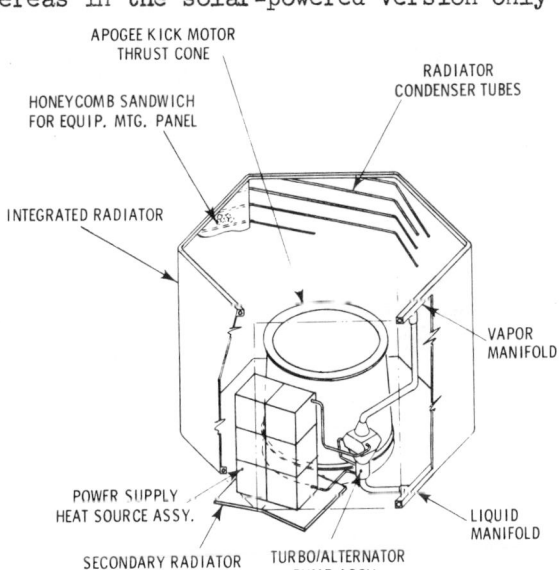

The arrangement of power system within the spacecraft envelope is shown in Fig. 12. In this case, the heat source fits inside the radiator, insertable from below. The secondary radiator, if required, faces downward from below the radiator.

Figure 13 shows a possible equipment mounting arrangement.

Fig. 12 Nuclear integrated reference design.

The weight of the heat source and converter equipment is balanced by most of the housekeeping electronic equipment on the opposite faces, similar to the arrangement used in the separated design.

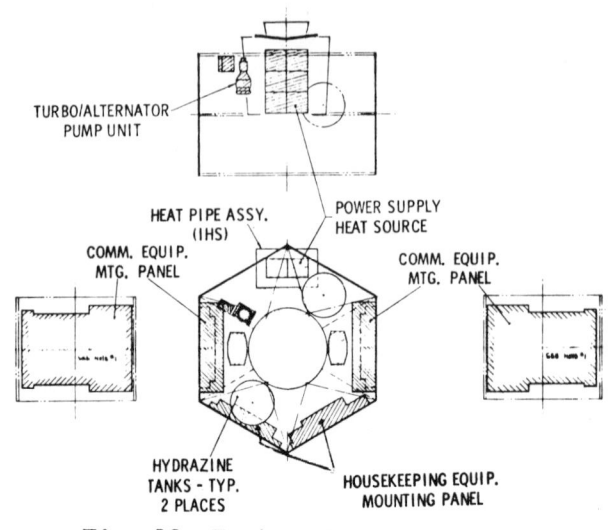

Fig. 13 Equipment mounting: integrated design.

Radiator Design

The radiator consists of spirally would aluminum condenser tubes adhesively bonded to external beryllium skins, forming a hexagonal box structure, as shown in Fig. 12. Where equipment support panels are required, aluminum honeycomb is bonded to the external skin between the tubes, and internal aluminum skins are bonded to the composite honeycomb/tube core to provide the internal equipment mounting surfaces (Fig. 14a). Threaded inserts are potted in the honeycomb, where required, to create equipment attachment points, and the internal skins are slotted near the edges to accommodate generous radiator-tube bend-radii in the six corners of the hexagonal structure. Vertical longerons are provided in the corners of the box to tie the panels together (Fig. 14b). The condenser tubes are welded

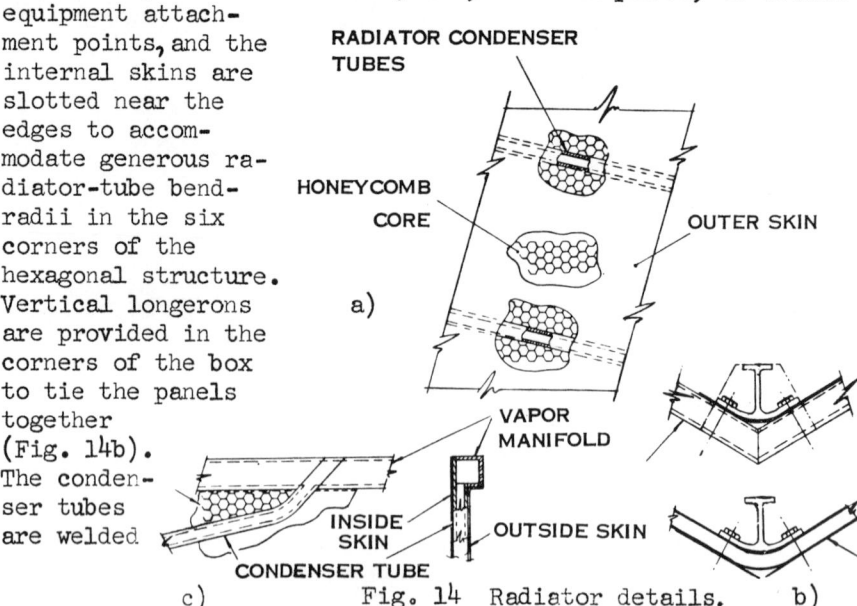

Fig. 14 Radiator details.

SOLAR VS NUCLEAR POWER: IS THERE A CHOICE?

to the liquid and vapor manifolds at the top and bottom surfaces of the hex (Fig. 14c).

The entire structure is quite similar to the honeycomb panels of the solar-powered design, except for the included radiator tubing and manifolds. The top and bottom manifolds add rigidity to the structure and contribute to its strength. Where equipment panels are not required, the combined tube/external skin structure is sufficiently strong, by itself, to withstand the boost flight loading environments.

Thermal Control

A major benefit of the integrated design is the achievement of automatic temperature regulation for any equipment contained within the radiator. Equipment mounted to the radiator interior walls will be constrained to follow the wall temperature closely. The spirally wound radiator tubes serve to distribute sun load or other unbalanced heat load uniformly over the radiator surface.

The maximum radiator temperature occurs with full sun load with a normal angle of incidence and a fully degraded thermal-control coating. This is the upper curve of Fig. 15; the maximum point on this curve is achieved at zero rf output, when the full heat source thermal power is dissipated through the radiator.

The minimum radiator temperature occurs with the solar-solstice sun load, nondegraded coatings, and full rf radiated power. As seen from Fig. 15, the maximum temperature variation is from 80° to 104°F between these extremes. By contrast, the solar-powered (and nuclear-separated) designs undergo a

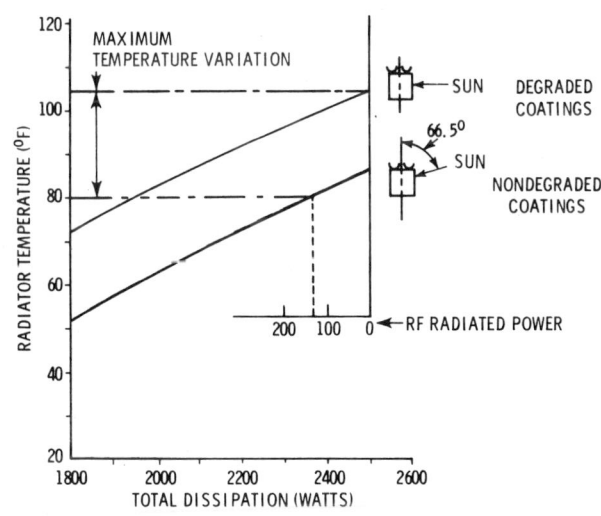

Fig.15 Temperature variation: integrated design.

gross temperature variation between 20° and 110°F, with the assistance of heaters at various locations.

VI. Solar vs Nuclear Spacecraft: Power and Weight Comparisons.

The power requirements and weight of the nuclear spacecraft have been determined mainly by comparison with the solar design, adding or subtracting components where indicated, and generally accepting the same assumptions. Therefore, the comparative values are likely to be valid to a greater accuracy than the absolute values of subsystem weights or power consumption in either of the systems. Nevertheless, the absolute values are based on detailed considerations and are the best estimates of technical specialists in each of the applicable disciplines.

The power requirements and weight breakdowns for the solar, nuclear-separated, and nuclear-integrated designs are shown in Tables 6 and 7, respectively. The communications, TT&C, and reaction control systems are unchanged in each of the designs. The attitude control system is affected by removal of the 14-lb solar-array drive assembly.

A significant reduction in the power consumption of the nuclear satellites is achieved by elimination of the batteries, even though no battery-charging power is budgeted at solar solstice. This comes about from elimination of the battery heaters. In the integrated version, most other heaters are dispensed with as well (Table 6).

Although the power consumption of the thermal control system is reduced by elimination of heaters, the weight of these heaters is small and compensated by the weight of added thermal control surface coatings in the nuclear versions. In addition, the separated design contains an added AKM plume shield, which was included in the weight of this subsystem.

In the power distribution assembly (PDA), most power is consumed by the voltage regulators needed for the heaters, battery-charge controllers, and the regulated bus. However, the output of the organic-Rankine generator is regulated naturally by the speed controller included in the electric power system. Therefore, most of the power consumption and a good deal of the weight of the PDA system are eliminated by either of the nuclear designs. Conservatively, however, only the weight of the 28-v heater regulator and ordnance controls used for solar array deployment (see Table 4) were removed from this subsys-

SOLAR VS NUCLEAR POWER: IS THERE A CHOICE?

Table 6 Relative power budgets

Subsystem	Power, w(e)		
	Solar (solstice)	Nuclear separated	Nuclear integrated
Communications	424.1	424.1	424.1
Telemetry and control	21.3	21.3	21.3
Attitude control	27.8	27.0	27.0
Reaction control	0.5	0.5	0.5
Electrical power and distribution	35.8	11.0	11.0
Power distribution	23.6	5.0	5.0
Harness	12.2	6.0	6.0
Thermal control heaters	69.5	39.5	7.0
E-W panels	18.5	18.5	0
Batteries	30.0	0	0
RCS tanks	4.0	4.0	0
Thrusters and lines	12.5	12.5	7.0
Transponders	4.5	4.5	0
Total	579.0	523.4	490.9

tem in Table 7. Further reduction in power consumption and weight is achieved by a reduction in the harness length, which results from elimination of the deployed solar arrays and the battery bus (Tables 6 and 7).

The net result is to establish the significantly different power consumption requirements for the nuclear and solar designs, as shown in Table 6, which then were used to determine the weights of the respective electric power systems. These are based on the conventional solar array/battery system

Table 7 Spacecraft weight comparisons

Subsystem	Solar	Nuclear separated	Nuclear integrated
Communications	186	186	186
TT&C	27	27	27
Reaction control (dry)	58	58	58
Apogee kick motor (dry)	56	56	56
Attitude control	114	100	100
Thermal control	42	53	42
Structure and integration	169	161	208
Thrust cone and trusses	82	75	75
Equipment panels and frames	87	87	133[a]
Electric power system	225	197	149
Array/converter	56	113	105
Battery/radiator	96	40	a
Power/dist. and control	25	15	15
Harness	48	29	29
Subtotal	877	839	826
Balance for RCS expendables	138	176	189
Total	1015	1015	1015

[a] Radiator is integrated into structural frame.

for the solar-powered spacecraft, and on the organic-Rankine system with unvented fuel for the nuclear designs.

The structural and integration hardware is broken down into two weight segments. The basic truss structure is reduced by the weight of the solar array support trusses (13 lb) and

SOLAR VS NUCLEAR POWER: IS THERE A CHOICE?

increased by the radiator support trusses (6 lb). (The supporting structure for converter components is included within the converter weights.) The equipment panels are essentially unchanged for the solar and nuclear-separated designs. The panels of the integrated version contain the radiator hardware, leading to this apparent weight penalty. Clearly, this is accompanied by a concommitant reduction in power-system weight.

Finally, these establish the overall system weight totals shown in Table 7. This illustrates a net weight advantage for the nuclear separated and integrated designs of 38 and 51 lb, respectively. This added weight margin can be used to extend system lifetime by adding propellant for stationkeeping and attitude control, as shown. The economic impact of this lifetime extension is discussed in the next section, where the comparison is extended to include the more advanced solar-battery technology and other versions of the nuclear systems.

VII. Economic Impact and Conclusions

In Ref. 1, it is concluded that, for this reference spacecraft, the most economically effective utilization of excess weight is in additional propellant for stationkeeping and attitude control. Propellant is consumed at a rate varying from 29 to 22 lb/yr for the first 10 years, mainly for north-south stationkeeping. Further weight savings in the power systems can, therefore, be used effectively by adding to the propellant supply. This permits a direct comparison of the systems effect of these power system variations, assuming that none of the other spacecraft subsystems will be affected further by changes to the power systems within each of the designs.

Figure 16 illustrates this comparison, in which mission lifetime is plotted as a function of power system specific power, assuming uniform propellant consumption of 25 lb/yr. The three curves, for solar, nuclear-integrated, and nuclear-separated systems, express the different weight balances and power requirements of the three spacecraft designs. Thus, as was illustrated already in Table 7, a nuclear system of a lower specific power can result in a significantly longer mission lifetime than that achievable by a comparable solar-powered spacecraft. For example, the reference organic-Rankine system (unvented heat source, beryllium radiator), although approximately 15% heavier than a state-of-the-art solar system of the same power level (Fig.3), can result in a net spacecraft weight reduction of approximately 50 lb and, therefore, a mission lifetime extension of about 2 yr.

A similar conclusion is seen to obtain in comparing the organic-Rankine system with a vented heat source (either curium- or plutonium-fueled) with the most advanced solar system combination now envisioned: violet-cell array with NiH_2 batteries. In this case, the solar system is over 40% higher in specific power; yet the nuclear system can achieve comparable lifetime in the separated configuration and increased lifetime in the integrated configuration. (It has been assumed for the purpose of this comparison that specific power will not change significantly with lifetime for the

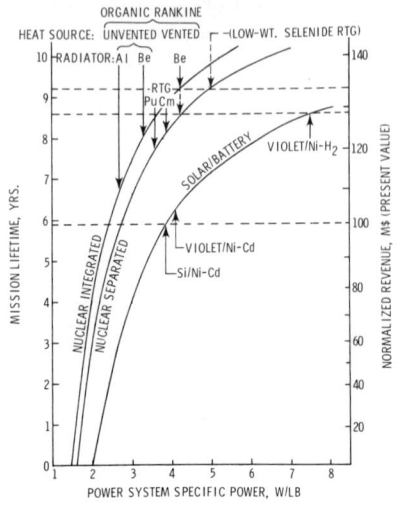

Fig. 16 Comparisons of nuclear and solar-powered versions of reference spacecraft.

period between 5 and 9 yr. This clearly is only an approximation both for solar-battery systems and for radioisotope-fueled nuclear systems.)

This increased lifetime capability can be translated directly into increased revenue produced by the satellite, against which subsystem cost increments can be compared. The right-hand scale of Fig. 16 shows the present value of expected satellite revenue, derived from Ref. 1. This includes the effects of predicted transponder failure rates and a 10%/yr discount rate. The scale reads directly in millions of dollars of total satellite revenue if it is assumed conservatively that the average income per operating transponder is 10^6/yr. From Fig. 16, therefore, we can derive a number of conclusions:

1) Either of the advanced nuclear systems clearly would be preferable to current solar systems, including violet cells.

2) The revenue impact of increased lifetime can justify costly design changes that reduce weight. For example, the transition from an aluminum to a beryllium radiator skin in the organic-Rankine system can result in a revenue increase of approximately $15x10^6$, orders of magnitude more than the cost differential between these radiators.

SOLAR VS NUCLEAR POWER: IS THERE A CHOICE?

3) The potential revenue advantage of the best of the nuclear systems is such as to exceed the predicted cost disadvantage, when compared to any of the solar systems.

This last point is illustrated in Table 8 by combining the information contained in Figs. 4 and 16. For this table, only recurring costs from Fig. 4 are used for the solar and nuclear systems. It is assumed further that the cost per watt of solar power will remain unchanged with the development of the advanced solar systems, and that no reduction in the cost of plutonium fuel will be achieved.

Although the nuclear-integrated design appears slightly superior to the separated version, the difference in weight is small and within the uncertainties. The advantage of separability for development and testing, and the greater ease of assembly of a separated design, could well result in a practical preference for this design. On the other hand, the tighter and simpler temperature control of the integrated version could tip the choice that way for many designers.

In summary, it has been illustrated that solar vs nuclear power, if compared on an overall systems basis, can lead to conclusions wholly different from those that would be drawn from a cost and weight comparison of the power systems alone.

Table 8 Revenue and cost comparisons, nuclear vs solar satellites

Solar \ Nuclear	Curium fuel		Plutonium fuel	
	Sel. RTG	Org. Rank.	Sel. RTG	Org. Rank.
	Difference in revenue (nucl.-solar, 10^6)			
Si array/Ni-Cd batt.	+25	+33	+21	+33
Violet/Ni-Cd	+20	+28	+16	+28
Violet/Ni-H2	-3	+5	-7	+5
	Difference in cost (nucl.-solar, 10^6)			
	+1.4	+0.4	+3.9	+1.1

Savings in the weight and power requirements of other satellite subsystems can result in a net weight advantage for the nuclear-powered design, even compared to the lowest-weight solar-battery systems now envisioned. This weight advantage, if used to extend satellite useful lifetime, can result in a revenue advantage that exceeds the possible cost disadvantage of the nuclear systems.

The advanced nuclear systems are seen to be commercially superior to present-day solar systems by a significant margin and potentially competitive with the violet-cell/nickel-hydrogen system for future applications.

References

[1] Raab, R. and Friedrich, S., "Design Optimization for Profit in Commercial Communications Satellites," AIAA Paper 74-467, April 1974, Los Angeles, Calif.

[2] "SNAP-27: Summary Report," GEMS-402, July 15, 1970, General Electric Co., Space Div., Philadelphia, Pa.

[3] "SNAP-19 Pioneer F&G Final Report," IESD 2873-172, June 1973, Teledyne/Isotopes, Timonium, Md.

[4] Bradshaw, G.B. and Postula, F.D., "Beginning of Mission Flight Data on the Transit RTG," 8th Intersociety Energy Conversion Engineering Conference, Proceedings, Aug. 1973.

[5] Caputo, R.S. and Truscello, V.C., "Long Life Performance Predictions of SiGe MHW-RTG," 8th Intersociety Energy Conversion Engineering Conference, Proceedings, Aug. 1973.

[6] "Economic Radioisotope Thermoelectric Generator Study Program, Final Technical Report," Appendix C, IESD 3112-2, Oct. 1973, Teledyne/Isotopes, Timonium, Md.

[7] "Economic Radioisotope Thermoelectric Generator Study Program, Final Technical Report," IESD 3112-1, Oct. 1973, Teledyne/Isotopes, Timonium, Md.

[8] "Economic Radioisotope Thermoelectric Generator Study, Final Briefing," AEC Cont. AT(04-3)-943, Aug. 1973, Gulf Energy and Environmental System, San Diego, Calif.

[9] "Mini-Brayton Economic RTG Study, Final Report," GEMS-417, Jan. 1974, General Electric Co., Philadelphia, Pa.

[10] "Organic Rankine Cycle Economic Radioisotope Thermodynamic Generator Study, Phase I Final Report," AEC Contract AT(04-3)-941, Jan. 1974, TRW Systems, Redondo Beach, Calif.

[11] Schock, A., "Light-Weight RTG Design," FSEC-NSG-217-74/31, Feb. 28, 1974, Fairchild Space & Electronics Co., Germantown, Md.

[12] "Standardization and Economics of Nuclear Spacecraft, Final Report-Phase I, SENSE Study," AEC SNSO 3063-2, March 1973, Fairchild Space & Electronics Co., Germantown, Md.

CHAPTER II—TERMINAL TECHNOLOGY

This chapter focuses on developments in the terminal subsystem of the satellite communications system and is generally considered to encompass the equipment from the intermediate frequency (typically 70 MHz) through radiation, or reception, of the desired signals at the radio-frequency.

The three papers selected give a good cross-section of current areas of technological interest in their field: unique terminal applications; the design of small, low-cost terminals; and unattended terminals.

The paper by Hooper et al., addresses the design of one of the "hot-line" terminals that constitute the new satellite-established direct communications link between Washington and Moscow. there are two parallel satellite paths employed: one is through the INTELSAT system and the other through the Soviet Molniya system. Because the Molniya space segment is characterized by highly elliptical orbits, rather than by the stable, circlular, geostationary orbits of most commercial and military satcom systems to date, the geometry and dynamics of the Molniya tracking task, differ significantly from the conventional, and it is the influence of these parameters that is the main thrust of this paper.

There are an increasing number of applications for satellite communications to be used in networks involving hundreds of potential terminals. Therefore, it is necessary for operators of these networks to keep the initial cost of each terminal as low as possible and to have, inherent in the basic terminal design, those features that will minimize recurring costs of operation and maintenance. The next two papers deal with this problem. The first one by Petrick and Abrahamson, investigates ways to reduce the initial cost of the terminal against the requisite terminal availability, whereas the second paper, by Miller, reports the results of the operational experience of Telesat Canada in utilizing small unattended receive-only Earth stations. The costs of operating personnel are a major element in the life-cycle cost equations, and both papers will be of great interest to those involved in the development of cost-effective satellite communications networks in the future.

EARTH STATION FOR THE U.S.-USSR DIRECT COMMUNICATIONS LINK

W. P. Hooper,[*] W. M. Rogers,[*] and J. G. Whitman Jr.[+]

HARRIS Electronic Systems Division, Melbourne, Fla.

Abstract

The Molniya Earth Station for the U.S.-USSR Direct Communications Link (DCL) is part of a new, intercontinental satellite relay system scheduled to replace the terrestrial "hot line" that has linked Washington and Moscow since its installation in 1963. The Earth Station, located at Fort Detrick, Md., will operate at C-band frequencies through the Molniya USSR communication satellites. The geometry and dynamics of the Molniya tracking task differ substantially from those that characterize present U.S. military and commercial satellite communications applications. The influence of these differences upon the configuration of the Earth Station around dual, independent communication systems is discussed. Following an introduction to the physical layout of the site and operations building, the design of each subsystem of the communication systems and of ancillary equipment that supports station operation is reviewed. Link parameters and major performance characteristics of the Earth Station and each of its subsystems are presented in summary form.

Presented as Paper 74-424 at the AIAA 5th Communications Satellite Systems Conference, Los Angeles, Calif., April 22-24, 1974.
*Associate Principal Engineer.
[+]Principal Engineer.

Fig. 1   DCL Earth Station at Fort Detrick, Md.

EARTH STATION FOR US-USSR COMMUNICATIONS LINK

Introduction

In 1963, a "hot line" was established between Washington and Moscow to permit instant communication between the heads of state of the U.S. and the USSR. The system has functioned well, providing insurance against the risk of international misunderstanding in times of crisis from its inauguration up to the present time. The communication link carries teleprinter traffic on two alternate paths: a primary cable circuit via London, Copenhagen, Stockholm, and Helsinki, and a backup radio circuit. The operational use of the "hot line" is infrequent, but regular testing of the link is conducted to verify its readiness. To date, no simultaneous outage of both the primary and backup links has been reported.

In 1971, with the potential of satellite relay links for highly reliable intercontinental communication well established, the leaders of the U.S. and USSR authorized the development and implementation of a new system that uses satellites to provide a direct link between their two nations. As before, two separate links are to be available, one through an Intelsat IV satellite and the other through a group of Soviet satellites of the Molniya II series. In October 1972, following a competitive procurement, the U.S. Army Satellite Communications Agency (USASATCOMA) awarded to Harris ESD a contract for the design, fabrication, installation, and 3-yr O&M support of a Molniya Earth Station. This Earth Station, Fig. 1, located at Fort Detrick, Md., is the U.S. terminus of the Molniya Direct Communications Link. The design and major performance characteristics of the Earth Station and its equipment, now undergoing final onsite testing, are described in this paper.

Communication Via Molniya Satellites

The USSR places its satellites of the Molniya series in elliptical orbits whose plane is inclined nominally 65° with respect to Earth's equator. Because of the large eccentricity (approximately 0.74), the satellite dwells in the region of apogee over the northern hemisphere for a surprisingly large fraction of its orbital period, then sweeps rapidly through a southern hemisphere perigee to reappear once again at northern latitudes. Through the selection of an orbit period of exactly 12 solar hours, the path of the subsatellite point on Earth's surface is made to retrace itself periodically. Stated differently, the Earth-fixed longitude over which the apogee occurs is kept fixed in a geographical area in which the satellite is suitably positioned to provide a desired communication relay geometry. The communication needs of the Soviet Union are met by placing the satellite apogee at a longitude centrally over the Eurasian land mass.

Since the 12-hr orbit passes through its apogee twice each day whereas Earth rotates on its axis only once during that time, a second point of apogee occurs over the western hemisphere. Because Moscow's 56° northern latitude enables it to look across the pole and view the satellite near its North American apogee for extended periods of time, satellites in this orbit provide a geometry that is well suited for communication between the Soviet Union and the western hemisphere. This is the geometry that has been selected for communication between Moscow and the Fort Detrick DCL Earth Station.

Molniya Orbit Configuration

A distant observer who maintained a geostationary position on the equator above the Atlantic Ocean would see the motion of the Molniya satellites as it has been illustrated in Fig. 2. The choice of an Earth-fixed coordinate frame for this illustration distorts the orbital figure, which would be a planar ellipse in inertial coordinates, into the looping trajectory that is pictured. Points along this trajectory which will be occupied by the satellite at successive 1-hr intervals are indicated. The satellite is over the northern hemisphere for almost exactly 11 hr of its 12-hr orbit. Near its North American apogee, it is mutually visible between Washington and Moscow for slightly more that 8 hr.

Because the satellite does not remain over the northern hemisphere indefinitely, a single satellite could not provide continuous communication capability. A system of at least three satellites must be used in order to guarantee that one of the satellites will be mutually visible between Washington and Moscow at all times. If three orbits are oriented such that all satellites follow the same trajectory at 8-hr intervals, then the achievement of continuous mutual visibility can be verified by means of Fig. 2. In actuality, the satellites making up the orbital configuration follow similar, but not necessarily identical, trajectories.

The Mercator projection in Fig. 3 illustrates the ground trace of the subsatellite point for a Molniya orbit with a 110° W equator crossing. Movement of the satellite slows dramatically as it moves into the northern hemisphere, climbing away from Earth. Following an extended period of mutual visibility while over the North American continent, its southward and Earthward motion makes it disappear, first

EARTH STATION FOR US-USSR COMMUNICATIONS LINK

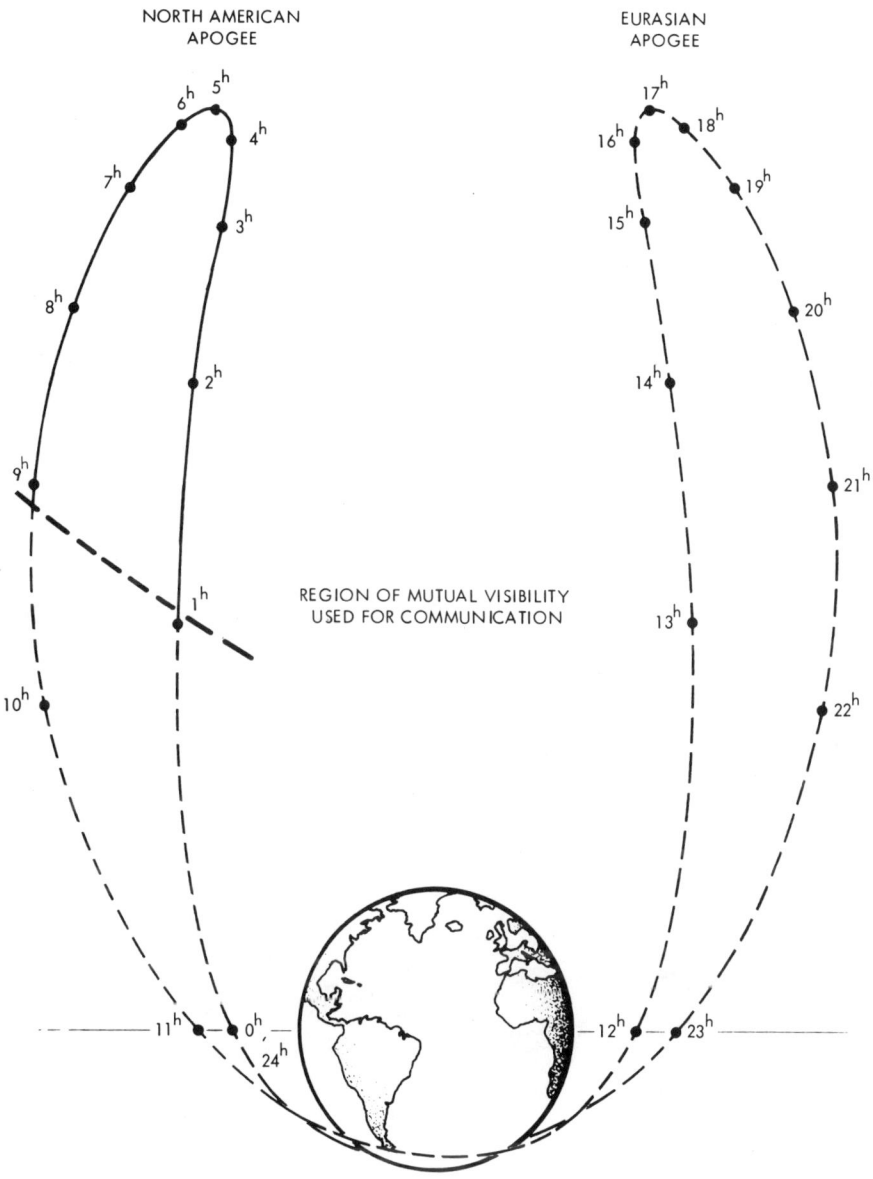

Fig. 2 Orthographic projection of Molniya satellite motion in Earth-fixed coordinate frame.

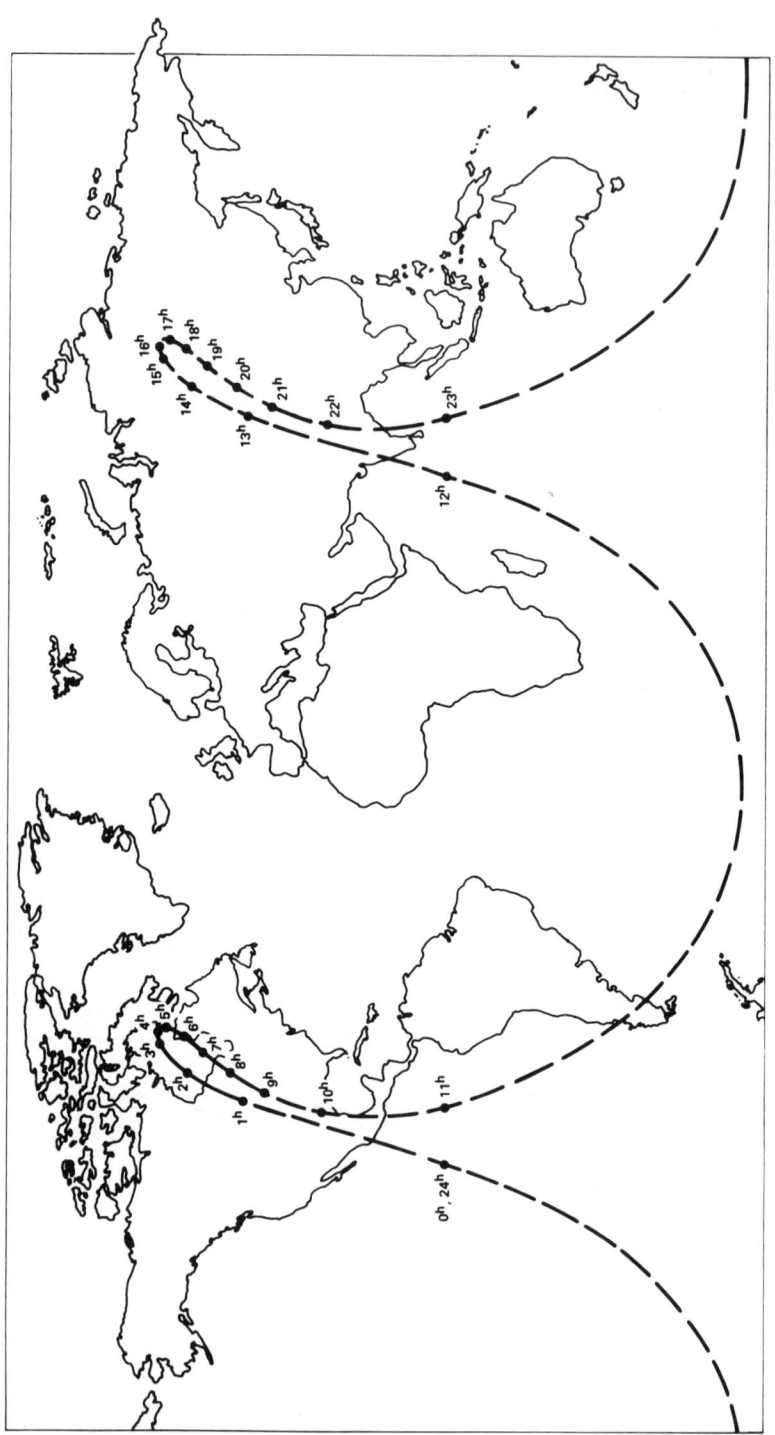

Fig. 3 Molniya satellite ground trace for 110° W equator crossing.

EARTH STATION FOR US-USSR COMMUNICATIONS LINK   121

from Moscow's view and then from Washington's. After a rapid trip through perigee in the southern hemisphere, the satellite again climbs toward apogee. Although the Eurasian apogee is sufficiently high to be mutually visible, this apogee does not provide favorable communications geometry for the DCL. Because of its moderate latitude, the Washington site is less well able to look over the pole at the opposite apogee than is the Moscow Earth Station. Thus, the Eurasian apogee is seen at a very low elevation angle from Washington and is not planned to be used for communication.

Tracking Geometry

Although the ground trace of the subsatellite point crosses the U.S. well to the west of the Fort Detrick DCL Earth Station, the great height of the satellite as it climbs toward and returns from its 40,000-km apogee makes it appear high in the Washington sky. An appreciation for the tracking geometry from the Earth Station may be gained by a review of Fig. 4. It can be seen that the satellite becomes visible south-southwest of Fort Detrick and climbs to 70° elevation in 0.5 hr. For the next 10 hr, the azimuth changes from southwest to north and back again, whereas the elevation varies only ±10° from 70°. Approximately 11 hr after it first became visible, the satellite vanishes beyond the horizon south of the station.

Because a large fraction of the satellite tracking, including all of the region of mutual visibility, occurs at high elevation angles, the use of an Az-El antenna positioner would not be ideal for the DCL application. Even a small shift eastward in the longitude of the orbital plane could bring the satellite directly overhead at Fort Detrick. If an Az-El positioner were used, the dynamic singularity that these positioners have at the zenith position would preclude the tracking of even the low angular dynamics presented by the satellite. Accordingly, a less conventional X-Y axis configuration was selected by USASATCOMA as optimum for the DCL Earth Station. The X-Y configuration also has points of dynamic singularity, but these lie on the horizon in the azimuth directions that are the extensions of the X axis in space. By locating these "key holes" at an azimuth of nominally 38.5^6 relative to north, Harris ESD placed them far from the region of mutual visibility for any Molniya satellite orbit.

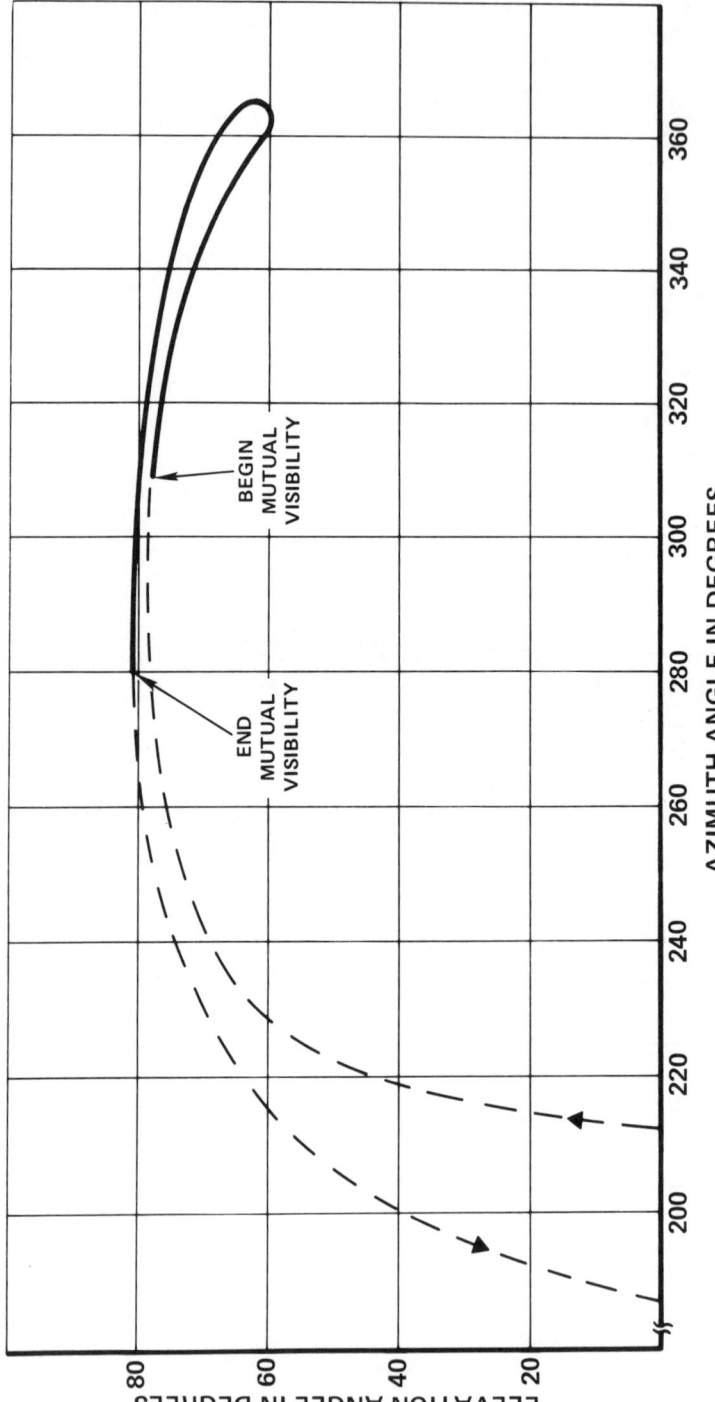

Fig. 4 Typical elevation vs azimuth trace at DCL Earth Station.

EARTH STATION FOR US-USSR COMMUNICATIONS LINK

Satellite Handover

Because no single satellite remains continuously within the region of mutual visibility, it is necessary at intervals of approximately 8 hr to hand over the communication link from a satellite that is leaving that region to another satellite that has just entered it. In order to permit acquisition, the establishment of tracking, and the adjustment of link parameters before the transfer of the actual communications traffic takes place, the DCL Earth Station is provided with two independent communications systems. For a large fraction of each day, both of these systems track a single satellite as it moves through the region of mutual visibility, thus providing backup to each other. As the satellite through which the communication link has been closed approaches the boundary of the region of mutual visibility, a second satellite becomes mutually visible. At a prearranged time, the backup antenna breaks track on the first satellite and moves through a short arc across the sky to acquire the second.

When the new link has been satisfactorily closed and tested, communications traffic is transferred to it. Shortly thereafter, the antenna that had originally supported the prime link disengages from the first satellite and slews to acquire the second. Because, as illustrated in the preceding figures, the two satellites are close together in the sky at the time that handover occurs, the entire process takes just a few minutes. Once handover is complete, both antennas track the active satellite. Except during the brief period of handover, incoming communications traffic is received by both of the communications systems of the DCL Earth Station. Should one of these systems fail, and should the extensive redundancy within the failed system not suffice to maintain uninterrupted communication through it, then communication would be restored immediately by transferring the traffic to the backup system.

Link Characteristics

The DCL Earth Station was designed with the capability of operating anywhere within a 500-MHz band for both transmission and reception. Transmit capability exists between 5725-6225 MHz, whereas the receive band is 3400-3900 MHz. In addition, dynamic characteristics are such as to permit operations with range to the satellite varying from 15,000 to 40,000 km, while maintaining EIRP control accuracy to 0.5 dB and holding frequency variation due to doppler to a minimum. Communication signals pass through a transponder in the satellite which translates the frequencies to the receive band. The link operates with

Table 1 Link characteristics

	Satellite	DCL Earth Station
Frequency band		
Transmit, MHz	3400-3900	5725-6225
Receive, MHz	5725-6225	3400-3900
EIRP per carrier, dBw	+2.5	45-70
G/T, dB	...	31
Loop control accuracy		
EIRP, dB	...	±0.5
Frequency, Hz	...	±500
Receive signal levels, dBm	...	-114 to -84

one communication channel from the USSR and one from the U.S. It also utilizes a pilot signal originating in the USSR. In the DCL Earth Station, this pilot signal is used for initial spatial and frequency acquisition, for autotracking, and for EIRP and frequency control.

The frequency uncertainty due to doppler would tend to make initial acquisition difficult without frequency search and a means of positively identifying the pilot signal. The DCL Earth Station uses a frequency sweep and recognizes the pilot signal by its special 4000-Hz tone modulation. Once acquisition has been accomplished, the receiver tracks the pilot throughout the pass. Control of EIRP and frequency of the DCL transmitter is accomplished with a closed loop through the satellite, as demonstrated in Fig. 5. The signal strength and frequency of the monitor signal are compared with the pilot signal, and errors cause the level and frequency of the uplink carrier to be moved accordingly.

A summary of the link characteristics is shown in Table I. Although the nominal frequencies for the transmitters and receivers are controlled by synthesizers locked to station standards, transmit frequencies can be changed up to ±200 kHz using a VCO in the modulator. The demodulators have frequency search capability of ±300 kHz. With 30-kHz peak deviation for a voice bandwidth of 3400 Hz, the spectral occupancy is less than 100 kHz. The present plans are to use subcarrier teleprinter signals within the 300-3400-Hz baseband.

Station Description

The DCL Earth Station is located on a 15-acre site at Fort Detrick, Md., approximately 50 miles northwest of Washington, D.C. Two 60-ft reflectors, each on X-Y pedestals, are adjacent to the operations building as shown in Fig. 6. Covered passageways are provided from

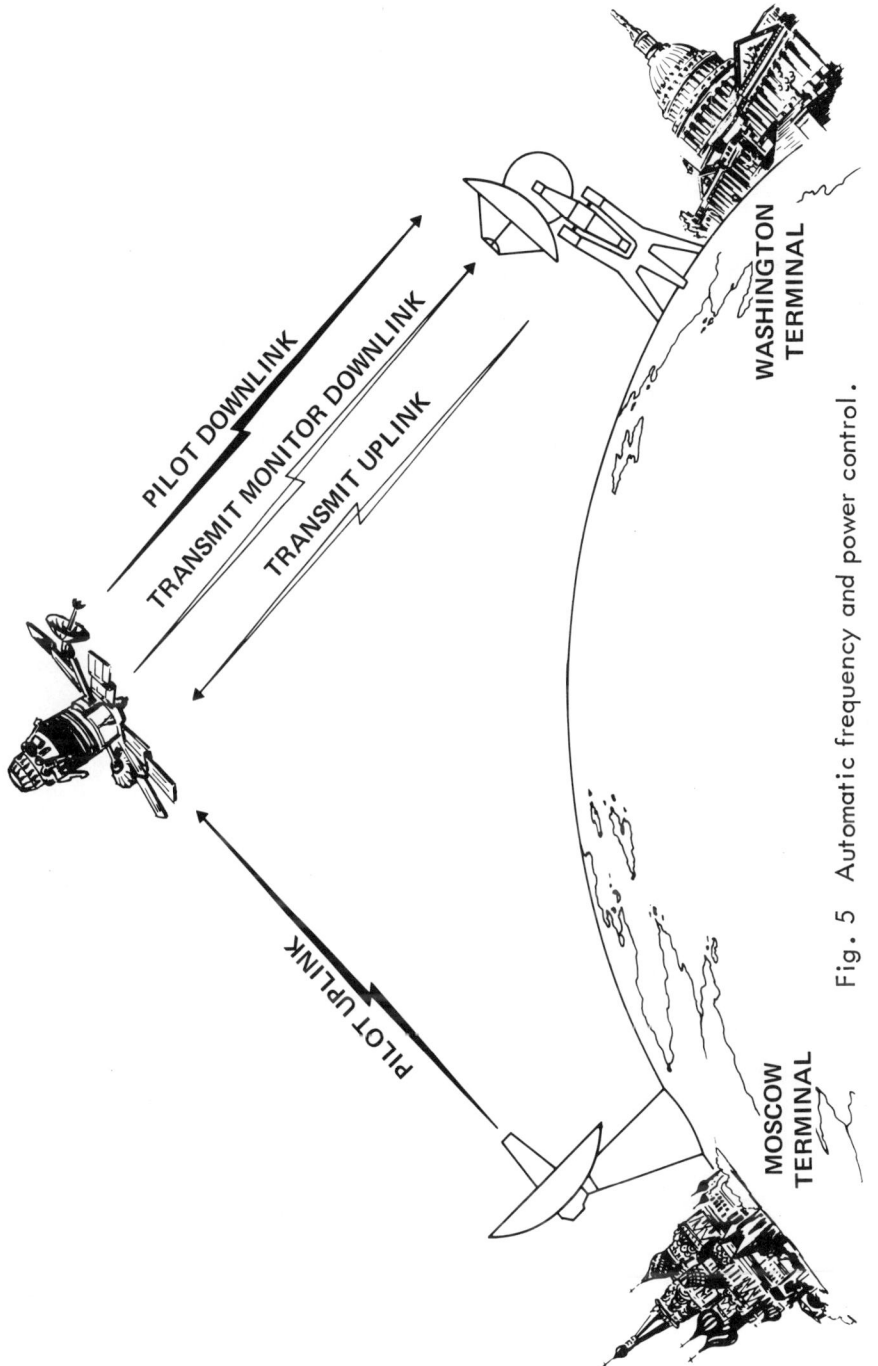

Fig. 5 Automatic frequency and power control.

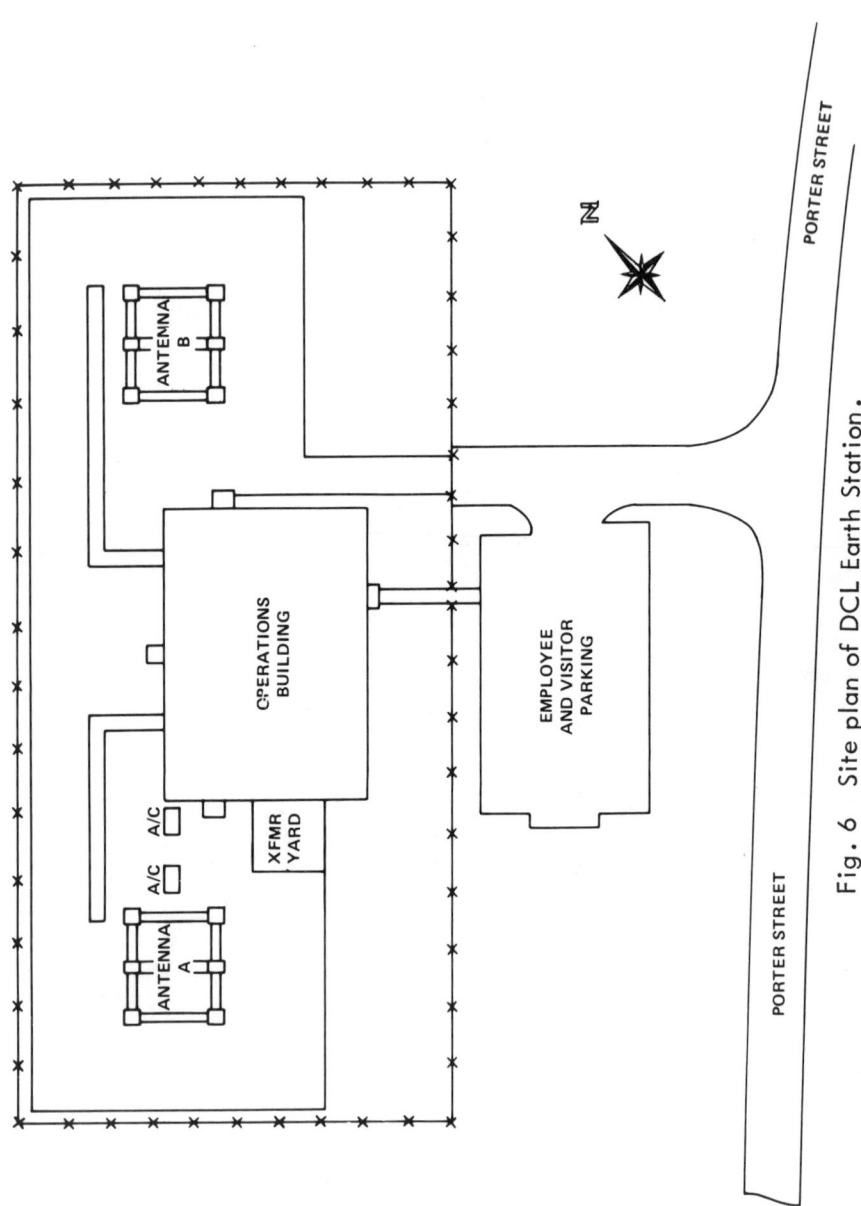

Fig. 6 Site plan of DCL Earth Station.

the building to each antenna to permit walkup access to rf equipment rooms on the back of the reflectors when the antennas are rotated to a maintenance position. The prefabricated steel building provides ample space for a self-contained operation including prime power distribution, building environmental control system, communication systems, management, and support activities. The operations building layout is shown in Fig. 7. Because of the critical mission requirements for DCL, high link availability is a prime goal. To accomplish this goal, redundancy exists throughout the facility. Dual 4160-v prime power lines are routed to two sets of transformers and switch gear. From there, distribution is made to the two communication systems and support equipments. The station load can be placed on one prime power input or split between them. Maximum station load is estimated at 800 kva when both antennas are under accelerated drive and reflector deicers are turned on.

Two turbine generators are located near the DCL facility, with each having the capability of supplying the full DCL load. These turbines supplement the existing power generation capacity at Fort Detrick. Either of the heat exchangers located at the left end of the operations building can control the building environment except on very hot days when both units can be operated in parallel. Air handlers 2 and 3 control the operations room environment; air handler 1 controls the other portions of the building. The power room contains two switch gears and two uninterrupted power systems (UPS), one for each communication system. These units will permit continuous operation of the communication equipment for 10 min with full load if prime power to the station is lost. Portions of the electronic equipment (i.e., programer subsystem, console, etc.) must operate when either system is in use, and, to accomplish this, a third critical power bus is formed by automatic switch selection of either UPS unit. Power is routed through the UPS at all times, battery drain being continuously replenished except during prime power outages. The telephone and fault alarm equipments are located in the telephone room. Twenty-five direct-dial phones are located throughout the station, providing necessary point-to-point communications for operation and maintenance and including conference capability. Any time a major or minor fault occurs in the communication systems or support equipments, an appropriate alarm will sound throughout the station on the public-address equipment.

Most of the electronic equipment is in the operations room and the hallway behind the operations room as shown in Fig. 8. Operational

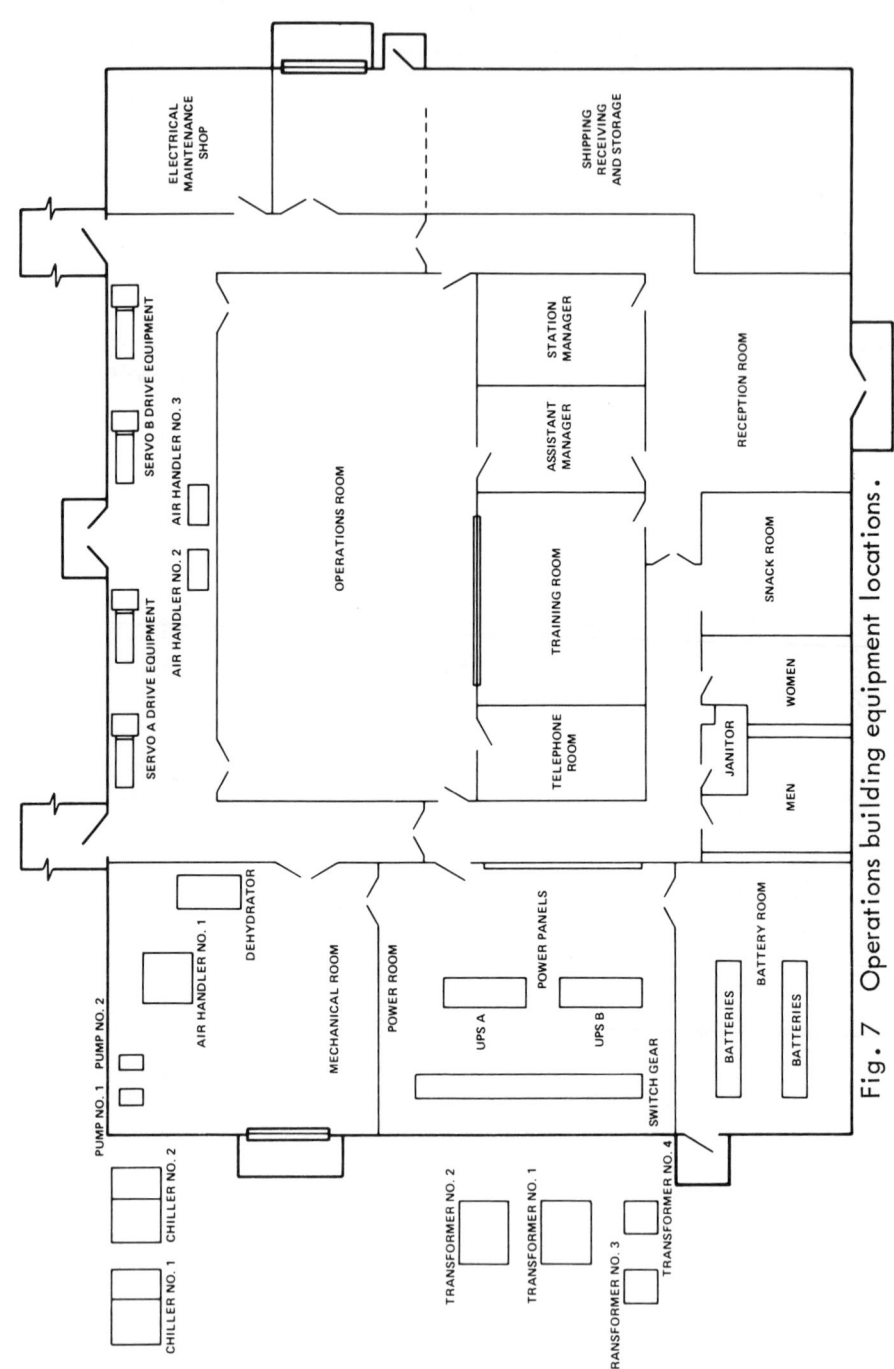

Fig. 7 Operations building equipment locations.

EARTH STATION FOR US-USSR COMMUNICATIONS LINK

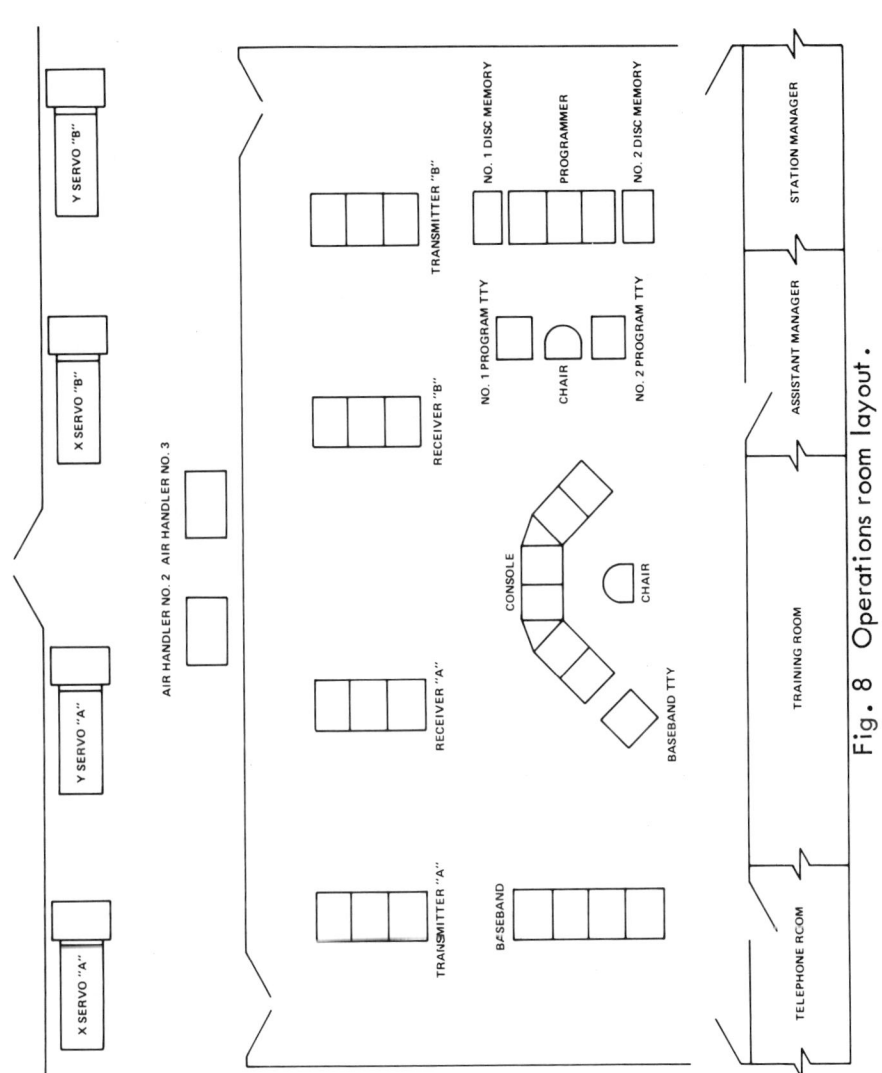

Fig. 8 Operations room layout.

control is centralized at the console flanked by the programer and baseband subsystems, with the electronic equipments for systems A and B located to the left and right, respectively, as viewed through the training-room window. All of the servo drive equipment is in the building except the motors. A raised floor is used in the operations room to form an air plenum for conditioned air to cool the electronics. Interconnecting cabling is routed beneath this floor. Defective equipment can be repaired in the electronic maintenance shop with replenishment parts from the bonded storage area as required. Depot-level repair is performed at the DCL facility, with vendor repair used only as required for subassemblies.

Operating Equipment

Dual Communication Systems

It has been seen that the maintenance of uninterrupted communication during the period of handover from one satellite to the next requires the Molniya Earth Station to utilize two parallel communication systems. These two systems are identical and independent, each consisting of an antenna, a servo subsystem, and receiving and transmitting subsystems. Within each communication system, additional equipment redundancy is provided. Critical items of equipment are monitored continuously to verify proper operation, and switchover from a failed unit to a redundant unit capable of maintaining the uninterrupted flow of communications traffic is automatic. Each communication system thereby achieves a reliability characterized by an MTBF predicted to be in excess of 835 hr, considering the outage of a signal route for more than 5 msec to be a failure. Availability of the entire Earth Station will be greater than 0.999.

Antenna and Servo. The antenna of each communication system consists of an X-Y positioner supporting a 60-ft parabolic reflector. A pseudomonopulse, autotracking feed illuminates a hyperboloidal subreflector held at the prime focus in the conventional Cassegrain geometry used in many satellite communication terminals throughout the world. Low microwave losses are achieved by locating the receiving subsystem low-noise amplifiers in an rf equipment room located immediately behind the reflector vertex. Convenient access to these amplifiers and other equipment in the rf equipment room is provided via a unique enclosed stairway carried by the antenna which interfaces with the covered walkways extending from the operations building to base of each antenna.

EARTH STATION FOR US-USSR COMMUNICATIONS LINK

The X-Y axis configuration illustrated in Fig. 9 provides near-hemisphere angular coverage, excluding only 10° conical keyholes that lie on the horizon along the extensions of the X axis. These are oriented far from the regions of mutual visibility and 90° away from the position of the boresite tower, which is located on a ridge northwest of the Earth Station. They, thus, are operationally of no consequence and their location relative to the boresite antenna permits a convenience in antenna measurements which is not normally achieved with the X-Y configuration. The servo subsystem of each antenna provides independent control of position and motion in the X and Y axes. Control is exercised by any one of three inputs, operator selected by positioning a mode select switch on the antenna control panel: 1) an autotrack error signal from the receiver, 2) a program track error signal from the programer, or 3) a slew input from the antenna control panel.

In the autotrack mode of operation, the error signal is an AM output from the pilot demodulator, modulated by a ferrite scanner in the feed with both X and Y axis error information. A phase demodulator in the servo further processes this error signal to separate the X and Y components. It then feeds the individual error signals to position loops that act to null the errors and bring the antenna beam into alignment with the satellite. In the program track mode, the position loops are closed through the use of a computer that compares the command pointing angles with the actual antenna angles sensed by X and Y axis synchro-to-digital converters that measure the position of each axis with an over-all accuracy of better than 0.01°. Pointing errors are converted into analog error signal inputs to the position loops. The manual mode utilizes X and Y axis slew controls for operator positioning of the antenna. A pair of d.c. motors connected in a counter-torque configuration is used to drive each antenna axis. The X axis uses 15-hp motors, and 10-hp motors are employed for the Y axis. Identical solid-state 6-phase SCR drives are used in each axis.

The antenna and its servo are capable of pointing and tracking to within a small fraction of the rf beam width without need for radome protection. A deicing system prevents the accumulation of ice or snow on the reflector and subreflector surfaces. Antenna and servo characteristics are summarized in Table 2.

Receiver. Figure 10 is a functional diagram of one of the receiving subsystems. The signal input from the feed passes through a bandpass

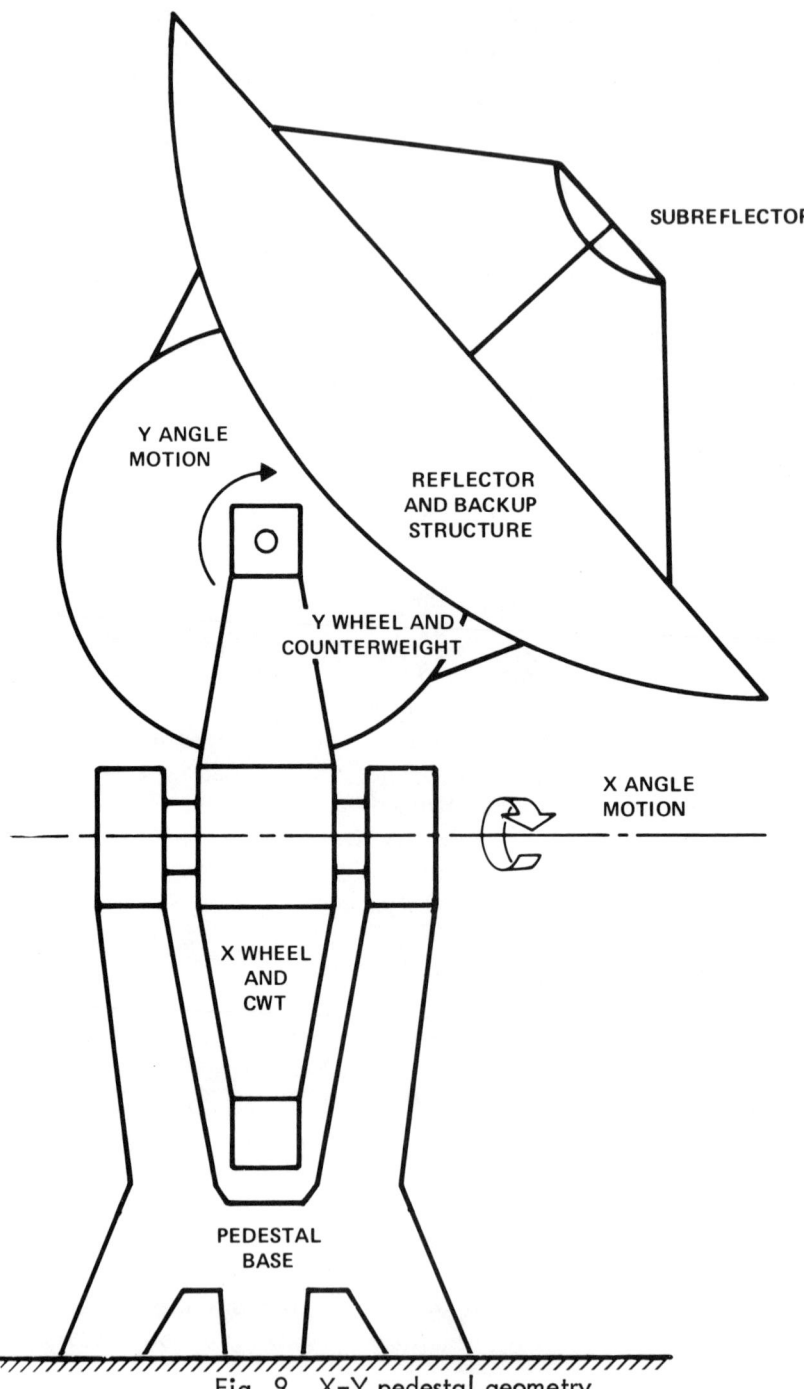

Fig. 9 X-Y pedestal geometry.

Table 2 DCL antenna and servo characteristics

Type: 60-ft reflectors on X-Y positioners

Frequency, MHz

    Transmit band: 5725-6225
    Receive band: 3400-3900

Gain, dB

    Transmit band: 57.5
    Receive band: 53.6

Beam width, deg

    Transmit band: 0.2
    Receive band: 0.3

Angular accuracy (3σ) in 30-mph winds, deg

    Autotrack: 0.020
    Program pointing: 0.035

Angular coverage, deg

    X axis: ±90
    Y axis: ±80

Angular dynamics

    Rates: ±0.3°/sec
    Accelerations: ±0.3°/sec^2

Acquisition capability: Spatial scan of 4° x 20° ellipse

filter that protects the receiver input from the transmitter output and other out-of-band signals, then goes through a waveguide transfer switch to the input of one of a pair of redundant low-noise amplifiers

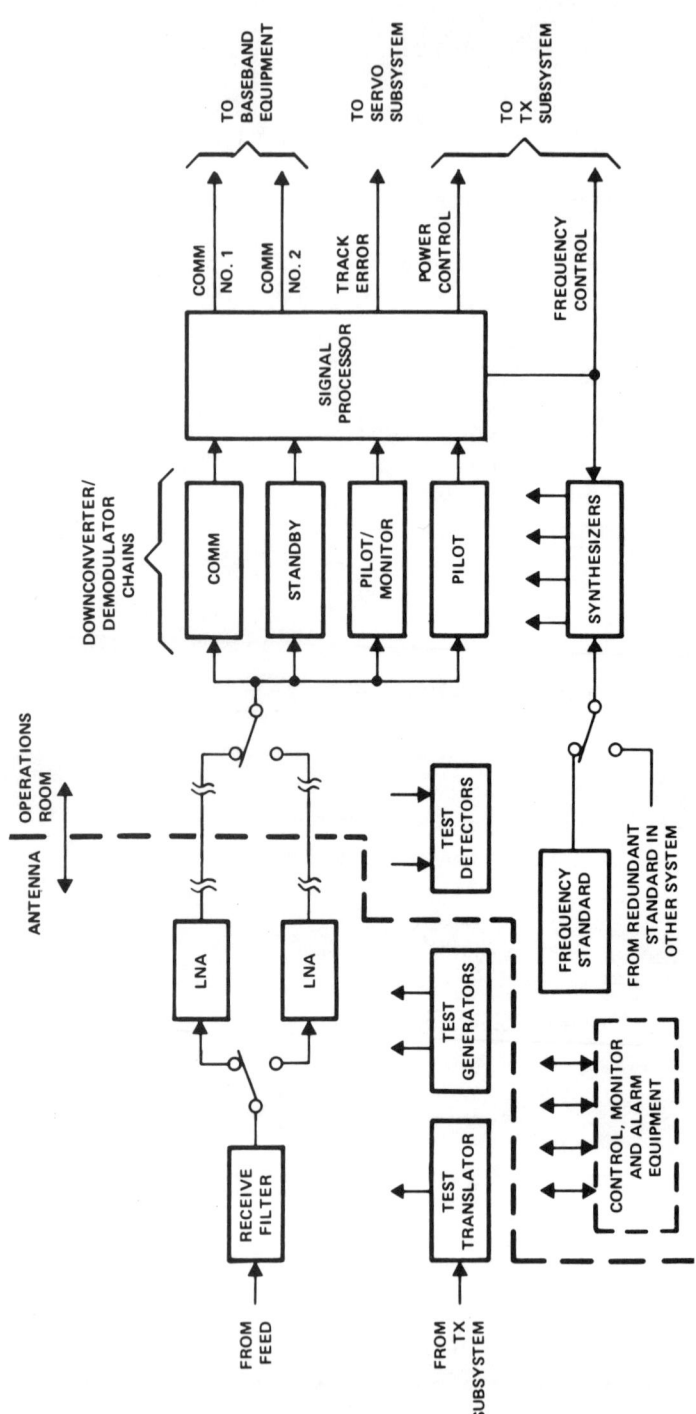

Fig. 10 DCL receiving subsystem.

EARTH STATION FOR US-USSR COMMUNICATIONS LINK 135

(LNA's). These LNA's are thermoelectrically cooled to provide a noise temperature of 70°K. The outputs of the LNA's are brought down from the antenna to the operations room in a pair of waveguides, where they are fed to a second waveguide transfer switch that selects the on-line LNA output. Directional couplers are located at various points before and after the LNA's for injection and recovery of test signals.

The output of the second transfer switch feeds a four-way power divider, which, in turn, drives four identical dual-conversion downconverters. A frequency synthesizer, one for each downconverter, is the source for a variable-frequency local oscillator that permits tuning to any channel in the receive band. Each downconverter has an output at 70 MHz and drives an FM demodulator. Of the four downconverter/demodulators, three are on-line and the fourth is a standby that can be switched in if a failure occurs in one of the on-line units.

Although the downconverter/demodulator chains serve different functions in the system, they are all identical and interchangeable. Operations that are unique to each chain are performed by the signal processor. The functions of the various downconverter/demodulator chains and their relationships to the signal processor are given in the following paragraphs.

Chain No. 1. This chain is tuned to the communication channel: the "hot-line" channel from the USSR. Only its baseband output is used. It is routed directly to the baseband equipment and is designated COMM NO. 1 on the block diagram.

Chain No. 2. This is the standby chain. If all of the other chains are operational, it is also tuned to the communication channel. Its baseband output is routed to the baseband equipment and is designated COMM NO. 2.

Chain No. 3. This chain is alternately switched electronically between the pilot channel and a monitor channel that monitors the transmitted signal as translated and retransmitted by the satellite. The automatic gain control (AGC) and automatic frequency control (AFC) voltages from the demodulator are used by the signal processor for automatic power control and automatic frequency control, respectively, of the transmitter.

Chain No. 4. This chain is tuned to the pilot channel and is the one used for autotrack. Its AM output is routed through the signal processor to the servo and is designated TRACK ERROR. A baseband output is also supplied to the processor. The pilot modulation is a 4-kHz tone, and the processor recognizes this tone in the baseband. However, if the tone is not present upon signal acquisition, the antenna is inhibited from going into autotrack and frequency search continues. This feature is incorporated so that the system cannot operate if it is tuned erroneously to another channel or if the pilot signal disappears.

The signal processor is the focal point for many functions of the communications equipment. Its major contribution, aside from some fault monitoring and control functions that will be left undescribed for the sake of brevity, is in frequency and power control. It performs that function by commutating the pilot/monitor receiver chain between the pilot and monitor channels and looking at the demodulator AGC and AFC as described previously. The commutation cycle has a period of 1 sec, during which the frequency synthesizer is tuned to the pilot channel for 500 msec and to the monitor channel for the other 500 msec. Toward the end of each 500-msec half-period, after all switching transients have died out and the AGC and AFC have had time to settle to essentially final values, the signal processor samples the demodulator AGC and AFC voltages. The monitor voltage samples are compared against the pilot samples, and error signals are developed to control the transmitter.

In the case of automatic power control, the monitor channel can be preset to have a signal level range of ±5.0 dB with respect to the pilot channel by means of a control on the processor panel. The quantity that is monitored is the demodulator AGC voltage, which has a constant slope of 100 mv/dB over the full signal dynamic range. For a programed offset of, say, +3.5 dB, the action of the control loop is to maintain a fixed voltage differential (in this case, +350 mv) between monitor and pilot AGC. This differential is maintained over the full satellite pass, despite the fact that the absolute received signal levels vary over a considerable range. The element controlled by the power control loop is an electronic attenuator that varies the drive to the power amplifier.

For frequency control, the transmitter output frequency is varied by electronically tuning the modulator. The action of the loop is such that the monitor AFC is forced to be equal to the pilot AFC, thus holding a constant frequency difference at the input to the receiver. This control is maintained over a total pilot frequency uncertainty range as great as

±300 kHz. Precise frequency control for the transmitter and receiver is obtained through the use of a precision quartz oscillator located in each receiver subsystem. Provisions are made to allow both systems to operate from one standard in the event of failure of the other standard. Completing the receiver subsystem are test generators and test detectors for LNA on-line gain monitoring, and a test translator that converts the transmitter signal down to the receive band for off-line testing.

Salient performance characteristics of the receiver subsystem are as follows: over-all noise temperature, 95°K; rf bandwidth, 500 MHz; tuning resolution, 40 Hz; demodulator bandwidths, selectable for peak deviations of 30, 42.5, and 65 kHz; signal dynamic range, -114 to -84 dBm; and spurious responses, less than -62 dB.

Transmitter. One of the transmitting subsystems is shown in Fig. 11. Baseband information is supplied to two frequency modulators, which impose this information on 70-MHz carriers. Each modulator output is passed through a dual-conversion upconverter that translates the frequency to the desired C-band transmit frequency, and then through a power amplifier. The parallel-redundant power amplifier outputs then pass through a crosspatch waveguide transfer switch and then to the antenna via parallel waveguide runs. A second transfer switch on the antenna selects one waveguide output to be on-line and directs it to the feed, connecting the standby output to a dummy load. Frequency selection for the transmitter is performed by two frequency synthesizers, one for each of the redundant transmitter chains. The synthesizers are slaved to the system frequency standard and have output frequencies in the vhf range. Frequency multipliers in the upconverters multiply the synthesizer output frequencies by 44, giving a frequency coverage of 500 MHz centered near 5 GHz which serves as the local oscillator to convert the signal to the final C-band transmit frequency. To set up the nominal operating frequency, the operator selects the synthesizer frequency that corresponds to the assigned transmit frequency and dials it with thumbwheel switches.

The nominal operating frequency is the frequency at which the transmitter would operate with no correction for doppler shift. With the system operating through the satellite, however, doppler correction is performed automatically by the signal processor, as previously described, which sends an error signal to the modulator to offset its output frequency by the correct amount. Provisions also are made for manual frequency control from the console. This permits the operator to

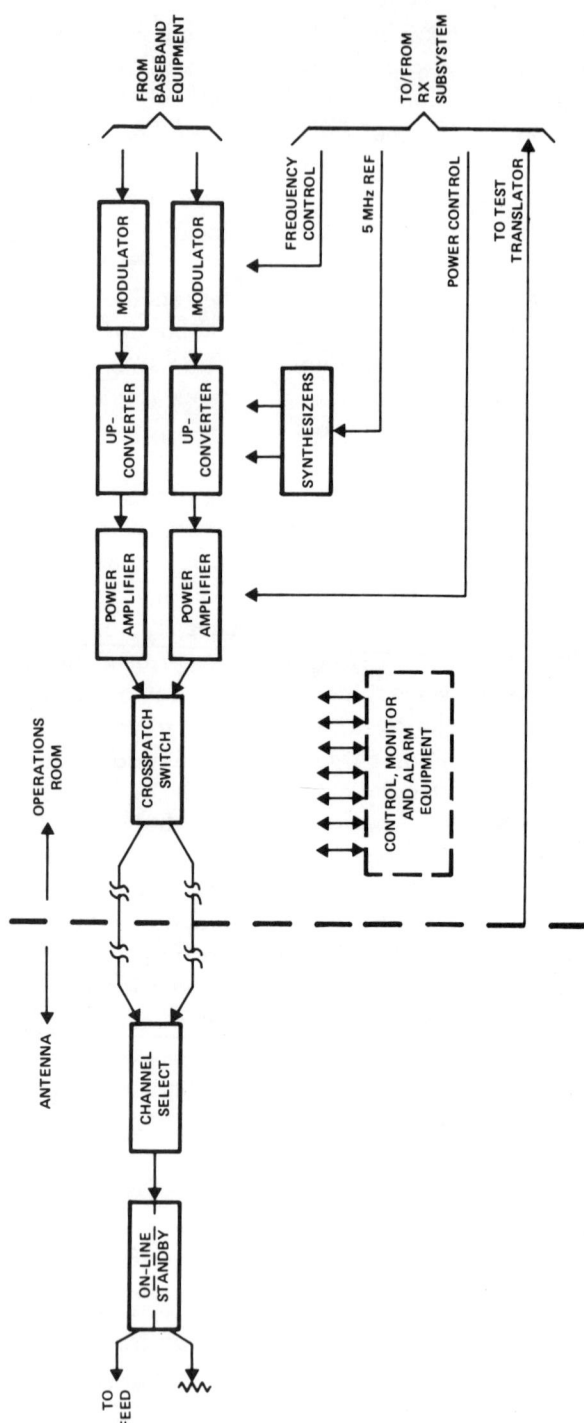

Fig. 11 DCL transmitting subsystem.

preset an estimated initial frequency offset computed within the programer subsystem before placing the system in the automatic mode.

The transmitter power output is controlled via the signal processor to give an EIRP range from 45 to 70 dBw. The signal processor output is a control voltage that goes to an electronic attenuator at the input of the power amplifier and establishes the proper drive level to give the commanded EIRP. The power control loop can be operated in two modes: 1) a "short-loop" mode, wherein the EIRP is controlled locally and a power sample from a directional coupler at the antenna feed is used to generate the error signal for the control loop; and 2) a "long-loop" mode, wherein the EIRP is controlled to maintain the transmitter monitor signal received from the satellite at a preset ratio in comparison with the pilot signal; the error signal in this case is derived by comparing the signal strengths of the monitor and pilot signals.

Broadband TWT power amplifiers, rated at 300 w, are used. They are operated well below saturation even at maximum EIRP, helping to insure trouble-free linear operation of the power control loop over its full dynamic range. The over-all transmitter performance characteristics are summarized as follows: power output, -12.5 to +12.5 dBw; rf bandwidth, 500 MHz; tuning resolution, 44 Hz; deviation capability, 10 to 100 kHz peak; switchover time, 200 msec; and spurious outputs, less than -80 dB.

<u>Loop Performance Characteristics.</u> Some of the transmitter and receiver performance characteristics can be specified and measured only when the two are tied together in a loop configuration, through the test translator or an equivalent frequency converter. These characteristics are summarized as follows: baseband test tone-to-noise ratio, 45 dB minimum with $C/kT = 62.5$ dB-Hz; baseband intermodulation products, less than -45 dB; baseband harmonic distortion products, less than -50 dB; EIRP control accuracy, ±0.5 dB; and frequency control accuracy, ±500 Hz.

Programer Subsystem

The principal function of the programer subsystem is the accurate direction of the pointing of the X-Y antennas in the two communication systems. The programer uses Datacraft Model 6024/5, 24-bit, general purpose processors to generate pointing information using orbital elements of the Molniya satellites as input data. As illustrated in Fig. 12, the programer subsystem is implemented with a pair of

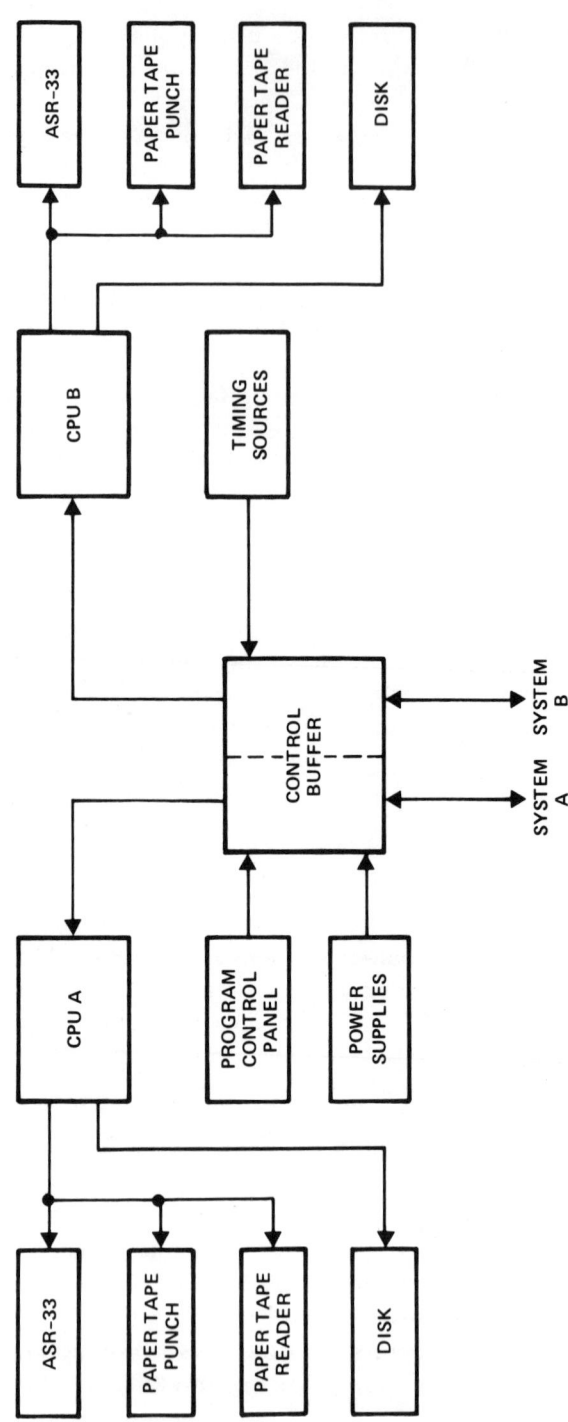

Fig. 12 Programer subsystem block diagram.

independent central processing units (CPU's) plus their peripherals. These interface through a hardware control buffer communicating by means of a 24-bit I/O channel and a priority interrupt bus. Both programers are controlled from a common panel at the Earth Station console. In normal operation, each programer directs a single antenna. When desired, either programer is capable of directing both of the antenna subsystems toward a common satellite.

Ancillary functions provided by the programer subsystem include the generation of spatial scans for acquisition, the adjustment of program pointing commands to reflect the known characteristics of antenna systematic errors, link computations that permit the manual setup of receiver and transmitter parameters so that radiated power and frequencies can be set properly at the time of satellite acquisition, and the comparison of predicted and observed acquisition look angles so that angle offsets may be added to the ephemeris generator output to facilitate subsequent acquisitions. Orbital elements and other input data are entered via the ASR-33 keyboard and may be recalled as hard copy for editing or verification.

Ephemeris Generation. Each programer has the capability of accepting and storing sets of orbital elements provided by the Soviet Union. To provide adequate ephemeris accuracy over periods between element updates which may be as long as 50 days, Harris ESD has incorporated in the DCL ephemeris generation software, a semianalytical perturbation theory that was developed specifically to accommodate highly eccentric 12-hr resonant satellite orbits such as those of the Molniya satellites. Using a numerical integration with a 1-day time step to avoid the time-consuming special perturbation techniques normally necessary to provide high-accuracy ephemeris generation, this theory includes lunisolar gravitational perturbation, the geopotential longitude resonance perturbation, the solar radiation pressure perturbation, and the effects of atmospheric drag near perigee.

Systematic Error Correction. Using analytical predictions confirmed by actual field measurements, Harris ESD has determined the parameters of a software model that represents the sources of systematic errors associated with each of the antenna subsystems of the DCL Earth Station. Specific error contributors that are included within this model are atmospheric refraction at rf frequencies, azimuthal misalignment and mislevel of the X axis of each antenna, X-to-Y and rf-to-Y axis non-orthogonality, and dead load deflections of each antenna structure as a function of pointing attitude.

Operations and Controls

As previously shown, redundancy exists throughout the DCL Earth Station, via both the two independent communication systems and the redundancy encompassed within each system. Key parameters of each system are monitored to provide information and control signals designed to maintain uninterrupted communication despite unavoidable equipment malfunctions. Failures within a unit will produce a control signal used for automatic switchover while supplying appropriate status information at the unit level and to the operator at the console. Based on the information displayed at the console, the console operator can direct other members of the team in the repair and restoration of the equipments. The console design permits a single operator to control both communication systems even during the handover period when two satellites are being utilized.

Occupying two bays each on the left and right sides of the console are the controls and status information for both communication systems. The servo control panel provides mode select capability, slew control, commanded and actual antenna position displays, and servo fault status. Adjacent to the servo panel, the communication control panel affords control and signal information for the transmitter and receiver. Receiver carrier-to-noise meters, baseband level meters, EIRP and frequency level and controls, and automatic/manual switchover controls are available to the operator on the communication control panel. Immediately above these two panels is located a communication status panel with sufficient information for the operator to identify faulty units readily.

The center portion of the console, two bays wide, is dedicated to station controls and status. The baseband communication panel and station control panel are mounted below the station status panel. Station status includes items such as structural interlocks, UPS status, reflector deicing status, and waveguide pressurization status. The station control panel includes controls for both CPU's in the programer. From this panel, the operator selects either computer to operate with either of the communication systems. Based on information received from the USSR, the operator will select an orbital element set for one of those sets available in the CPU. After a few seconds delay, commanded angles, range, range rate, EIRP, and frequency offset information is displayed for operator use. Indicators on this panel are illuminated when the elements in use predict that the satellite is above the horizon at the U.S. and/or USSR Earth Stations.

Program track selection on the servo panel will result in the antenna being driven to the predicted satellite location. During the acquisition process, frequency sweep and a nominally elliptical spatial scan whose major axis is oriented along the satellite track are available to aid acquisition. Again at the operator's choosing, when signal conditions are properly satisifed in the receiver, the servo will go to auto-track and the transmitter will be turned on, initiating the sequence for automatic frequency and EIRP control through the satellite. All of this is accomplished without operator intervention. After performing link checkout test with the new satellite, the communication link will be transferred. The programer command angles then will be used to slew the second antenna to the new satellite for communications backup. The selection of either communication system as prime for the link is accomplished by a single switch on the station control panel.

ECONOMIC CONSIDERATIONS FOR LOW-CAPACITY SHF SATELLITE COMMUNICATIONS EARTH TERMINALS

G. P. Petrick[*] and C. M. Abrahamson[+]

HARRIS Electronic Systems Division, Melbourne, Fla.

Abstract

Increasing requirements for low-capacity terminals are anticipated as the number of communications satellites increases and the cost per transponder channel decreases. The trend expected for these terminals is toward minimizing the life-cycle cost to achieve a specified availability. This paper examines cost tradeoffs for modest G/T and EIRP yet high-availability terminals. The tradeoffs indicate that, for a given availability, lowest cost is obtained with antenna diameters large enough to permit use of nonparametric amplifiers and low-power transmitters. An example of an Earth terminal with modest communications capacity, high availability, and minimum cost is described.

Introduction

A growing requirement for low-capacity Earth terminals is anticipated in the future as the number of communication satellites increases and satellite costs per channel decrease to the point that the space subsystem segment is not a disproportionate percentage of the total link cost. Such satellite communication links then can be competitive with

---

Presented as Paper 74-459 at the AIAA 5th Communications Satellite Systems Conference, Los Angeles, Calif., April 22-24, 1974.
[*]Presently employed by Electronic Communications, Inc., Marketing, St. Petersburg, Fla.
[+]Associate Principal Engineer.

land lines, TROPO, and high-frequency links. Although many military requirements such as tactical considerations, link availability, and communications flexibility practically will dictate the use of satellite communications, it is still extremely important to minimize the cost of these terminals. The major limitations of present low-capacity terminals are the following: 1) they utilize too much satellite power; and 2) they are too costly.

By using the candidate terminals discussed in this paper at the data rates defined and the operational configurations discussed, a minimal amount of satellite power is utilized. In addition, by implementing the equipments and techniques discussed, a terminal can be produced for approximately $100,000-150,000 in moderate quantities.

This paper defines the major cost elements of a terminal and discusses potential cost reduction techniques that can be implemented using the present state of the art. The low-capacity terminals considered are defined to be capable of handling digital data at rates of 75 baud to 16 kbit/sec, which includes TTY and digital voice. Two candidate terminals with values of 14/23-db G/T, 55/64-dbw EIRP, and constant availability are discussed which will provide low error rate communication while utilizing a small percentage of the power of a satellite such as the DSCS II.

The low-capacity terminals considered can be developed for tactical applications, but prime emphasis of this paper addresses a nontactical and special user requirement. Low-cost, low-capacity terminals are especially economical when used in a low-duty-cycle demand assignment mode such as in a polling net and in a broadcast mode. Since unattended operation is often a requirement, low-cost terminals of the future must exhibit high availability and low life cycle as well as low initial cost. Therefore, utilization of all solid-state components with maximum use of microwave integrated circuits (MIC) and large-scale integration (LSI) is mandatory for such a terminal. The quantities required for the low-capacity terminals will be sufficient to realize the cost-savings benefits of these techniques.

Terminal Costs

Major Subsystem Costs

The total Earth terminal hardware costs are concentrated in a grouping of five major subsystems: 1) antenna, 2) transmitter power amplifier and driver, 3) low-noise amplifier, 4) up and down converters, and 5) modem (including coding/decoding equipment if appropriate for the desired service). The individual performance requirements for the first three are adjusted to achieve the desired EIRP and G/T at minimum cost. The cost influence of the last two is then additive, depending upon the number of channels and type of service required.

It is easy to appreciate that, for "small" antenna diameters, a more powerful transmitter and a quieter receiver are needed than for "large" antenna diameters. Although antenna cost increases with diameter, transmitter and receiver costs decrease as less power is required and more noise is permitted. Intuitively, it is suspected that a minimum total cost will be obtained for some combination of antenna diameter, transmitter power, and receiver noise temperature which satisfies the EIRP and G/T objectives. To find this minimum cost combination, it is necessary first to determine, on an individual basis, just how expensive antenna diameter, transmitter power, and receiver noise temperature are.

Cost vs Performance

Before discussing the actual cost vs performance data, it is well to re-emphasize just what kind of terminals and what kind of costs are under study. In this paper, attention is directed toward a modest performance configuration operating in a modest environment. Factors such as size, weight, and setup time, which would be important parameters in a tactical system, are not considered in this paper. It is required further that the total cost of such a terminal be low enough to attract 10's to 100's of users. Therefore, it is recurring costs only that are of concern in this section of the paper. It also is important to realize that the costs presented are the present and near-future cost. Changing technology requires periodic upgrading.

The relationship between antenna diameter and cost has been treated by others; one in particular was found useful for the special category of smaller tracking antennas (under 30 ft in diameter) designed for a quantity market which is the subject of this paper.[1] The separate antenna and feed relationships given in the referenced paper have been combined to give the following expressions, plotted in Fig. 1, where D is the antenna diameter in feet: 1) minimum cost ($10^3) = $1 + 0.8D + 0.016D^2$;   2) average cost (10^3) = $30.4 + 2.36D +

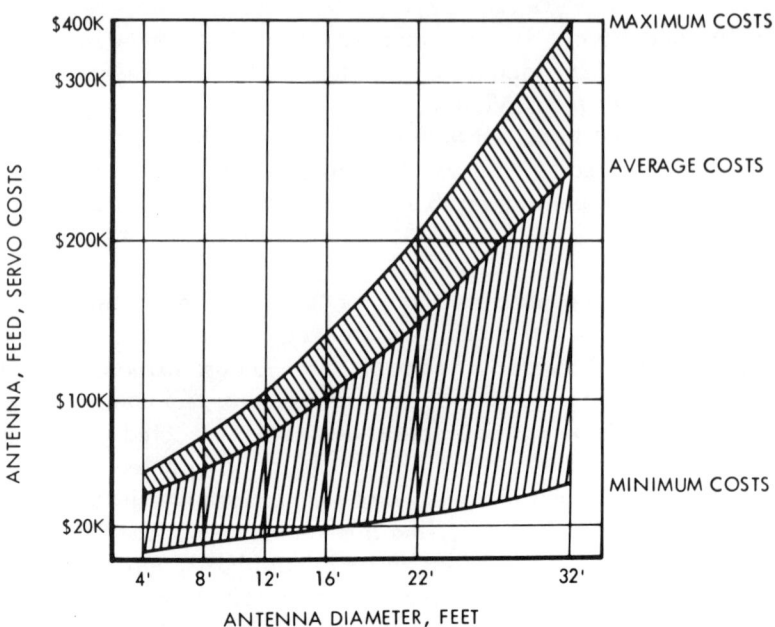

Fig. 1 Antenna, feed, servo cost vs antenna diameter.

$0.134D^2$; and 3) maximum cost (10^3) = $42.0 + 3.13D + 0.191D^2$. These costs include antenna, feed, and servo equipment. As defined in the referenced paper, maximum costs represent military systems, average costs represent revenue-oriented commercial systems (e.g., COMSAT and AT&T), and minimum costs represent commercial systems that would be the outgrowth of a highly competitive and mass production market. It is expected that the costs for the type of terminal addressed in this paper will be in between average and minimum, with a bias toward the minimum cost curve. The justification for this low-cost estimate is the implementation of a low-cost tracking technique such as step track and the use of a limited sector coverage open-frame antenna pedestal.

Figure 2 represents the anticipated cost picture for transmitter power over a three-decade range from 1 w to 1 kw. These costs are for complete subsystems capable of operating over the entire 7.9- to 8.4-GHz band, and assume a nominal -10 dbm available from the up converter.

ECONOMIC CONSIDERATIONS FOR EARTH TERMINALS

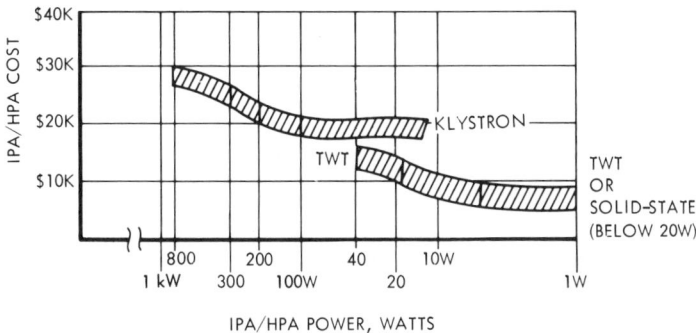

Fig. 2 IPA/HPA cost vs HPA power (7.9/8.4 GHz).

In the power range from 500 to 1000 w, klystrons are the most cost-effective, since the beam power supply is appropriately sized for the output power required. The power supply and the protective and monitoring circuitry required greatly influence the cost in this power range. The range between 200 and 500 w is achieved either by reduced rf drive using the 500-w klystron transmitter or, more economically, with a TWT HPA. In the range between approximately 20 and 200 w, the TWT is the prime candidate. At power levels below 20 w are IMPATT diode amplifiers, which are available presently with 5-w output and, assuming appropriate product engineering and quantity production, will be competitively priced with TWT's.

Figure 3 indicates the cost associated with the range of noise temperatures achieved by devices from cryogenically cooled parametric amplifiers to Schottky barrier mixers. Between the two cost extremes are thermoelectrically cooled paramps, room-temperature paramps, tunnel diode amplifiers, and, ultimately competing with tunnel diode amplifiers, the gallium arsenide field effect transistor amplifier. All of these devices are capable of operating over the full instantaneous bandwidth from 7.25 to 7.75 GHz.

Figures 1-3 indicate the expected cost ranges as a function of antenna diameter (terminal size), transmitter power, and receiver noise temperature. These are the prime variable cost elements; tradeoffs to achieve a minimum cost terminal for a specified EIRP and G/T now can be conducted.

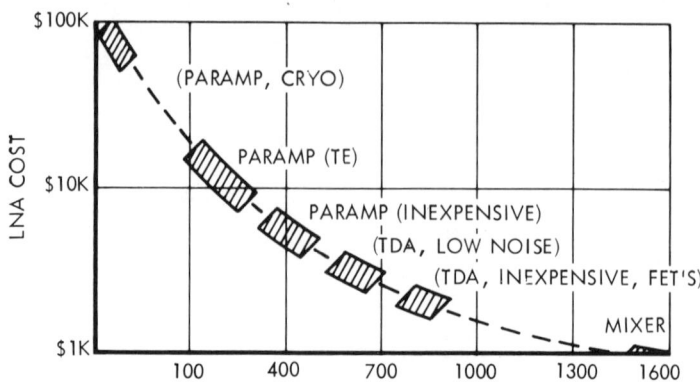

Fig. 3 LNA noise temperature, °K (7.25/7.75 GHz).

Candidate Terminals

Definition of a modest communications capacity terminal with a low acquisition and life-cycle cost is the major goal for this study. The communications capacity of 75-baud teletype and 16-kbit/sec digital voice is realized with nominal EIRP values of 55-64 dbw and G/T value of 14-23 db/°K at each end of a duplex link. These major system parameters are, therefore, those for which the antenna diameter, transmitter power, and receiver noise temperature tradeoffs will be made in order to define two minimum-cost candidate terminals.

The basic elements of the candidate terminals are shown in the block diagram of Fig. 4. It is a single-thread (i.e., nonredundant) configuration, permitting full duplex operation, tunable over the entire uplink 7.9- to 8.4-GHz range and over the entire downlink 7.25- to 7.75-GHz range. One communications carrier for the uplink and one for the downlink are accommodated, with antenna tracking, via step track, accomplished using either the downlink communications carrier or the satellite beacon. With the basic configuration defined, it is possible now to develop the final information required for the terminal cost curves. This is done by first examining the influence of antenna diameter on transmitter power and receiver noise temperature for a constant EIRP and G/T.

Tradeoffs for 64-dbw EIRP, 23-db G/T Terminal

Figure 5 indicates the basic relationships assumed in calculating the receiver noise temperature and antenna diameter combinations for

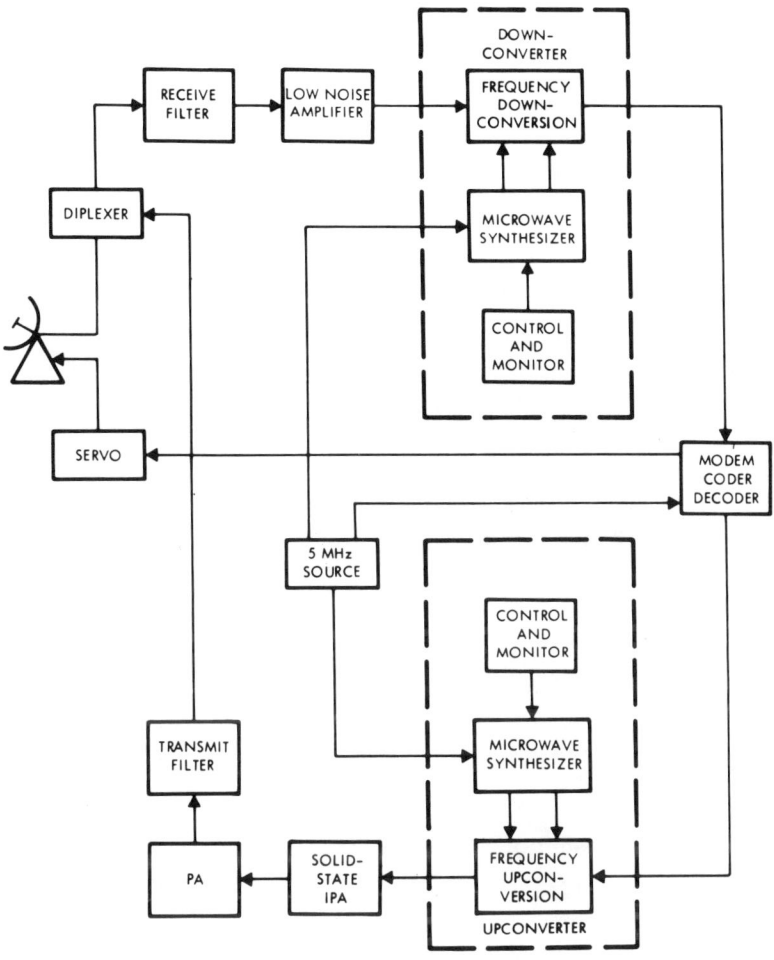

Fig. 4 Basic Earth terminal block diagram.

constant G/T. Table 1 summarizes the values calculated for antenna diameters from 6 to 32 ft. The antenna gain was calculated assuming an efficiency of 55%. The required transmit power level assumes that the HPA is mounted on or near the antenna structure with a loss of 0.5 db between it and the feed.

Table 2 summarizes the cost information from Figs. 1-3 for the range of parameters listed in Table 1. Also shown is the class of device which would be used to obtain the required transmitter power and receiver noise temperature. The table indicates that the minimum-

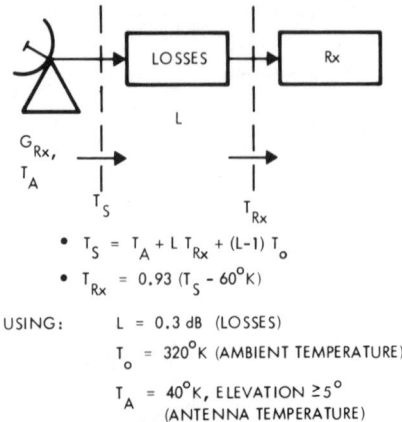

Fig. 5 Receiver noise temperature relationship. $T_S = T_A + L T_{Rx} + (L-1) T_o$; $T_{Rx} = 0.93 (T_S - 60° K)$; using $L = 0.3$ db (losses), $T_o = 320°K$ (ambient temperature), and $T_A = 40°K$, elevation $\geq 5°$ (antenna temperature).

Table 1 Antenna diameter (D), transmitter power (P_{Tx}), and receiver noise temperature (T_{Rx}) for EIRP = 64 dbw and $G/T_S = 23$ db/°K

D, ft	G_{TX}, db	Required P_{TX}, dbw	Required P_{TX}, w	G_{RX}, db	Required T_S, db/°K	Required T_{RX}, °K (NF, db)
6	41.0	25.0	320.0	40.5	17.5	...
8	43.5	22.5	180.0	43.0	20.0	37 (0.5 db)
12	47.2	18.8	77.0	46.7	23.7	158 (1.8 db)
16	49.7	16.3	43.0	49.2	26.2	355 (3.1 db)
22	52.5	12.0	16.0	52.0	29.0	688 (5.0 db)
32	55.7	9.2	8.3	55.2	32.2	1488 (7.5 db)

ECONOMIC CONSIDERATIONS FOR EARTH TERMINALS 153

cost system would utilize an antenna in the 12- to 16-ft-diam range. However, it is judged essential that, in addition to constant EIRP and G/T, the cost comparisons should be made on the basis of constant availability. Availability is defined simply as the ratio of (uptime) to (uptime plus downtime) and involves not only the expected failure rates (λ_i) but the mean-time-to-restore an equipment to operation after the failure (MTR$_i$). If the ratio MTR/λ is small (say, less than 0.01 with both quantities expressed in hours), total terminal availability, A_{total}, may be expressed very simply as

$$A_{total} = 1 - U_{total} = 1 - \sum_i \lambda_i \, MTR_i$$

where U_{total} is the total terminal unavailability expressed as a summation of the individual products of failure rate and mean-time-to-restore for each equipment. Figure 6 indicates the general approach used in calculating availability.

Table 2 Total single-thread terminal variable costs vs antenna diameter (cost, 10^3; EIRP = 64 dbw; G/T = 23 db/°K)

Diameter	Antenna		HPA		LNA		Total	
D, ft	Min. cost	Avg. cost	Min. cost	Avg. cost	Min. cost	Avg. cost	Min. cost	Avg. cost
			(Klystron)		(Cryocooled paramp)			
8	10	60	20	25	60	100	90	185
			(Klystron; TWT)		(TE-cooled paramp)			
12	14	78	15	20	8	10	37	108
			(TWT)		(Room temp. paramp)			
16	19	103	12	16	4	6	35	125
			(Solid-state; TWT)		(TDA; FET amp)			
22	28	147	10	15	2	3	40	165
			(Solid-state; TWT)		(TDA; FET amp)			
32	45	243	5	10	2	3	52	256

Using this procedure for each terminal size, it was found that the availability for the single-thread terminals requiring the higher-power transmitters was 0.9989 for the 8- and 12-ft and 0.9992 for the 16-ft, vs the 0.9994 and 0.9995 calculated for the terminals utilizing the lower-power transmitters. Making the HPA redundant was the most cost-effective method found for improving the availability of the 8-, 12-, and 16-ft terminals to 0.9994. The cost data shown in Fig. 7, therefore, represent a constant inherent availability of approximately 0.9994 for all terminal configuration and, it is believed, present a more accurate cost comparison.

It is noted that with a 22-ft antenna a complete solid-state terminal can be implemented, a TDA (or an FET amplifier in the not-too-far-distant future) in lieu of a parametric amplifier, and a solid-state HPA instead of a tube-type HPA. Factors that make this an attractive configuration (factors *not* included in the cost or inherent availability anal-

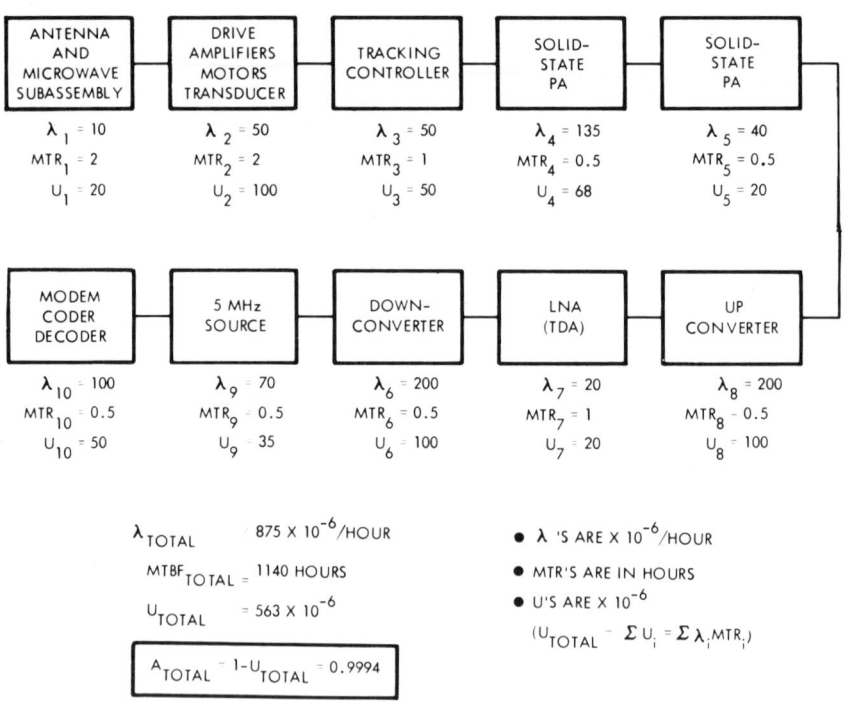

Fig. 6 Earth terminal inherent availability diagram (22-ft antenna). λ's are $\times 10^{-6}$/hr; MTR's are in hours; and U's are $\times 10^{-6}$ ($U_{total} = \sum U_i = \sum \lambda_i \cdot MTR_i$).

yses) are lower life cycle cost due to reduced maintenance, and higher operational readiness due to reduced complexity and few adjustments.

Table 3 Antenna diameter (D), transmitter power (P_{Tx}), and receiver noise temperature (T_{Rx}) for EIRP=55 dbw and G/T_S= 14 db/°K

D, ft	G_{TX}, db	Required P_{TX}, dbw	Required P_{TX}, w	G_{RX}, db	Required T_S, db/°K	Required T_{RX}, °K (NF, db)
4	37.5	19.5	90.0	37.0	23.0	130 (1.5 db)
6	41.0	16.0	40.0	40.5	26.5	363 (3.3 db)
8	43.5	12.0	16.0	43.0	29.0	688 (5.0 db)
12	47.2	9.3	8.5	46.7	32.7	1674 (7.9 db)

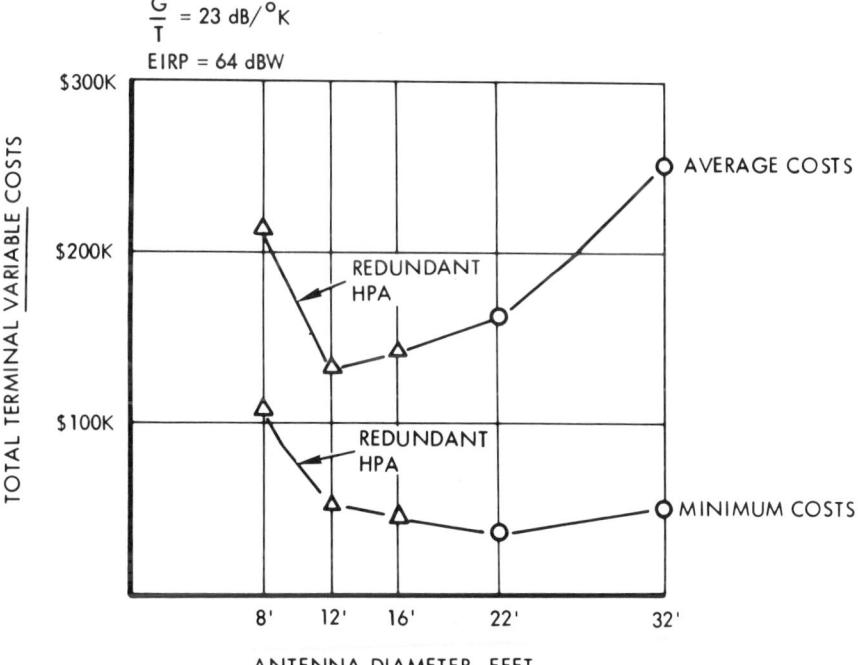

Fig. 7 Terminal variable costs vs antenna diameter (64-dbw EIRP).

Tradeoffs for 55-dbw EIRP, 14-db G/T Terminal

Table 3 lists the parameters necessary to attain 55-dbw EIRP and 14-db G/T for antenna diameters from 4 to 12 ft, using the relationships of Fig. 5, assuming an antenna efficiency of 55%. Table 4 indicates the total costs for a single-thread terminal using the data from Figs. 1-3. Figure 8 presents the cost data in graphic form for terminals with constant availability and indicates than an 8-ft diam will yield minimum cost. This terminal requires no redundancy to achieve an inherent availability of 0.9994, and it utilizes a TDA (or FET) and a low-power TWT (or solid-state) HPA.

Hardware Implementation

An engineering model of the 14-db G/T candidate terminal was assembled by Harris ESD, utilizing equipment developed for the U.S. Army SATCOM Agency. This terminal, shown in Fig. 9, was demonstrated successfully at the 1973 AFCEA Convention linking 75-baud TTY through the IDCSP and DSCS II satellites.

The up and down converters utilized were developed by Harris ESD on an exploratory development contract with SATCOM. The tunnel diode amplifier and IMPATT diode IPA were purchased items. The IMPATT diode HPA was developed by Raytheon for the SATCOM Agency, whereas the modem, coder/decoder, servo control, and antenna are Harris ESD product-line items.

Antenna

The step-track antenna shown in Fig. 10 is one of a family of open-space, frame antenna terminals developed especially for use as communication satellite terminals. In addition to providing transmit and receive capability in the 7.25- to 8.4-GHz frequency range, it is capable of autotracking through the use of a unique step track algorithm and motor-driven linear actuators in two orthogonal axes. Utilization of step track and the limited sector coverage of the pedestal permit lower implementation cost than conventional techniques.

The antenna consists of a four-section parabolic reflector, a Cassegrain self-supporting feed system, and an rf equipment enclosure in the dish center hub for mounting preamplification components. The two-axis mount supplies sky coverage by movement in the traverse (limited azimuth sector) axis and in elevation. The below-axis pedestal accommodates initial siting and antenna ground clearance.

Table 4 Total single-thread terminal variable costs vs antenna diameter (cost, 10^3; EIRP = 55 dbw; G/T = 14 db/°K)

Diameter	Antenna		HPA		LNA		Total	
D, ft	Min. cost	Avg. cost	Min. cost	Avg. cost	Min. cost	Avg. cost	Min. cost	Avg. cost
			(Klystron)		(TE-cooled paramp)			
4	6	42	15	20	8	10	29	72
			(TWT)		(Room-temp. paramp)			
6	8	50	12	16	4	6	24	72
			(Solid-state; TWT)		(TDA; FET amp)			
8	10	60	10	15	2	3	22	78
			(Solid-state; TWT)		(TDA; FET amp)			
12	14	78	5	10	2	3	21	91

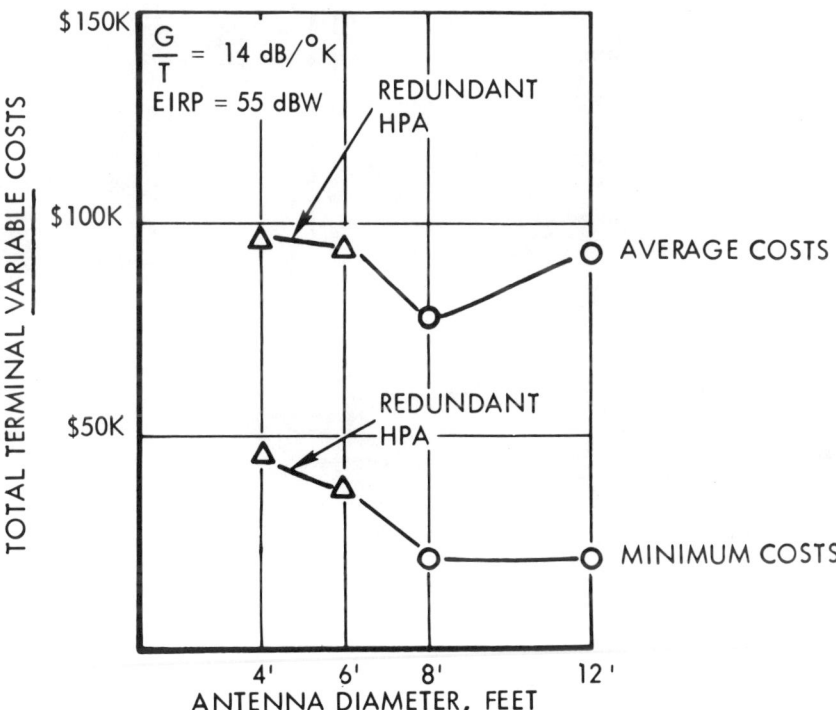

Fig. 8 Terminal variable costs vs antenna diameter (55-dbw EIRP).

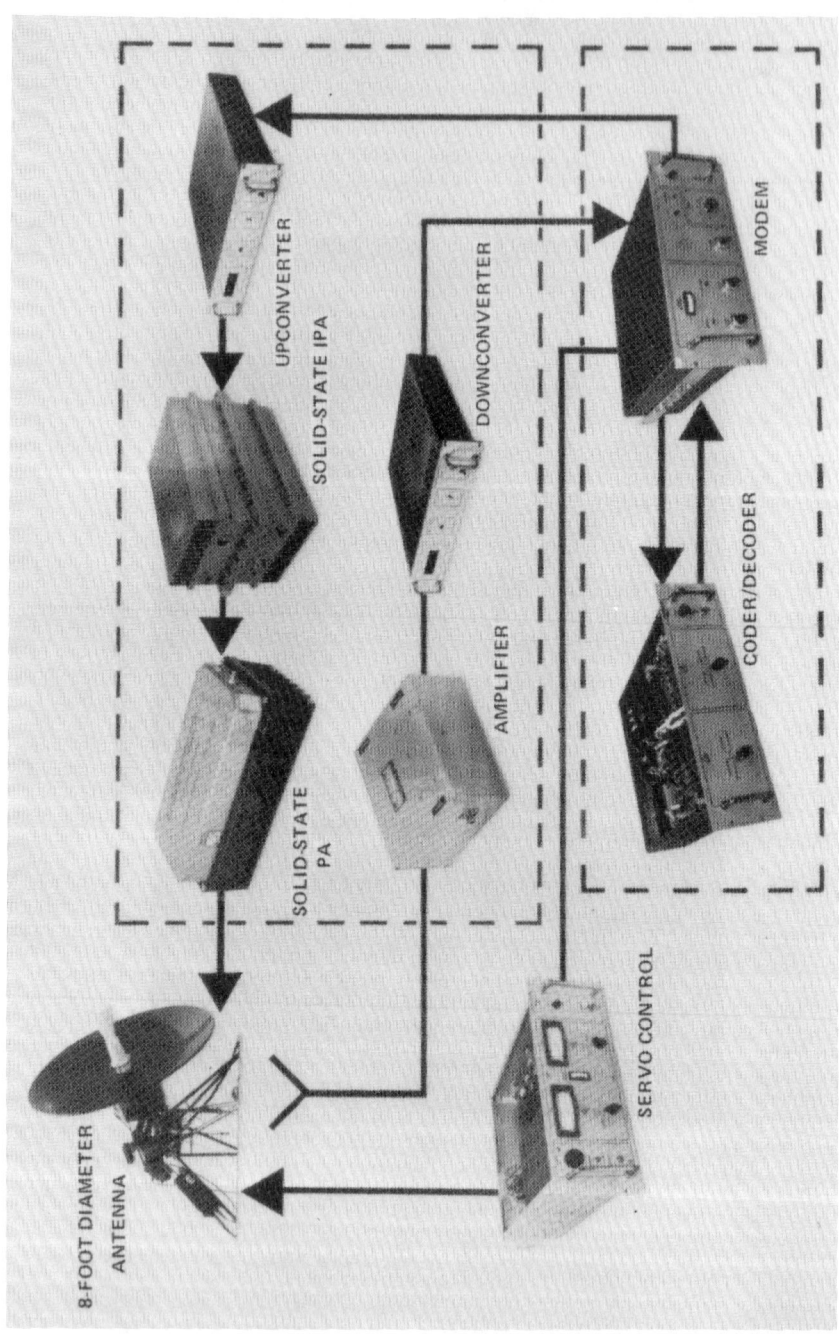

Fig. 9 X-band solid-state satellite communications terminal hardware.

ECONOMIC CONSIDERATIONS FOR EARTH TERMINALS

The reflector is a 4-section, 8-ft-diam parabolic dish. The reflector panels are fabricated by a proprietary procedure from honeycomb core covered by a fiberglass skin and then laminated with a conductive reflector surface. Each quarter-section of the dish weighs approximately 18 lb. The Cassegrain feed is self-supporting from the reflector vertex. The antenna system utilizes step-track servo control.

In manual tracking, an operator watches the received signal level meter and, alternating the axes, moves the antenna to see if the level can be increased. The operator periodically repeats this process as the satellite moves. In step-track, the antenna is moved automatically (stepped) a given position increment angle ($\Delta\theta$) in each axis, alternately. The signal level is sampled (or integrated), and the antenna beam is stepped in one axis. After the step is made, the new level is integrated and compared with the preceding sample from that

Fig. 10 Economical step-track antenna subsystem.

axis to determine the direction of the next angular step. Thus, the step-track servo control has shifted the burden of constant attention and subjective judgment from the operator to an algorithm designed to seek the peak of the beam by observing changes in the received signal as shown by the step-track motion and making decisions from that information. Table 5 summarizes the antenna system specification.

Up and Down Converters

Although not involved in the tradeoffs for minimum <u>relative</u> cost, the up and down converters shown in Fig. 11 contribute to the <u>absolute</u> cost. These converters utilize hybrid microwave integrated circuits (HMIC) to assist in achieving a design that will have low cost production. The converters are rack-mountable, incorporating synthesizers, LO circuitry, rf/IF circuitry, 5-MHz reference oscillator, and power supplies in a completely self-contained drawer, ready for system integration. The desired X-band frequency is selected on the front panel thumbwheel switches, which read out directly, a valuable operational convenience. The synthesizer-LO chain is frequency agile, requiring no manual tuning.

<u>Solid-State Power Amplifier</u>. This unit is a 5-w, developmental IMPATT diode amplifier developed for the Army SATCOM Agency by Raytheon. Four of these units could be combined and driven by a solid-state IPA to provide the 16 w for the two candidate terminals. With the current state of the art of IMPATT diode technology, it is believed that a 20-w amplifier could be available for terminal use in the 1975 to 1980 time frame.

Other Equipments

The low-noise receive amplifier utilized was a tunnel diode amplifier with a noise figure of 4.7 db. The transmitter intermediate power amplifier provided a power output of 0.75 w and 40 db of gain.

The modem utilized employed biphase PSK and was used in conjunction with a convolutional encoder, Viterbi decoder that provided a bit error rate of 10^{-5} for an E_b/N_o of 6 db.

Terminal Communications Capability

The communication capability of the candidate terminals is evaluated for data rates of 75 baud to 16 kbit/sec in the following

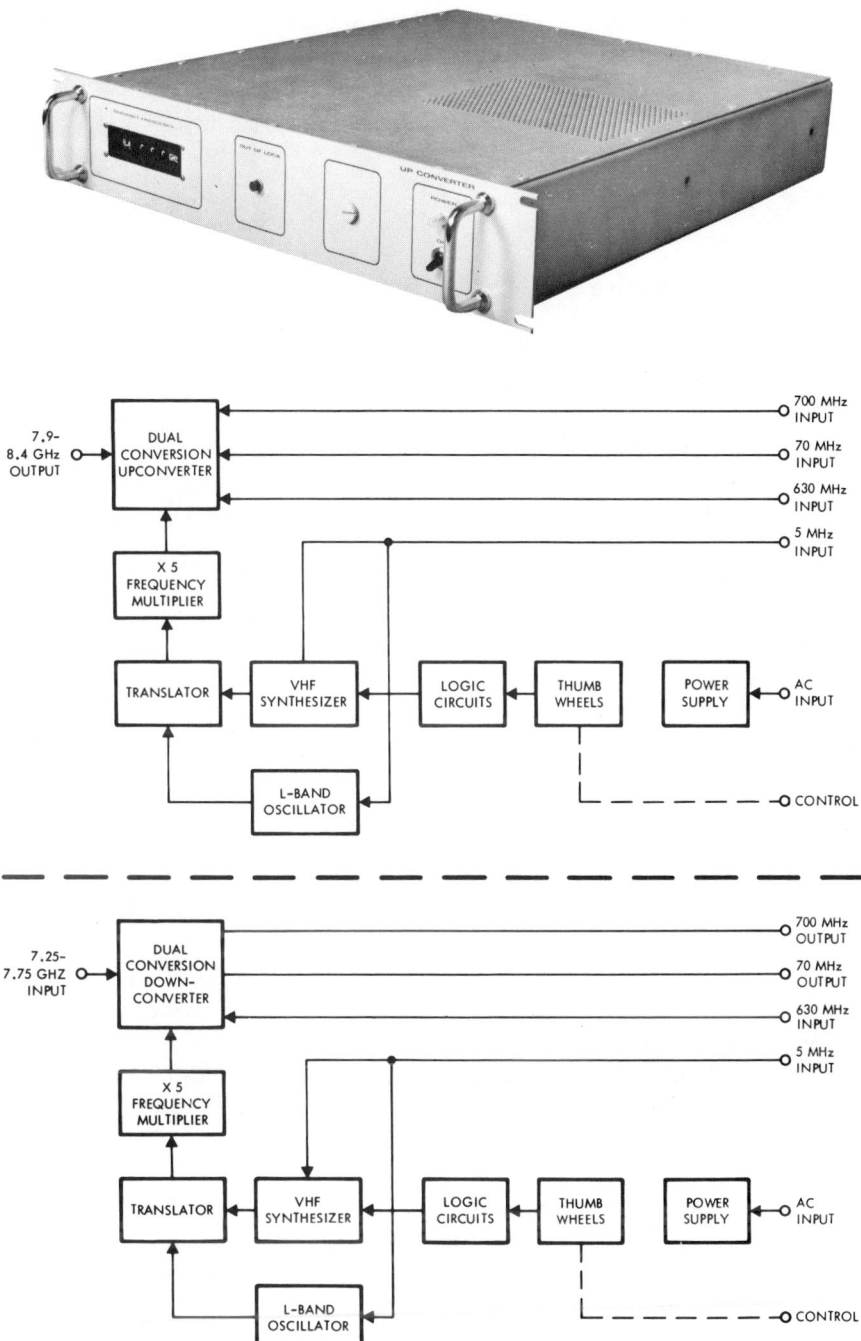

Fig. 11 Solid-state up and down converters.

Table 5 Summary of step-track antenna specifications

Frequency	7.25 to 7.75 GHz, receive 7.9 to 8.4 GHz, transmit
Gain (nominal)	43 db, receive 43.7 db, transmit
Transmit/receive isolation	15 db, minimum
Side lobes	-15 db
Axial ratio	2 db, maximum
Tracking accuracy	Typical 0.3 db (1σ) with 38 db C/kT
Size	8-ft-diam reflector
Weight	381 lb
Wind environment	45 mph, with gusts to 60 mph, operating; 60 to 120 mph slow
Sky coverage	$30° \times 30°$ sector

modes: 1) back-to-back through the DSCS II in the Earth coverage mode; 2) back-to-back through the DSCS II in the narrow coverage mode; 3) a duplex link through the DSCS II in the Earth coverage mode to a 39-db G/T terminal such as the AN/MSC-60; and 4) a duplex link through the DSCS II in the narrow coverage mode to a 39-db G/T terminal such as the AN/MSC-60.

The terminal communication capability will be defined as a percentage of satellite power utilized. This is done since, all too often, a terminal cost is considered independent of the space subsystem cost. The motive behind the definition is valid, since the prime power of a satellite constitutes the largest fraction of its weight, and both launch cost and satellite cost have been found to be generally proportional to satellite weight.[2] Therefore, it is the weight that

ECONOMIC CONSIDERATIONS FOR EARTH TERMINALS

determines the cost, the rf power being directly proportional to the weight because of the prime power required.

The required signal power at the terminal is determined based on the following assumptions: 1) digital data with PSK modulation; 2) a bit error rate (BER) of 10^{-5}; 3) utilization of forward error correction coding/decoding; 4) $E_b/N_o = 6$ db for 10^{-5} BER; and 5) noise contribution from the satellite is negligible.

Therefore, the required C/kT is 24.8 db for 75 bit/sec and 48.5 for 16 kbit/sec, where C/kT is carrier power-to-noise density ratio, and E_b/N_o is the carrier power-to-noise power ratio in the bit rate bandwidth.

The required EIRP from the satellite is expressed as follows:

$$EIRP = (C/kT)_{required} + L_{FS} + L_O - (G/T_R) + k + M$$

where EIRP is satellite effective isotropic radiated power, L_{FS} is free space loss = 202.6 db, L_O is miscellaneous losses = 1.0 db, (G/T_R) is the terminal antenna gain to system noise temperature ratio, k is -228.6 db (Boltzman constant), and M is margin = 6 db.

For the two candidate terminals discussed previously, the required satellite power is shown in Table 6. The percentage of total transponder power required is shown in Table 6. Analysis of Table 7 indicates that, at TTY data rates, neither the 14-db G/T nor 23-db G/T terminal utilizes an excessive percentage of the satellite power. However, at the digital voice rate, the smaller terminal requires one-fifth of the entire Earth coverage transponder and over 1% of the narrow-beam transponder.

However, in order to size the terminal for a given link requirement, the total system cost must be considered for both satellite and ground terminal. The monthly DSCS II space subsystem cost based on development, procurement, launch, and O&M costs, averaged over a 10-yr life cycle, is given in Table 8. If the candidate terminals were fully developed and produced in quantities of 100 to 500, the total terminal cost would be approximately as follows:

14-db G/T terminal recurring cost	$ 90,000
Maintenance over 10-yr life (based on 10% maintenance cost per year)	90,000
	$180,000

Table 6 Required satellite EIRP for single link

Terminal	Data rate	Satellite EIRP, dbw	Required ground terminal EIRP with transponder at nominal gain setting, dbw	
			Earth coverage	Narrow beam
14-db G/T, 55-dbw EIRP (8-ft antenna)	75 bit/sec	-7.8	53.2	43.2
	16 kbit/sec	15.5	76.5*	66.5*
23-db G/T, 64-dbw EIRP (22-ft antenna)	75 bit/sec	-16.8	44.2	34.2
	16 kbit/sec	6.5	67.5*	57.5

*Unattainable with satellite nominal gain setting.

Table 7 Percentage of satellite power required

Terminal	Data rate	Percentage of DSCS II satellite power required			
		Full duplex with like terminal		Full duplex with 39-db G/T terminal	
		EC mode	NC mode	EC mode	NC mode
14-db G/T (8-ft antenna)	75 bit/sec	0.1	0.006	0.05	0.003
	16 kbit/sec	20.0	1.25	10.0	0.62
23-db G/T (22-ft antenna)	75 bit/sec	0.012	0.00075	0.006	0.00038
	16 kbit/sec	2.5	0.158	1.25	0.079

Table 8 Monthly space subsystem cost allocation[3]

Terminal	Data rate	EC monthly cost, $10^3	NC monthly cost, 10^3
14-db G/T (8-ft antenna)	75 bit/sec	0.375	0.024
	2.4 kbit/sec	12.0	0.76
	16 kbit/sec	80.0	5.05
23-db G/T (22-ft antenna)	75 bit/sec	0.047	0.003
	2.4 kbit/sec	1.5	0.095
	16 kbit/sec	10.0	0.63

ECONOMIC CONSIDERATIONS FOR EARTH TERMINALS

Over a 10-yr life span, terminal monthly cost is approximately $1500/month exclusive of operation costs. In a similar manner, the approximate monthly cost for the 23-db G/T terminal is $2000/month.

Therefore, if a 16-kbit/sec link were to be established, utilizing the narrow-beam mode, the total link cost utilizing the 14-db G/T terminal would be as follows:

$5050/month for transponder
1500/month for terminal
6550/month total

If the 23-db G/T terminal were utilized, the total link cost would be as follows:

$ 630/month for transponder
2000/month for terminal
2630/month total

Therefore, the total system cost would be less, utilizing the more expensive terminal. However, if the data rates were limited to TTY, the system utilizing the 14-db G/T terminal would be more economical.

Consider a system utilizing 10 terminals, operating at 16 kbit/sec in which the terminals are polled or utilized in the broadcast mode. In either case, the transponder cost would be time shared effectively. The total system cost utilizing the 14-db G/T terminals would be as follows:

$ 5050/month for transponder
15,000/month for 10 terminals
20,050/month total

If the 23-db G/T terminals were utilized, the total system cost would be as follows:

$ 630/month for transponder
20,000/month for 10 terminals
20,630/month total

Therefore, for such a system, use of the lower-cost terminals would result in a lower over-all system cost. This clearly indicates the importance of considering the space subsystem cost, data rates, and network configuration in the design of a cost-effective system.

The cost estimates in this paper are those of the authors, with informal contribution from suppliers of terminal subsystems. The space segment costs are taken directly from Ref. 3.

References

[1] Cuccia, C. L. and Teicher, S., "The Economics of Antenna Receiving Systems," June 1969, Philco-Ford Corp., WDL Div., Microwaves.

[2] Hadfield, B. M., "Satellite Systems Planning Cost," Rept. MTP-138, Oct. 1972, Mitre Corp.

[3] "Satellite Communication Reference Data Handbook," July 1972, Defense Communications Agency, Appendix F.

OPERATIONAL EXPERIENCE WITH
SMALL UNATTENDED TELEVISION RECEIVE EARTH STATIONS

A.D.D. Miller*

Telesat Canada, Ottawa, Ontario, Canada

Abstract

This paper describes the initial basic operational performance of a satellite remote television service receive system, and assesses the extent to which the initial objectives have been achieved over the first year of operation. The Earth stations provide high-quality live color television service to 24 remote Northern Canadian communities with a basic service quality in excess of 48 db. The unattended equipment, housed in a climate-controlled fiberglass shelter, operates with outside temperatures ranging from -70° to $105^\circ F$, with icing conditions. A 26.5-ft antenna and $100^\circ K$ low-noise amplifier are used. In addition to television, the equipment provides a second audio circuit and automatic control of the associated unattended TV transmitter. By careful planning of spares deployment and with refined maintenance procedures, during the first year of operation the design availability of 99.2% was exceeded.

Presented as Paper 74-454 at the AIAA 5th Communications Satellite Systems Conference, Los Angeles, Calif., April 22-24, 1974. The assistance received from many members of the Engineering and Operations staff of Telesat Canada in providing data for this paper is gratefully acknowledged.

* Formerly Project Engineer, Earth Station Engineering, Telesat Canada; currently President, Miller Communication Systems Limited.

I. Introduction

Remote television distribution via satellite into the vast expanse of the Canadian North commenced on Feb. 5, 1973 when residents of Goose Bay and Frobisher Bay received live programing for the first time. Later that day, several communities in the Pacific Time Zone also received programing via satellite. These events were preceded by a great deal of planning and a significant implementation program both by Telesat and Canadian Broadcasting Corporation. A coordinated approach was maintained throughout the project to insure the orderly provision of an acceptable and reliable service via satellite. Where possible, receiving Earth stations were colocated with CBC transmitter facilities, thereby easing the problem of connection. In many locations, the broadcaster housed some of his equipment in the Earth station shelter. Since the broadcaster himself effects some operational control on the facility to increase the flexibility and utility of the satellite service, an overall network control facility was designed as a joint venture.

This paper deals with the remote television stations (RTV) and seeks to identify some of the factors contributing to high service performance in a facility of this type. An analysis of service performance is taken from well-validated service records, which have been compiled from the first year of operation of remote television stations.

II. Basic Operational Requirements

The basic requirement was for the provision of color television programing, together with two audio channels, to be fed to 24 remote locations (listed in Table 1) for full-time continuous service. Unavailability objectives of less than 0.8% for each transmit-receive link were postulated as a service objective. The service quality required by the broadcaster, the Canadian Broadcasting Corporation, is identified, and the system design implemented to achieve the desired performance is described in a companion paper.[1]

An additional operational requirement was that any location should be capable of being commanded to receive one of several television feeds under the control of the broadcaster. This necessitated the provision of a frequency agile receiver and led to the design of the network control system, which allows commands to be passed from television operating centers

Table 1 Remote television stations

Cassiar,	British Columbia
Churchill,	Manitoba
Clinton Creek,	Yukon Territory
Dawson,	Yukon Territory
Elsa Mayo,	Yukon Territory
Faro,	Yukon Territory
Fort Chimo,	Quebec
Fort George,	Quebec
Fort Nelson,	British Columbia
Fort Simpson,	Northwest Territories
Fort Smith,	Northwest Territories
Frobisher,	Northwest Territories
Goose Bay,	Newfoundland
Great Whale,	Quebec
Inuvik,	Northwest Territories
Magdalen Islands,	Quebec
Norman Wells,	Northwest Territories
Pine Point,	Northwest Territories
Port-au-Port,	Newfoundland
Uranium City,	Saskatchewan
Watson Lake,	Yukon Territory
Whitehorse,	Yukon Territory
Yellowknife,	Northwest Territories
Rankin Inlet,	Northwest Territories
Sept-Iles,	Quebec

in Montreal and Toronto via an additional audio channel multiplexed on the same carrier with the television and program channels.

III. Remote Television Earth Station Design

The Earth station design implemented to meet the system and operational requirements is as indicated in the simplified block diagram, Fig. 1. The specification summary is shown in Table 2.

Antenna

The antenna is a shaped, dual reflector system of Gregorian configuration equipped with a broadband feed horn capable of transmitting and receiving in the 5.925-6.425-GHz and 3.7-4.2-GHz bands, respectively. The addition of the transmit capability was deemed prudent to permit future service

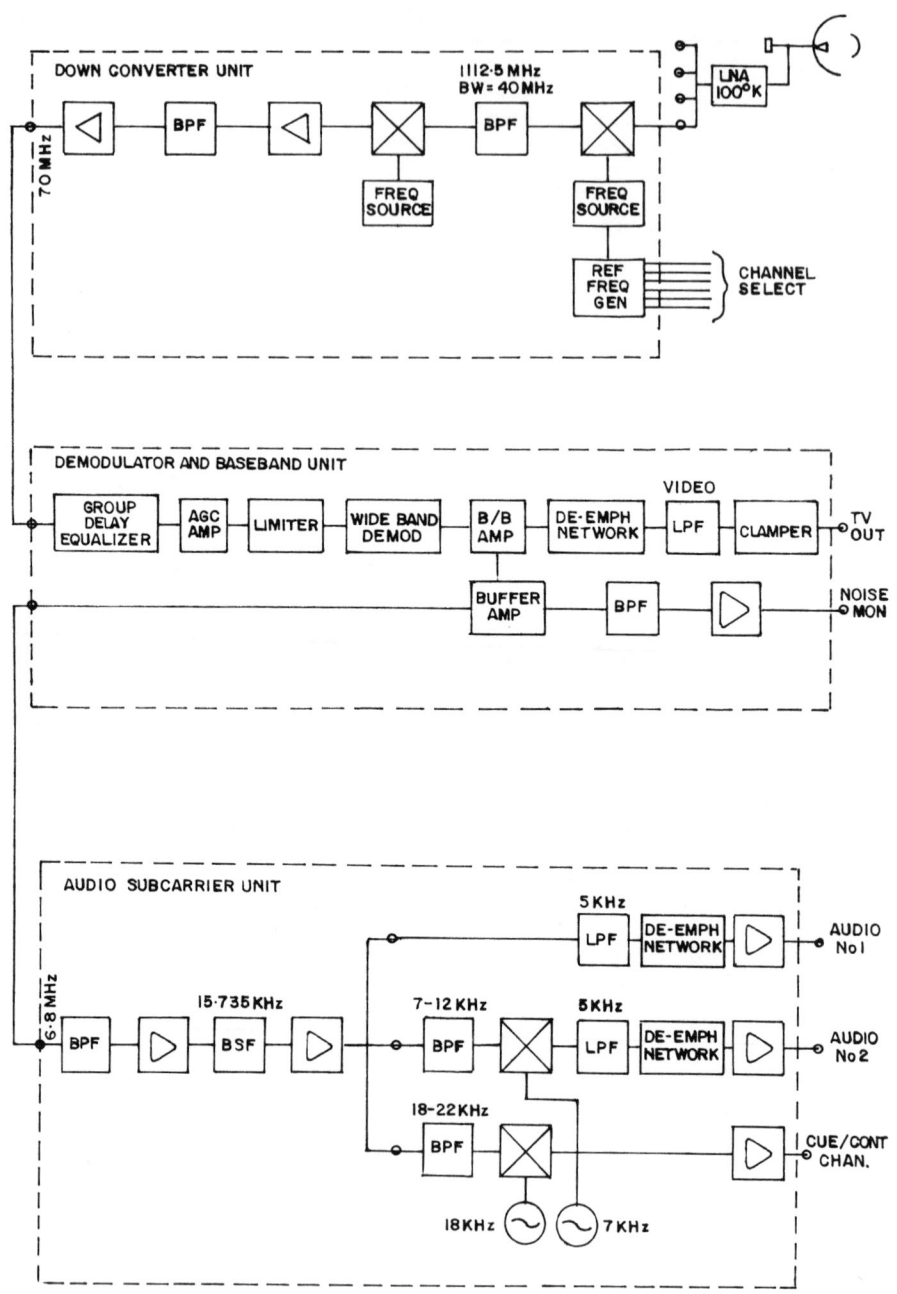

Fig. 1 Simplified block diagram/RTV.

Table 2 Basic Earth station characteristics

Figure of merit, G/T db/°K	> 26.0 + 20 log (f/4) db
System noise temperature	< 160°K
Antenna gain (receive)	> 48.5 + 20 log (f/4) db
Antenna gain (transmit)	> 50.5 + 20 log (f/4) db
Low-noise amplifier	
Frequency band	3.7-4.2 GHz
Noise temperature	< 100°K
Gain	50 db
Frequency agile receiver	
Frequency band	3.7-4.2 GHz
Number of frequencies	6 equipped (12 total capability)
Noise figure	< 15 db
L.O. stability	Better than 1 part in 10^6/hr 1 part in 10^5/day
Channel selection	Front panel switch or remote
Demodulator and baseband	
Linearity	< 1% over ±16 MHz
Center frequency	70 MHz
De-emphasis video	CCIR rec. 405 525 line (Oslo 1966)
De-emphasis audio	CCITT rec. J-21 (Geneva 1964)
Subcarrier frequency	6.8 MHz
Energy dispersal waveform	30 Hz
Video bandwidth	15 to 4.2 Hz
Out-of-band noise monitor	8.932 MHz
Video signal-to-noise ratio (weighted)	Typically 54 db
Gain frequency response	±0.5 db between 500 kHz and 4.2 MHz referred to line frequency
Differential gain	Typically 4% (10% APL) 2% (50% APL) 2% (90% APL)
Differential phase	Typically 1° (10% APL) 0.5° (50% APL) 0.5° (90% APL)

extensions requiring transmission to the satellite. Provision is made for manual polarization adjustment, which is set during installation to positions computed according to locations of ground station and satellite.

The subreflector and feed horn are protected against icing by a system of radiant lamps which can be changed readily from a maintenance platform behind the main reflector. The mount structure is designed to be repointed manually to cover any of the three orbital positions of Anik satellites: 114°W, 109°W, 104°W.

Based on environmental studies conducted at an earlier stage in the project, it was decided to install main reflector electric pad heating at four locations where heavy icing conditions were anticipated. Because of the high cost of connection and operation, the antennas at two of these locations are being operated without main reflector heating. To date no adverse effect on performance has been detected. It should be noted that little published information exists on the signal degradation effect of ice and snow buildup on antenna performance, although the cost of fitting such a deicing system in a small-aperture television receive station can be a substantial proportion of the cost of the total station.

Low-Noise Receiver

To meet the system performance, a wide-band low-noise amplifier is used. This amplifier consists of two stages of parametric amplification, followed by a low-noise transistor amplifier giving an overall gain of 50 db. The overall receiver noise temperature of less than 100°K is achieved by thermoelectric cooling of the parametric amplifiers and circulators. The unit contains monitoring functions allowing metering of bias voltages, currents, pump power, and the various power supply voltages. A transmit reject filter is fitted ahead of the amplifier to provide additional isolation of transmit frequencies from that inherent in the orthomode coupler design. Space provision has been allowed for retrofit of a redundant low-noise amplifier, if required.

Frequency Agile Receiver

The frequency agile receiver is designed to be tuned to any of 12 channel frequencies, either manually or by remote command. As the system is configured presently, only 6 of the 12 crystals are fitted. In current operation, up to four of these frequencies are exercised by remote command by the broadcaster in the daily routine of television and radio program distribution to remote television stations. The downconverter is followed by a wide-band FM demodulator and an audio subcarrier unit. Other features of the receiver include the provision of an energy dispersal waveform, carrier level detection, and out-of-band noise monitoring.

Audio Multiplexing

The system of multiplexing of the audio and cue/control channels is by normal frequency division multiplexing techniques. Two program audio channels, each of 5-kHz baseband, and one cue and control channel of nominally 3.0-kHz bandwidth are assembled on a common baseband. This multiplexed baseband signal is then frequency-modulated on a subcarrier of 6.8 MHz and inserted in the video baseband. At the remote television receiver, the process of demultiplexing is performed. Normally the first audio is used as the television audio, whereas the second audio frequently is used to distribute radio programing.

Power Supply Subsystem

In designing the power supply system, the variation of quality of commercial power at remote locations had to be taken into account. It was therefore decided to implement a -24-v-d.c. distribution system to all communications equipment utilizing a rectifier/floating battery uninterrupted supply, requiring little maintenance. This system also provides additional filtering of ripples and noise to insure a highly stable, clean power system for the communications equipment. Batteries supporting the full traffic load for more than 8 hr have been provided.

Built-In Test Equipment

In addition to the normal metering built into the various subsystems, the station is also provided with picture and waveform monitors.

Equipment Shelter

The equipment shelter structure is constructed in a unitized fashion using a polyurethane foam low-density core sandwiched between fiberglass laminates. This material has excellent structural and insulation characteristics. Additional structure support utilizing steel bracings and beams was included in the design to facilitate transportation and connection to concrete foundations on-site. The building is proportioned generously, allowing future plant extensions to be accommodated readily.

Since the overall design and construction are somewhat novel, particular attention was paid to wind, snow, and ice loadings during the design stage. An attractive feature of the design is that it requires no maintenance such as painting or structure renewal of any kind. The environment of the shelter is thermostatically controlled by a heater and constant-speed fan arrangement, operating in conjuction with a louver that adjusts its aperture automatically, using a temperature-sensitive expanding/contracting element. Some stations are also fitted with standard air-conditioning units.

IV. Project Implementation

The project implementation was characterized and, indeed, dominated by the extensive logistics problem. To minimize cost and maximize assurance that facilities would operate reliably when installed, a complete remote television Earth station was constructed in the prime contractor's plant in Waterloo, Ontario. This station fulfilled many roles, among which were system design confirmation, equipment design assurance, manufacturers's field staff training, Telesat operations staff training, maintenance agents training, and development of test routines. The testing conducted in the complete station environment complemented the normal engineering and quality assurance programs.

The field work schedule problem was eased by having three field crews working simultaneously. The field acceptance tests were witnessed by Telesat staff and demonstrated to Canadian Broadcasting Corporation staff on the same visit. This arrangement worked out most successfully and provided Telesat maintenance staff with invaluable experience. Performance tests were carried out both by the use of a special test set to provide a local test loop facility and also via satellite. Since telephone communication was difficult (sometimes impossible), a prearranged test program was developed utilizing standard TV test signals injected from a manned location with one-way communication via one of the associated audio channels. Later on in the program, even the one-way voice connection could be dispensed with.

V. Operation of Remote Television Facilities

The station is designed to operate in an automatic and unattended fashion with a minimum of maintenance. Initial design availability analysis indicated that an equipment design goal of 99.55% could be achieved with a nonredundant configuration and a travel plus repair time of 12 hr. This availability target left a small implementation margin from the operational system availability target of 99.2% for the first year of operation.

The only operational aspects requiring change of state of any element of the Earth station are 1) the on-off control of space heater, fans, and air-conditioning units, which is accomplished automatically; and 2) the changeover of channel select, effected by remote control by the broadcaster.

A unique custom-built network control system was designed and put into service as a joint effort by Telesat and Canadian Broadcasting Corporation. The network operation is controlled both manually and by utilizing small computers in Montreal and Toronto. These centers are tied together by a data transmission system for coordination and backup purposes. Appropriate commands are transmitted by frequency shift keying (FSK) tone generators in a prescribed code and received and detected at all Telesat Earth stations distributing television. Typical commands executed at the remote television stations are channel selection of receiver (TV feed selection); audio/video feed to broadcast transmitter (on/off); and audio 2-feed (on/off).

VI. Maintenance Strategy and Philosophy

The development and setting up of maintenance support to meet the service requirements presented a formidable challenge. Historically, commercial satellite communications of high-density Intelsat-type required a fairly large staff manning stations continuously. In essence, there was no historical base on which to draw, or to assist in the derivation of staffing, sparing test gear deployment, or logistics.

Staffing

All stations in the Telesat system, excluding the heavy-route stations at Allan Park and Lake Cowichan, are maintained under an operations manager of unattended stations with a staff of two supervisors and six technicians, four based in

Ottawa and two in Frobisher, Northwest Territories. The stations maintained by this staff include 1) six network-quality television transmit and receive stations located near major cities across Canada; 2) 24 remote television stations located in the far north; 3) two northern telecommunication stations transmitting and receiving frequency division multiplex (FDM) message service and television (Frobisher Bay and Resolute, which is 700 miles from the North Pole); and 4) nine thin-route Earth stations in the Eastern Arctic, with another eight similar stations due for service by the end of 1974.

To assist this skilled technical staff, arrangements have been made with various telecommunication entities operating in areas where Telesat Earth stations are located, to provide support in case of failure. These arrangements have provided about 11 man-weeks of assistance during a 1-yr period. To insure that these maintenance agents were trained adequately, Telesat supervisory staff set up on-location training courses in locations as distant as Whitehorse, Yukon Territory.

Spares and Repairs

Since field replacement rather than repair is a key philosophy, it was necessary to insure an adequate supply of spares strategically placed in depots chosen to optimize the total use of spares, maintenance agents, and scheduled transportation. The depots chosen were Hay River, Northwest Territories; Whitehorse, Yukon Territory; Ottawa, Ontario; and Allan Park, Ontario.

Four sets of each subsystem were provided, together with a selection of plug-in modules. Generally, passive components, such as antenna feed, are kept only at Allan Park to protect against catastrophic failure or damage. Each station was supplied with an expendable spares kit, including lamps and fuses. Repair and realignment of returned defective modules or subsystems is undertaken at the Allan Park heavy-route station, where a repair and test facility has been set up. Such arrangements have proved to be extremely successful, even during the first year of commercial system operation.

VII. Operational Fault Reporting

Two basic levels of fault reporting have been instituted: 1) outage and service degradation faults, which are collected at the network control center in Ottawa for the eventual purpose of quantifying performance for customer contractual reasons; and 2) detailed technical reports on equipment failures or service anomalies aimed at evaluating the adequacy of the facilities provided.

The network operations center is located at the Allan Park heavy-route station, which is staffed on a full-time basis. Engineering order wire facilities exist at Allan Park connecting to Lake Cowichan, Frobisher Bay, and Telesat sites, as well as the CBC television operating centers, Bell Canada Network Control Center at Barrie, Ontario, and the remote stations maintenance section in Telesat head office in Ottawa. The network operation center is, therefore, the day-to-day operation interface between Telesat and customers. A formal system of "ticketing" insures that faults are isolated, trouble-shooting is pursued, and the effect of faults on service performance is accurately recorded. These fault records are analyzed and summarized by head office staff in reports distributed monthly to Telesat management and design staff. This improves awareness of any equipment performance deficiencies and assists in the identification of areas requiring operational, procedural, or design attention. Thus, service faults are recorded and analyzed carefully, and the results used to determine appropriate action for system improvement.

VIII. Service Availability Performance

The remote television stations were placed successively into service commencing Feb. 5, 1973. By the end of February, some 15 stations were feeding TV to community TV transmitters. Although the remaining stations went into service in subsequent weeks, no detailed analysis of service record was undertaken. The results presented, therefore, summarize the service record of 15 of the 24 locations in service over a 12-month period. Exclusions from the availability calculations are 1) service degradation from impace of sun transit, and 2) extended commercial power outage.

Of the 15 locations, six had perfect availability records, i.e., 100% availability. Five of these perfect six also had perfect audio channel records. Only one station failed to meet the service availability objective of 99.2%. The lowest availability was recorded on a station where the environmental control system (heater) failed in the same period as an extended commercial power outage. Even after power was restored, low-temperature conditions prevailed, which adversely affected equipment performance.

Diagrams indicating equipment failure analysis and service outage impact are shown in Figs. 2a and 2b. A further analysis of the receiver and audio demultiplexing equipment faults shows failures of the normal type: 1) three phase-lock oscillator failures; 2) three control-system failures; 3) two demultiplexing equipment failures; 4) one voltage regulator failure; and 5) six degraded service faults, which were unlocated, nonrecurring, but attributed to the receiver subsystem.

It is worthy of comment that, although much difficulty was expected with ice/snow contributing to signal degradation, the first two winters have shown only a few incidents. The service availability performance over the first year of operation has exceeded predicted performance.

IX. Future Design Trends

With the establishment of commercial regional satellite communications systems on the currently proposed scale, it is inevitable that more and more operating companies will seek designs capable of unattended operation. Significant operational cost savings can accrue to a company prepared to put engineering planning effort into such concepts. The success of such an operation rests with proper choice of system protection margins, coupled with good system and equipment designs.

Future trends will have to concentrate on reduction of parts counts, easily replaced modules, improved environmental control systems, and design capable of being serviced locally by relatively untrained staff. Significant cost reductions are anticipated with the introduction of miniaturiation of equipments housed in smaller equipment shelters. The accent must be on reliability; good reliability programs are likely to distinguish between good and excellent equipments for television applications.

X. Summary

Operational performance of the RTV stations over the first year met the service availability objectives of 99.2% with a handsome margin to spare. This has clearly demonstrated the

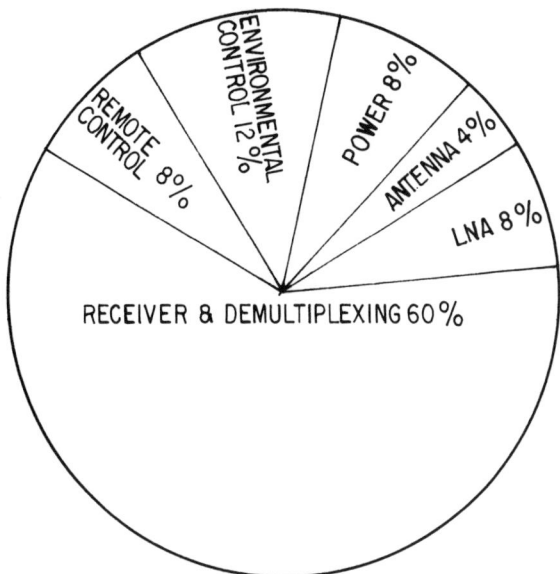

Fig. 2a Equipment failure analysis.

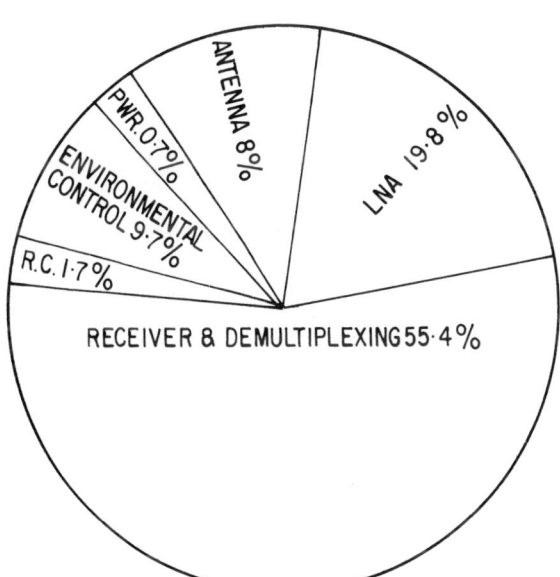

Fig. 2b Service outage impact.

viability of operating satellite communications ground stations in an unattended mode. Technological improvements to communications equipment and improved attention to the design of peripheral equipment will lead to even further improvement in reliability. This conclusion can be extrapolated to include more complex facilities operating in both transmit and receive modes.

References

1 Weese, D.E. and Smart, F.S., "Measured Communication Performance of the Telesat Satellite System," published elsewhere in this volume.

CHAPTER III—CONTROL TECHNOLOGY

The third major element of a satcom system is the control subsystem. This is the most all-pervasive of the three major elements because it encompasses every user, every terminal, and every satellite. As such, it must be integral to the total system design and operation and be responsive in time to the demands placed upon it.

There are a variety of types of control; system, satellites, terminals, operational modes, links, and channels are all subject to control procedures. The first two of the three papers of this Chapter have to do with the implementation of ground terminals for satellite control, and the remaining concerns itself with link control.

In general, a satellite control terminal is to monitor the status of the onboard systems via the command link. By tracking the satellite signals, the relative position of the spacecraft is known. During the launch sequence, the ground terminal also may be responsible for orbital injection. The Canadian domestic satellite system is operated by Telesat Canada, and its ground control terminal is described by Kowalik in the first paper of this group.

A vital role is played by computers in these control stations. It is underlined at the outset of that discussion, and it is further emphasized in the second paper. The design of a ground control complex, which controls satellites in two distinct systems that have a common owner and operator, is the topic of the next paper. The COMSAT General Corporation is both the operator of the space segment to be used by the domestic ATT/GTE systems, as well as the major owner and operator of the MARISAT system, which is a maritime services satellite. van Hover and Gribbin describe the east cost and west coast stations and their interaction with the Control Center in Washington, D.C.

In frequency division multiple access systems, the power of each carrier is carefully calculated to yield an optimum trade-off between the constant bound of total repeater output power and the power link losses and availability associated with the individual carriers. In general, because of the complexities involved with these link calculations, the procedure has been to keep the carrier power constant and to include a somewhat exorbitant amount of power to cover temporal variations in path loss. The direction of the paper by Ince et al. is to design, and validate by simulation, an adaptive link power control technique which permits greater satellite capacity by replacing the fixed margins allocated per link.

TELESAT SATELLITE CONTROL SYSTEM

H. Kowalik[*]

Telesat Canada, Ottawa, Ontario, Canada

Abstract

The Telesat satellite control system is required to perform the mission functions for the Telesat geostationary communications satellites, beginning with the injection of the satellite into its inclined transfer orbit by the Thor Delta launch vehicle, through apogee motor firing and station acquisition. Orbit control tolerances for commercial service are ±0.05° in latitude and longitude, and to meet these requirements the system has some particular features. This paper is a general description of the hardware and software system, with particular reference to the special features and performance of the system.

I. Introduction

The space segment of the Telesat baseline system consists of two satellites in geostationary orbit.[1,2] The satellites were launched from Cape Kennedy by NASA on Thor Delta 1914 vehicles, with the first satellite, Anik I, launched on Nov. 9, 1972, followed by Anik II on April 20, 1973. Arrangements with NASA provided for the injection of the spin-stabilized satellites into the highly elliptical and inclined transfer orbits, with parameters defined by Telesat. Anik I and Anik II are stationed at 114° and 109°W long, respectively. Telesat assumes responsibility for the mission following spacecraft and launch vehicle third-stage separation at the first equator

---

Presented as Paper 74-451 at the AIAA 5th Communications Satellite Systems Conference, Los Angeles, Calif., April 22-24, 1974. The author is grateful to members of the Satellite Control Division and to all those who contributed to the design and implementation of this system.
[*]Director, Satellite Control Division.

crossing. To perform the mission operations throughout the transfer orbit, the apogee motor firing, and station acquisition, and to conduct stationkeeping during commercial service, Telesat designed and implemented a satellite control system and trained the required staff to carry out the objectives.

Insofar as a spin-stabilized satellite launched by a Thor Delta through a transfer orbit phase requires certain basic elements in the mission control system, Telesat was required to implement such a system with tracking, telemetry, and command (TT&C) facilities in both the eastern and western hemispheres connected to a mission control center. In the western hemisphere, a permanent TT&C facility is collocated with the eastern heavy route communications Earth station, at Allan Park, Ontario. It performs TT&C functions both for mission operations and during commercial service. In the eastern hemisphere, Telesat leased a transportable TT&C station (TTS) for the missions from the Hughes Aircraft Company; the station is located on the island of Guam. The Satellite Control Centre (SCC), from which all mission operations are conducted in transfer orbit through arrival on station, and from which all satellite operations are carried out during commercial service, is located at Telesat headquarters in Ottawa.

The TT&C functions are controlled from the SCC using a network of minicomputers connected via high-speed (2400 bits/sec) full-period-private-line data circuits (FPPL); during the transfer orbit operations, this includes the TTS on the island of Guam. The TT&C baseband equipment is computer-controlled and is integrated with a local minicomputer at each station. The TT&C station computer is connected with an SCC computer for transmission of telemetry and tracking data to the SCC and for relaying satellite commands from the SCC. During transfer orbit operations and before the satellite arrives on station, as an extra precaution, satellite commands are carried out manually at the station. The computers at the SCC are connected to a computer utility via FPPL, which is accessed from the SCC via remote job entry software. The computer utility is used for processing the mission programs.

Except for two heavy route (HR) 98-ft-diam antennas, the Telesat communications antennas are fixed, with provision for initial adjustment manually, for any given satellite. To meet the performance criteria for the 33-ft-diam antennas in the Earth station system, it is desirable to control the orbit to within $\pm 0.05^\circ$ in both latitude and longitude. To meet this requirement, a particular tracking system configuration, together with certain minimum tracking accuracies, was needed.

Also, orbit determination and prediction programs, with accurate dynamical models of the orbital perturbing forces and ephemeris generators, were developed.

II. Spacecraft Telemetry and Command Subsystems

The essential features of the spacecraft telemetry and command subsystems are described as they interrelate with the satellite control system, and a simplified block diagram of the baseband characteristics is shown in Fig. 1.[3] The telemetry subsystem is divided into two independent sections for redundancy, with two phase-modulated transmitters permanently coupled to the shaped-beam antenna and a bicone antenna. The antennas are polarized orthogonally, with the bicone polarization parallel to the spin axis. The transmitters have low- and high-power modes, with the high-power mode employed with the bicone antenna.

As shown in Fig. 1, there are three types of data provided via telemetry, namely, satellite status (PAM), real-time data, and range measurement, with any two of the three data types provided at any given time. The two PAM encoders are redundant except for the thermal data, with each encoder providing 22 channels of satellite status information; the PAM frame period is 6 sec. The real-time data consist of the sun sensor, Earth sensor, and master index pulses; the master

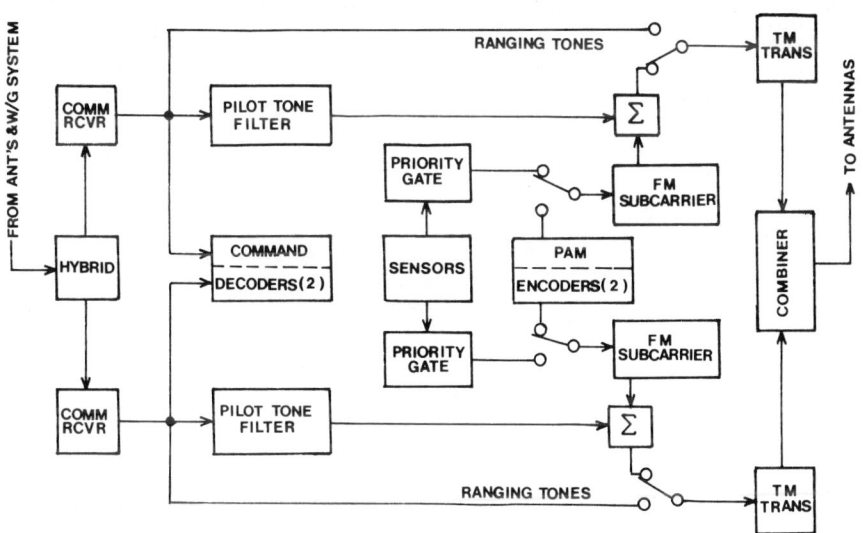

Fig. 1 Spacecraft baseband telemetry and command subsystems.

index pulse denotes the orientation of the shaped-beam antenna relative to the spinning section of the satellite. In the event of pulse coincidence, a priority gate selects, in order of priority, the master index, the sun sensor, and the Earth sensors. The north and south Earth sensors, squinted at $5°$ relative to the satellite equatorial plane, also are displaced in the equatorial plane in order to facilitate transmission via a single channel without overlap; there is no onboard Earth sensor processing.

The command subsystem is redundant, with the receivers permanently coupled to both an omnidirectional turnstile antenna and the shaped-beam antenna. The antennas are orthogonally polarized, with the shaped-beam antenna polarization parallel to the spin axis. The command subsystem is an audio tone to digital system comprising four audio tones, with a two-step process to allow verification of the command number via telemetry before executing. The satellite communications antenna is despun mechanically and is controlled in its Earth-pointing configuration via a ground-based pilot beacon.[3] The beacon is an integral part of the command subsystem, and the command carrier is modulated continuously by one of the four command tones, as required for the pilot beacon. Except during commanding, the system operates on one particular tone, appropriately denoted as the pilot tone. Range measurement is carried out through the T&C subsystems during the mission when the direction of the spin axis is unfavorable for acquisition of the shaped-beam antenna. With the antenna locked on the beacon, and throughout commercial service, ranging is performed through a communications channel.

III. Tracking, Telemetry, and Command System

The Allan Park main TT&C facility consists of a 36-ft TT&C antenna (TTAC) with a set of baseband ground control equipment (GCE) situated adjacent to the communications equipment in the station. The HR also is equipped with TT&C capability, and its baseband equipment is cross-linked with the TTAC GCE to provide system redundancy and high availability.

The basic system design philosophy is to conduct the missions with the TTAC system as the prime TT&C facility, with the HR system providing emergency backup. For the transfer orbit, the Allan Park complex is supported further for vital telemetry and command functions only, such as apogee motor firing, by temporary installation of T&C facilities on Telesat's other HR communications antenna at Lake Cowichan on Vancouver Island. This is mainly to provide geographical diversity for these vital functions with regard to factors such as weather, fire, communications links, commercial power, etc. The TTS on the island of Guam completes the TT&C facilities for the mission phase. Further to the basic design philosophy for the TT&C system, all TT&C functions for Anik I were switched from the TTAC to the HR system at the commencement of commercial service on Anik I, with the TTAC providing emergency backup. Additional details are given later on the tracking function. The TTAC then was available for the Anik II mission and for the control of that satellite on station. This fundamental design philosophy could be extrapolated for any growth in the commercial system comprising three satellites in orbit and a second communications antenna at the Allan Park Earth station.

With the provision of interim service to a U.S. customer on Anik II and with the TTAC required as a backup to the Anik I system, particularly for pilot beacon, a third antenna was needed to provide an uninterrupted and reliable pilot beacon for Anik II, even before a second communications antenna is constructed for Canadian customer growth. Thus, a 26-ft-diam antenna (TAC-2) and a set of T&C equipment were installed and placed into operation at the end of November 1973. The TTAC system provides emergency backup to both the HR and TAC-2 systems while it is available for support of another mission.

For high availability of the pilot beacon, the design philosophy is to have redundant command transmitters for each satellite system that is carrying commercial traffic, that is, on both the HR and TAC-2 command systems. Each system has automatic switching between its transmitters, and the HR is supported further by the TTAC system by means of a beacon interlock mode, which provides automatic switching to TTAC but which must be engaged manually. The TTAC system also has redundant uplink facilities with automatic switching between transmitters. The main characteristics of the antennas and rf system are listed in Table 1.

The TTAC antenna is a precision tracking antenna with azimuth and elevation (AZ,EL) data sufficiently accurate such that these data, together with range measurement, provide for orbit control to within $\pm 0.1°$ in latitude and longitude. In Table 1, it is shown that the tracking error is within $0.004°$ rms; this is for an LNA input signal of -150 db-w with average weather conditions. The antenna may be operated in manual, autotrack, or program track mode. The TTAC antenna has an operating frequency range spanning the 500-MHz communications bandwidth in the 4- and 6-GHz bands and is used for satellite transponder tests as well as for TT&C. Each transmitter has seven preset frequency bands, which are remotely selectable,

Table 1 Antennas and their rf characteristics

Parameter	TTAC	TAC-2	TTS	HR[a]
Reflector diameter, ft	36	26.5	24	98
Antenna feed	Cassegrain	Gregorian	Cassegrain	Cassegrain
Mount	EL/AZ	Fixed	EL/AZ	Wheel and track
Auto track	Monopulse	N.A.	Monopulse	Step track
Pointing error, rms, deg	0.015	±0.04 (manual)	0.04	0.03
Tracking error, rms, deg	0.004	N.A.	0.01	0.01
Tracking velocity, deg/sec	1.0	N.A.	1.0	0.3
Angular travel, deg				
Azimuth	±270 from S	±20 (manual)	±200	±170 from S
Elevation	0 to 92	±10 (manual)	0 to 90	0 to 90
Polarization rotation, deg	±360	±10 (manual)	±90	±360
GT at 4 GHz, db	28	20.5	29.7 max.	37
Low-noise amplifier	100°K paramp	4.5-db transistor amp	30°K paramp, 150°K paramp	100°K paramp
Effective isotropic radiated power, db-w, max.	85	76	92	87
High-power amplifier	3-kw klystron	300-w traveling-wave tube (TWT)	15-kw klystron	300-w TWT, 1.5-kw klystron

[a] Characteristics pertinent to TT&C functions.

with one tunable to the command and even transponder channels and the other to the command and odd transponder channels.

The telemetry receive system includes two LNA's (redundant), a three-channel tracking down-converter, a phase-lock monopulse receiver, and two telemetry receivers, one for each satellite beacon. The sum and azimuth difference channels of the tracking down-converter also serve as the main and standby telemetry down-converters. Input to the telemetry receivers is at 70 MHz. The HR and TAC-2 antenna systems are equipped with a complete set of uplink and downlink equipment, similar to the TTAC system (Table 1).

Interfacing between the TT&C GCE in the equipment room and the three antennas and their respective rf equipment is at 70 MHz, through the TT&C patch panel (Fig. 2), for both uplink and downlink. In accordance with the T&C system design requirement (Sec. II), all antennas, including the TTS, are linearly polarized, with transmit orthogonal to receive, and rotatable.

Ground Control Equipment

Further to the automatic switching provided for the command pilot transmitters on each antenna, there are two points in each set of GCE which are monitored and connected to switching logic for automatic switching of the command pilot tone to a spare. These are the command generator output and the FM modulator output. A redundant (dual) command generator and FM modulator on the TTAC system is made available via a power divider, as a redundant standby to both the HR and TAC-2 command systems, simultaneously if necessary (Fig. 2). The switching arrangement for the HR pilot source by switch S1 is shown as an example; the TAC-2 system (not shown) is identical.

As shown in Fig. 2, most of the GCE is interfaced with the computer. The command generator can be coded directly or can be operated through the station computer or remotely from the SCC in Ottawa. A synchronous controller provides timing to the command generator for operation of the spacecraft reaction control system (RCS). An antenna despin control unit provides for acquisition and east-west adjustment of the satellite shaped-beam antenna. With regard to telemetry, the PAM video is decommutated and fed to the station computer, an analog PAM bar chart display, and the command generator. The command generator displays the contents of the satellite command decoder in order to provide for verification of the

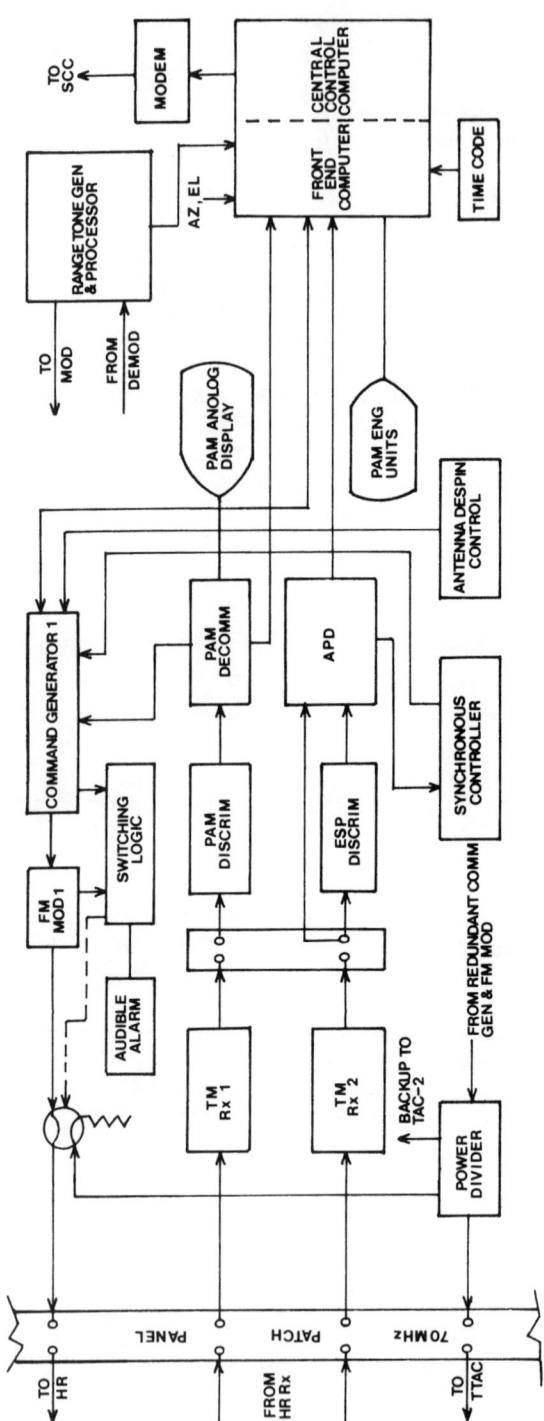

Fig. 2 A typical set of ground control equipment.

command number before transmission of the execute code. Video display of PAM in engineering units also is provided.

For sensor data, a custom-designed attitude pulse digitizer (APD) processes the sensor video into the timing events of the sensor pulses and digitizes the events for the computer input. Recently, a new processing technique has been devised for computerized Earth sensor waveform analysis. The technique involves analog-to-digital conversion of the Earth sensor waveform with high fidelity. Transformations representing the inverse transfer functions of the Earth sensor bolometer and electronics are applied to the digitized waveforms to obtain the input waveform characteristics of the infrared radiation intensity with respect to Earth scan. This technique, together with the very important factor of telemetering the Earth sensor output in its original analog waveform, enables attitude determination to an accuracy of about $\pm 0.01°$.

For range measurement, the well-established method of transponding a set of audio tones and measuring phase delays is employed. Four audio tones at frequencies ranging from 35.4 Hz to 27.7 kHz are used, with the three lowest tones employed for ambiguity resolution in the highest frequency tone. The range generator/processor is controlled by the station computer and in turn by the SCC computer (Fig. 2). During commercial service, with range measurement performed via a communications transponder, the ranging tones are multiplexed into unoccupied frequency slots of the baseband spectrum. Details on performance are given in Sec. V.

The transportable tracking station on the island of Guam is completely compatible with the Telesat satellite and is employed in the mission up to apogee motor firing. The characteristics of the antenna and rf system are summarized in Table 1. The GCE is the same as that shown in Fig. 2 except that automatic switching of uplink equipment is not required, and there are some differences in equipment detail with regard to recorders, message units, etc.

IV. Computer System

The Telesat computer system has been custom-designed for the data acquisition and processing requirements during the mission phases and for operational control during commercial service. The fundamental system requirements for the mission phases are the usual basic factors. The system must be capable of handling large quantities of data during high-activity periods in transfer orbit and in the drift orbit (period between apogee motor firing and arrival on station). System availability must be very high, with down-time periods of no more than a few minutes during certain critical periods. Clearly these basic requirements can be achieved in a number of ways, but the costs will vary.

Although it was recognized that a large, scientifically oriented computer is required for handling and processing the large mission programs and carrying the necessary precision in the data, it also was clear that Telesat engineering could never make full use of such a large machine, and it was never felt that subleasing the free time on the computer would be in the best interests of the Corporation. Thus, the fundamental approach taken was to design a system of minicomputers for processing data that does not require a large machine, with the capability of accessing a large machine for processing the large mission programs. In addition, with minicomputers a modular design and implementation approach was readily feasible, with computer system expansion carried out in accordance with expansion in the space segment as required in support of growth in commercial service. This has been achieved, and an illustration of the fundamental aspects of the hardware system is shown in Fig. 3.

The computer functions at the TT&C stations were described briefly in Sec. III and Fig. 2. The Guam and Allan Park station computers are connected via an FPPL circuit to a data control computer (DCC) at the SCC. The DCC is connected to a data storage computer (DSC), which in turn is connected via FPPL to a Univac 1108 system. The Univac 1108 is accessed via remote job entry (RJE) software. The TT&C station computers, DCC, and DSC are Hewlett-Packard (HP) machines owned by Telesat, with the exception of the TTS computer, which is part of that station. The Univac 1108 computers are owned and operated by a computer utility in Ottawa, Computel Systems Limited. Word size on the HP computers is 16 bits, with memory capacities on the units ranging from 4K to 32K, depending on individual machine requirements. The Univac 1108 operates with a 36-bit word, and each machine is equipped with a 65K memory.

The basic functions of the station and SCC computers are to handle tracking, PAM, and sensor data from the stations, transmit command codes from the SCC to the command generator, and relay ordinary teletype communications messages between the SCC and the stations. The DCC and DSC, with the peripheral video display and plotting equipment, provide the mission team with data in real time, in engineering units, for an

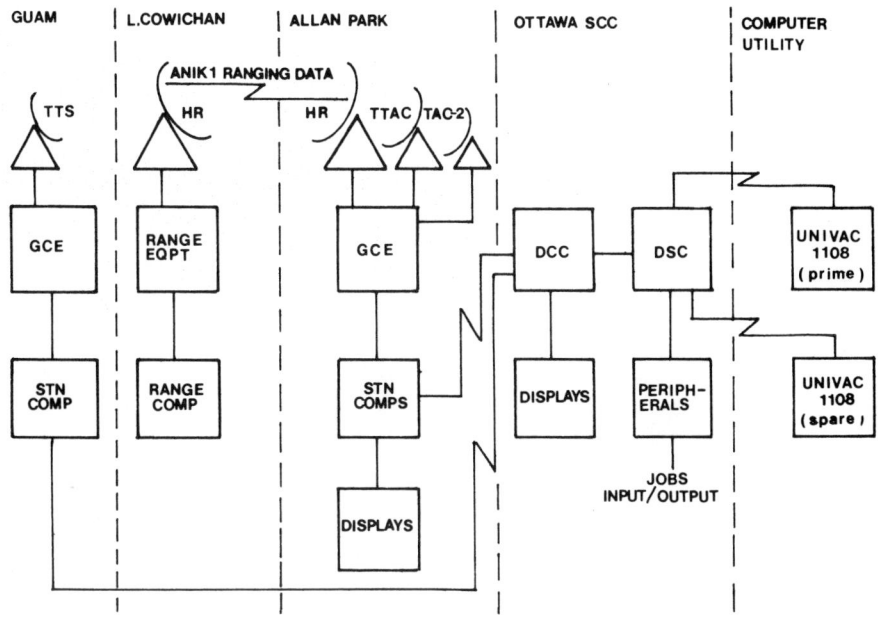

Fig. 3 Basic elements of data acquisition and processing system.

immediate appreciation of satellite status; immediate trends in critical parameters of interest can be observed on analog recorders, and historical plots showing long-term behavior are available on a digital plotter. During commercial operations, it is standard practice to plot PAM once per week and then dump the data from disk onto digital tape for permanent storage. Tracking and sensor data are retained on a DSC disk until required by the analysts for orbit and attitude determination on the Univac 1108. The data are transferred to the Univac 1108 via RJE by ordinary job control cards as part of the batch job. Remote control from the SCC of the TT&C station computers and certain GCE equipment is through the DCC from a console teletypewriter in the SCC; this includes satellite commanding, range measurement, and selection of data collection rates.

The basic function of the Univac 1108 system is to handle the large mission and stationkeeping programs. The programs are stored on the 1108, and all input/output functions are carried out on the DSC at the Telesat computer facility, with input data called from the DSC disks as required. Results generated by the 1108 are returned to the DSC disks for buffer storage and then output on the DSC peripheral line printer(s). Thus, in essence only the 1108 central processing unit is used. The operation is very economical, and costs for use of the 1108 have varied according to activity and have been very low in comparison to owning and operating an adequate facility.

As stated earlier, system availability must be very high, especially during the mission phases. For the in-house minicomputer system, extra units were procured to support the online system shown in Fig. 3. While serving as spares, the units are configured into a so-called off-line system, which represents the on-line system sufficiently well to provide a facility for system diagnostics and for software development in support of system expansion and improvements. Furthermore, with regard to availability, the 1108 facility in Ottawa is backed up by other 1108's outside the city, owned by the computer utility company. Very satisfactory contractual arrangements were made with regard to machine priority for the mission phases, and job turn-around has been very satisfactory. The system software for controlling the in-house HP system is written in Assembler language, and the system has been developed in multiprograming mode. The multiprograming feature provides for efficient use of machine time and for simultaneous operation of the various functions, including receiving and display of tracking and satellite data, commanding, range measurement, message service, and input and output of jobs on the 1108.[4]

V. Orbit Control

When Telesat began system design on the orbit control requirements, there was very limited information available with regard to controlling a geostationary satellite, in a steady-state operation, to tolerances in the order of $0.1°$ in inclination and longitude. Some studies were carried out at an early stage to determine the merits of orbit determination based on data derived from dual ranging from two points widely separated, in particular from Allan Park and Lake Cowichan. This was compared to orbit determination accuracies derived from azimuth, elevation, and range data from a single station, and it was shown readily that orbit determinations from dual ranging should be significantly more accurate. It also was recognized that biases in the data should be of lesser significance in ranging measurements than in azimuth and elevation data. Weather conditions, particularly wind, would affect azimuth and elevation data, whereas ranging data should be quite immune to environmental conditions. As communications systems design progressed, it was evident that it would be desirable to control the satellite orbit to within ±0.05° in latitude and longitude in order to satisfy the performance criteria for certain network television receive stations. In order to provide orbit control to within ±0.05°, as one of the analysis criteria, it was considered necessary to determine the position of the satellite to roughly one order of magnitude better, that is, to within ±0.005°. Simulation programs were developed to analyze and define the requirements utilizing simulated data with noise.

Two types of systems were considered as models, denoted, for discussion purposes, as systems A and B. System A comprised orbital data derived from a single station in the form of azimuth, elevation, and range, and system B comprised range measurements from Allan Park and Lake Cowichan. Although the stations and equipment were still in the design phase, the best values known at the time were used to simulate errors in the models. Noise error in the ranging data was assumed to be ±45 ft rms, and azimuth and elevation errors were considered to be ±0.013° rms. The results in Figs. 4 and 5 provide comparisons between the two systems for 3σ uncertainties in inclination and longitude determinations, with data rate as a

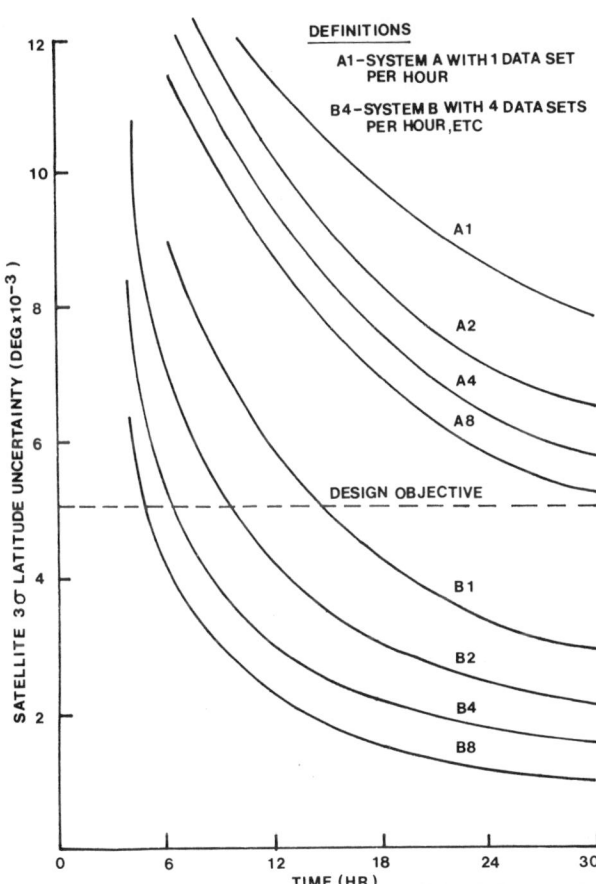

Fig. 4 Satellite latitude uncertainty vs time.

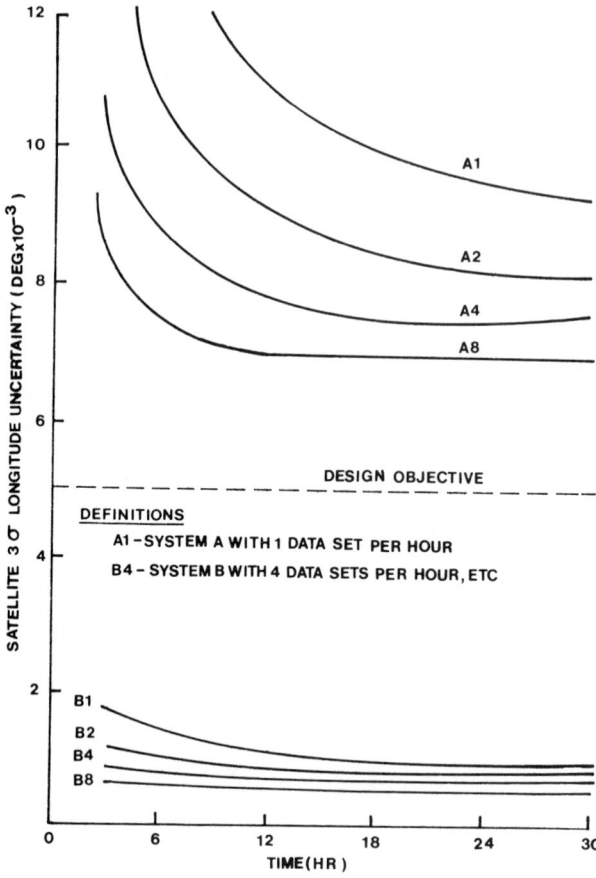

Fig. 5 Satellite longitude uncertainty vs time.

parameter. Figure 4 demonstrates that system A will not provide ±0.005° accuracy in latitude determination (3σ) after 30 hr of data collection for the data rates considered, whereas system B will provide the required accuracy after 18 hr for the lowest data rate, and after 6 hr for the highest data rate considered. Similarly, Fig. 5 compares systems A and B with respect to longitude uncertainty (3σ), and the superiority of system B over system A clearly is evident.

It also is apparent from Figs. 4 and 5 that latitude and longitude determination to within ±0.005° should be fairly straightforward with dual ranging, whereas system A is inadequate for this degree of control. On the basis of this analysis, it was decided to proceed with a dual-range measurement system and to work out the details in practice. In addition to developing a dual-range measurement system with high accuracy, it also was necessary to develop the pertinent orbit and prediction programs utilizing high-order terms for the tesseral and zonal harmonics. A precise solar radiation pressure model also has been developed from data taken over the past year, which has improved system accuracy further and has made orbit prediction and propagation more accurate than initially possible.

With ranging carried out via the satellite transponder, there is ample carrier-to-noise ratio, and range accuracy is limited by phase meter resolution; thus, in practice, range measurement accuracy is ±20 ft rms. The resultant latitude and longitude uncertainties for Anik I are very low, in the order of ±0.0002° and ±0.00005°, respectively. Although orbit determination is very accurate, stationkeeping is carried out only to the degree of tolerance required for commercial service. Other factors associated with stationkeeping tolerances are frequency of maneuvers, amount of data processing and analysis, accuracy in the reaction control system impulses, accuracy in the orbit propagations, longitudinal position of the satellite as it affects drift accelerations, and perhaps others. Thus stationkeeping costs generally increase with tighter control, but stationkeeping fuel consumption vs stationkeeping tolerances is not a significant factor provided that the tolerances are wide enough to avoid corrections for diurnal variations. Currently Anik I is controlled to within ±0.035° as a soft limit in order to guarantee ±0.05°. This is achieved with ranging data taken every 2 hr alternately from Allan Park and Lake Cowichan. Anik II is controlled to ±0.10° in latitude and longitude, with tracking data provided by the TTAC antenna in the form of azimuth, elevation, and range, with measurements every 2 hr.

VI. Mission Control Programs

The mission and stationkeeping programs may be described in four categories. One category consists of prediction programs, which include launch window analysis, shadow prediction, acquisition prediction, predictions of sun angle, Earth angle, signal strength, and other geometric factors. These programs are essential, not only for the planning of the mission profile, but also during the course of the mission to predict upcoming events. A second category consists of the determination-type programs, which include orbit and attitude determination programs. These programs accept unprocessed tracking and sensor data and determine statistically the state of the orbit and attitude of the satellite. A third category involves programs associated with a specific maneuver of the satellite. These are the reorientation program, hydrazine control system calibration program, apogee motor firing programs, thruster programs for various thruster geometries, and maneuver evaluation programs. The fourth category consists of stationkeeping programs, which predict the motion of the satellite in geostationary orbit and help to determine the optimum cycle and magnitude for orbit and attitude corrections required to maintain the satellite on station.

Each of the programs, written in Fortran IV, consists of a main calling program and a group of mathematical subroutines. All of the preceding programs are distinct in that they are used independently and require minimal derived input data. In most cases, input data consist of satellite orbit and attitude, velocity or time increments, and some spacecraft parameters. Program development is continuing to improve these mission and stationkeeping tools. There is an objective to reduce or minimize the extent of the high-level analytical effort required for routine operations during commercial service. The attitude determination technique described in Sec. III is an example.

VII. Concluding Remarks

With two successful missions and with over a year of stationkeeping experience, the satellite control system has been tested and proven adequately; few design changes have been required, and these have consisted mainly of minor circuit modifications. The major system design concepts have worked out very well, and in particular the computer system design and orbit control techniques have proved to be very satisfactory. Any extensions to the system will be based on the same approach.

An analysis has been carried out to determine the orbit control capability based on range data consisting of turn-around ranging, where turn-around range is twice the total path length through the satellite from a ranging station to a remote station. As an example, for which tests are being carried out, a ranging carrier from Allan Park is received at Lake Cowichan and is looped back to Allan Park. Variations in parasitic phase delays at Lake Cowichan are kept at an insignificant level by looping back at the 70-MHz IF. Analyses indicate that, for the Telesat configuration, where the satellite is situated in longitude between the two stations used for ranging, range plus turn-around range are required in order to compare favorably with dual range measurements. In this case, range plus turn-around range consist of a range measurement from Allan Park to the satellite, followed by a turn-around measurement from Allan Park to Lake Cowichan.

With regard to pilot beacon availability and reliability, the objective is such that the satellite communications antenna availability is specified at 99.998%, one order of magnitude higher than the highest figure of availability objectives in commercial service. For a 99.998% availability, the duration for satellite antenna outage works out to about 10.5

min/yr. Provided the uplink equipment is functioning, an average period for satellite antenna reacquisition is about 5 min; thus the availability objectives can tolerate about two pilot beacon interruptions per year. During the first half of 1973, there were several beacon interruptions on Anik I, with a cumulative period for satellite antenna outage of about 93 min, greatly exceeding the design figure. The outages from the beginning of service in January to about May were chiefly due to problems associated with a shakedown period, and there were intermittent problems with one of the 1500-w transmitters on the HR antenna (Table 1). The problem has since been resolved, and a dedicated spare transmitter also has been added to the pilot beacon system for Anik I. Thus, from May 1973 up to mid-February 1974, there was only one 3-min satellite antenna outage on Anik I. It is expected that this level of performance will be maintained.

References

[1] Chinnick, R.F., "The Canadian Telecommunications Satellite System," Journal of the British Interplanetary Society, Vol. 26, No. 4, April 1973, pp. 193-202.

[2] Almond, J. and Lester, R.M., "Communications Capability of the Canadian Domestic Satellite System," IEEE International Conference on Communications, June 1971.

[3] Harrison, L. et al., "Canadian Domestic Satellite (Telesat), A General Description," IEEE International Conference on Communications, June 1971.

[4] Domb, U., "Duplex Computer System for Control of Canada's Anik Satellites," Canadian Aeronautics and Space Journal, Vol. 21, No. 2, Feb. 1975, pp. 59-64.

DESIGN OF A GROUND CONTROL SYSTEM TO OPERATE DOMESTIC AND MARITIME SATELLITES

A. J. E. van Hover[*] and W. J. Gribbin[+]

Comsat General Corporation, Washington, D. C.

Abstract

An integrated and automated tracking, telemetry, and command (TT&C) ground system was designed for the Comsat General domestic and maritime satellite systems. The TT&C Earth stations, processing equipment, system control center, computer facilities, and communication links are described, as well as the system design criteria. Tradeoffs of manning vs automation, capital vs operating costs, and innovation vs use of known technology are analyzed. General telemetry data management and reduction philosophies are addressed.

I. Introduction

The time frame of both the Comsat General Marisat and Comstar I systems presented a unique opportunity to design an integrated ground-support system with cost savings for both programs. This comes about through both economy of scale and savings in number of operators required. The system design incorporated both the features learned from 10 years of operating Intelsat satellites and new

---

Presented as Paper 74-483 at the AIAA 5th Communications Satellite Systems Conference, Los Angeles, Calif., April 22-24, 1974.

[*]Maritime Satellite Office.
[+]Domestic Satellite Office.

technology, such as the use of automatic switching and minicomputers.

II. Communications Satellite System

The Comsat General TT&C ground segment is designed initially to support a system of five satellites, with later possible expansion to eight. The initial system comprises two Marisat satellites, one serving the Atlantic and the other serving the Pacific, and three Comstar I satellites designed to serve the continental U.S., Hawaii, Puerto Rico, and Alaska.

The Marisat has two communications packages onboard, one at uhf to be leased to the Navy and the other at L and C bands to provide communication services to commercial users. The Navy will provide its own ground segment, and ground communications for commercial users can be provided through the Comsat General TT&C antenna. The entire Comstar I capacity will be leased to AT&T, which will also provide their own communications Earth segment. The TT&C functions for both Marisat and Comstar I are under the control and operation of Comsat General, which is responsible for both providing a communications service and seeing that the satellites remain operational.

III. Space Segment

Several aspects of the satellite design impact the design of the ground-control and monitoring system. Both Marisat and Comstar I are being built by Hughes Aircraft Company and are, respectively, derivative of Anik and Intelsat IV. Both types of satellites are similar from a TT&C ground operator's point of view, except for some significant aspects. These differences are summarized in Table 1 and are discussed in the following paragraphs.

Comstar I is of the gyrostat class. This is significant from a TT&C point of view as a gyrostat is stable only when a given difference in spin speed exists between the despun platform and the spinning drum. If a failure occurs and the spin speed difference between the platform and drum drops below the stability threshold, the spacecraft would begin to nutate and probably would end up spinning around the transverse axis (flat spin). A flat spin might damage the spacecraft and render it unusable for further service when recovered. The Marisat, on the contrary, has a favorable inertia ratio and is stable even if the antenna should spin up. The operation of a gyrostat requires continuous telemetry monitoring and fast switching of critical subsystems in case of failure to avoid the flat-spin situation.

Another difference which impacts the TT&C is the spin speed: Comstar I normally will spin at 50 rpm and the Marisat at 100 rpm. This impacts clocks required to synchronize commands initiating firing of onboard thrusters. The other major difference is in the despin control systems. Both satellites use modified Intelsat IV digital despin controllers, but, in Comstar I, an analog system has been added to obtain finer pointing. This system is similar in performance to that used in Anik and might present special monitoring problems if the operational point exceeds the linear range of the control system. There, a reversal of polarity occurs, and the platform of the satellite can spin up, causing a flat-spin situation.

Table 1 Satellite specifications

	Marisat	Comstar I
Launcher	Delta 2914	Atlas-Centaur
Stabilization	Spin (100 rpm)	Spin (50 rpm)
Communication frequency	Uhf, L/C, C/L	4 and 6 GHz
Attitude control	Digital	Digital & analog
Inertia ratio	Favorable	Gyrostat
TT&C	Pulse code modulation Real-time FM	Pulse code modulation Real-time FM FM accelerometers
Other systems	Operationally identical	

IV. Ground Control System

A. Design Requirements

Extensive study of the user requirements for high reliability, observation of both the Intelsat[1] and the Telesat systems, and system considerations permitted certain criteria to be set. In other words, to operate a successful commercial satellite system, the following criteria should be met:

1) A centralized control center should be the heart of the system, and it should be located close to people knowledgeable of satellite subsystems.

2) Continuous real-time and historical data should be available readily for both day-to-day operational monitoring and for immediate analysis in case of failure.

3) Commands should be verified at several levels before being transmitted.

4) In case of critical failure, switching response time should be rapid to avoid the flat-spin situation.

5) Tracking and ranging information should be available to the celestial mechanics computer programs at predetermined intervals.

6) Antenna pointing data should be available from the celestial mechanics program to any user, including the TT&C Earth stations.

7) Launch operation: the control center should be able to process data from the Intelsat TT&C earth stations that support launch operations.

8) And, most important of all for a successful commercial venture, cost should be minimized and reliability maximized.

These criteria were subject to constraints because of the design of the satellites and their orbital positions. They also translate into specific requirements determined by the design of the satellite or ground system. The first two criteria dictate that the pulse code modulated (PCM) data stream from the satellite should be transmitted to the system control center (SCC) with minimum interruption. These data should be checked and stored for use in case of failures or for historical trend

GROUND CONTROL SYSTEM TO OPERATE SATELLITES

analysis. Other telemetry data needs such as real-time sensor pulses and the analog accelerometer signal are required only occasionally and should be available on an "as-needed" basis. Criteria 3 and 4 dictate that all commands should be sent from the system control center with backup capability from the Earth station. Special provision also is required to identify critical component failure and initiation of switching to avoid flat spin.

To insure correct commanding of the satellite, commands are looped back from both the Earth station and the satellite via telemetry. This provides two levels of verification, with additional checking via software in the control center processor. Ranging is initiated in the same manner as commanding at the SCC, and ranging, angle tracking, and time annotation are transmitted back to the SCC and to the celestial mechanics computer. Antenna pointing data are generated at the celestial mechanics computer after orbit determination and transmitted to the TT&C Earth stations and all customer stations as required.

B. Ground System Description

1) Antenna System. As one of the Marisats is placed over the Pacific and the other over the Atlantic, a TT&C antenna is required on each coast of the U.S. The Comstar I satellites are all visible to an Earth station placed anywhere in the continental U.S. The geographical separation of the stations required by Marisat also provides better transfer orbit coverage and redundancy for both systems.

To meet the criteria of full-time telemetry, each satellite requires a dedicated antenna; thus, for five satellites, two antennae are required at one site, and three at the other. These antennae need accurate tracking performance for orbit determination and are, therefore, expensive. To increase system reliability and to decrease cost by elimination of a tracking system, the fifth tracking antenna was replaced by two limited-motion antennae, one at each site. This makes both Earth stations identical, which also has the advantage of standardization.

As no angle tracking is available from the manual limited-motion antennae, a dual ranging mode was required, with both small antennae pointing at one satellite. Orbit determination results using dual ranging data from two geographically separated antennae were obtained from Intelsat and Telesat tests. The orbital determination accuracy was as good and probably better than angle tracking and ranging data obtained from one antenna only. This scheme also had the advantage of providing six antennae for five satellites, which gives redundancy. An antenna diameter of 13 m was required to provide sufficient margin in the Maritime system link budget. The same antenna size was also selected for the full tracking domestic antenna to provide backup for the Marisat antenna. The two limited-motion antennae, dedicated to Comstar I, would give satisfactory performance with a diameter of 10 m. The operational layout of antennae is shown in Fig. 1. It should be noted that, in case of failure, an insignificant amount of telemetry data will be lost while an antenna switches to another satellite to obtain tracking data for orbit determination.

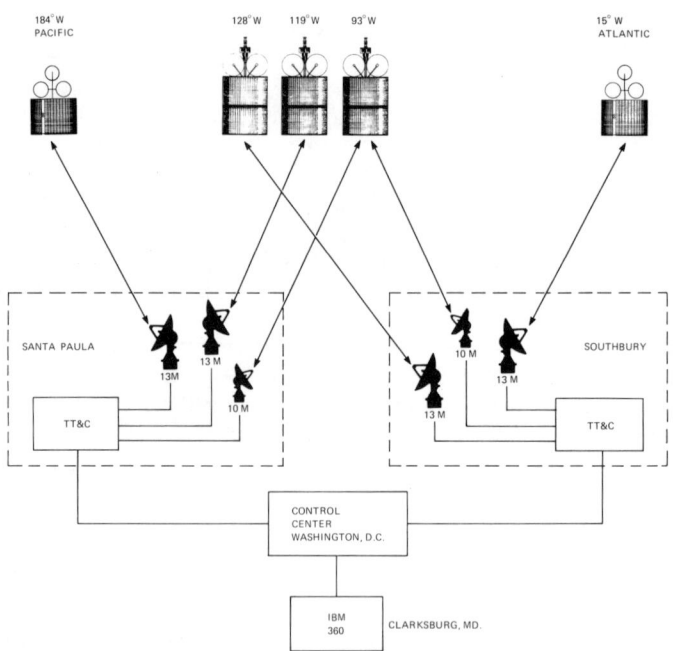

Fig. 1 Comsat General TT&C system.

GROUND CONTROL SYSTEM TO OPERATE SATELLITES

2) <u>Control System</u>. To meet the criteria of centralized operational control with all data and command capability available at the control center, a certain degree of automation was required in the system. The control center processor is designed to be the heart of the system, with the Earth station processors and the celestial mechanics computer (IBM 360) acting as remote peripherals (see Fig. 2). Communications links tie the Earth station processors and the IBM 360 to the control center processor.

In the prime day-to-day operational mode, assuming no failures or launches, the system will operate with a human operator monitoring at the control center only and with the personnel at the Earth stations acting primarily in a maintenance role. In case of failure, it was found simpler, more reliable, and cost-effective to initiate operator intervention at the Earth station to reconfigure the equipment instead of implementing a fully automated failure switching system. This aspect is thought to be important, as, during initial operation or during unexpected problems, a smooth transition can be made from the automatic operational mode to manual backup with operator intervention. In the backup mode, the system will have reduced capability but sufficient for monitoring and control of the satellites without undue load on the operators.

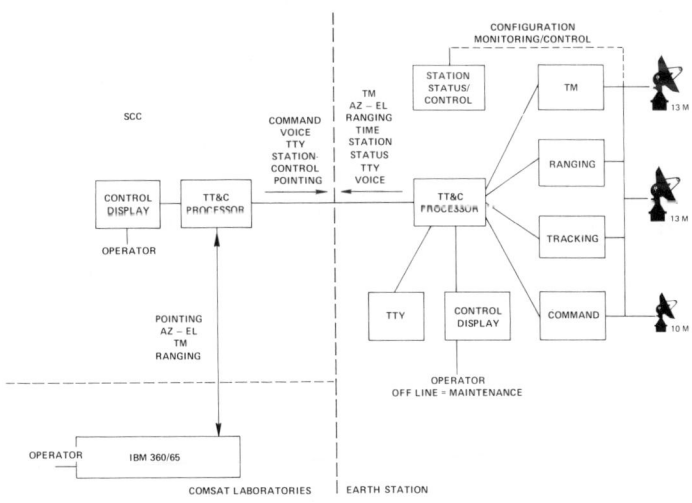

Fig. 2   TT&C functional configuration.

Launch is a peak load situation where manpower in the system must be increased. The design was not based on the launch load requirements, as this is an exceptional situation where manpower can be increased on a temporary basis.

V. Subsystems Description

The integrated ground support system (Figs. 3 and 4) is comprised of the following subsystems: angle tracking, command, telemetry, ranging, data processing and communications, station configuration monitoring and control, and radio frequency.

A. Angle Tracking

Three modes of spacecraft tracking are provided: manual, program, and autotrack. In the manual mode, Earth station operators enter demanded angle data via a thumbwheel entry system. Program tracking provides a means of automating the acquisition of a spacecraft through local processor control using pointing parameters generated by the celestial mechanics computer. Monopulse tracking systems permit close-loop autotracking capability. Antenna positions are routed to the celestial mechanics computer facility for spacecraft ephemeris generation.

B. Command

Commands may be initiated and transmitted to the spacecraft by either the Earth station or the system control center (SCC). In the normal operating mode, command messages are entered via interactive keyboard entry system at the SCC and transmitted via the interfacility data link to the Earth station computer. Error correcting/detecting block coding is used in this transmission to virtually eliminate data transmission errors. The command messages are logged in the Earth station computer and forwarded to the command generator. The command generator, in turn, encodes the message into a serial RZ FSK baseband for transmission to the satellite network.

Spacecraft positioning commands require synchronization of execution with the spacecraft spin rate and angle of roll. This is accomplished through

GROUND CONTROL SYSTEM TO OPERATE SATELLITES

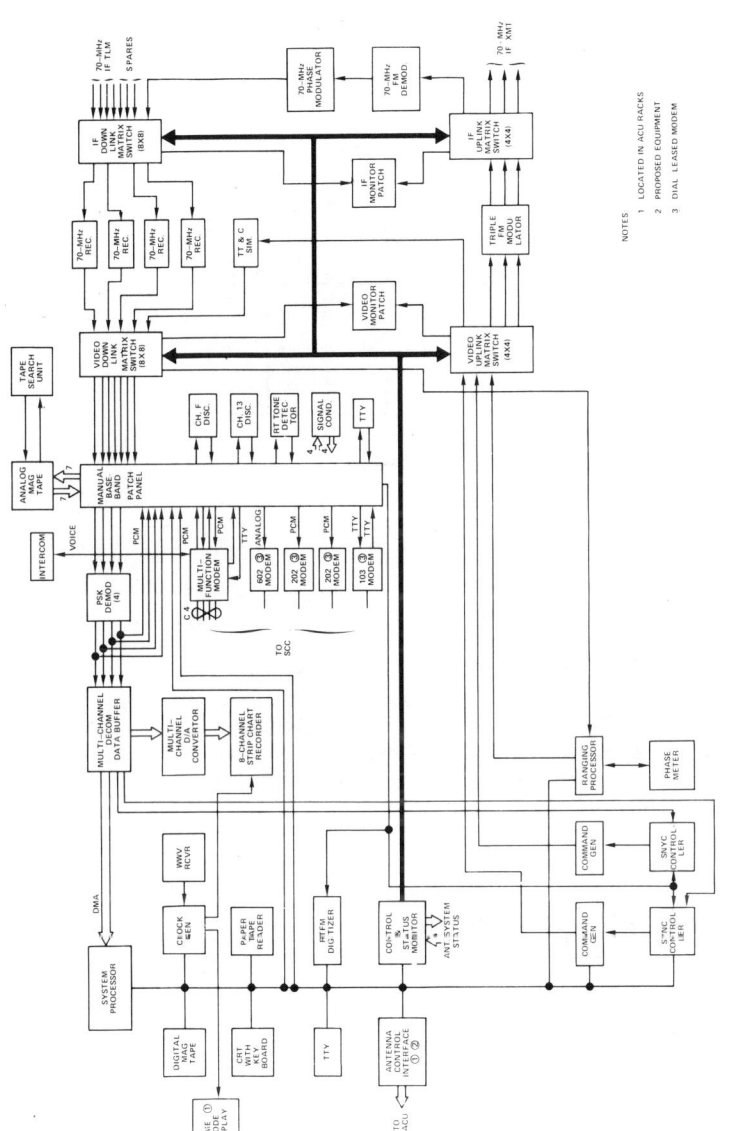

Fig. 3 Earth station processing equipment configuration.

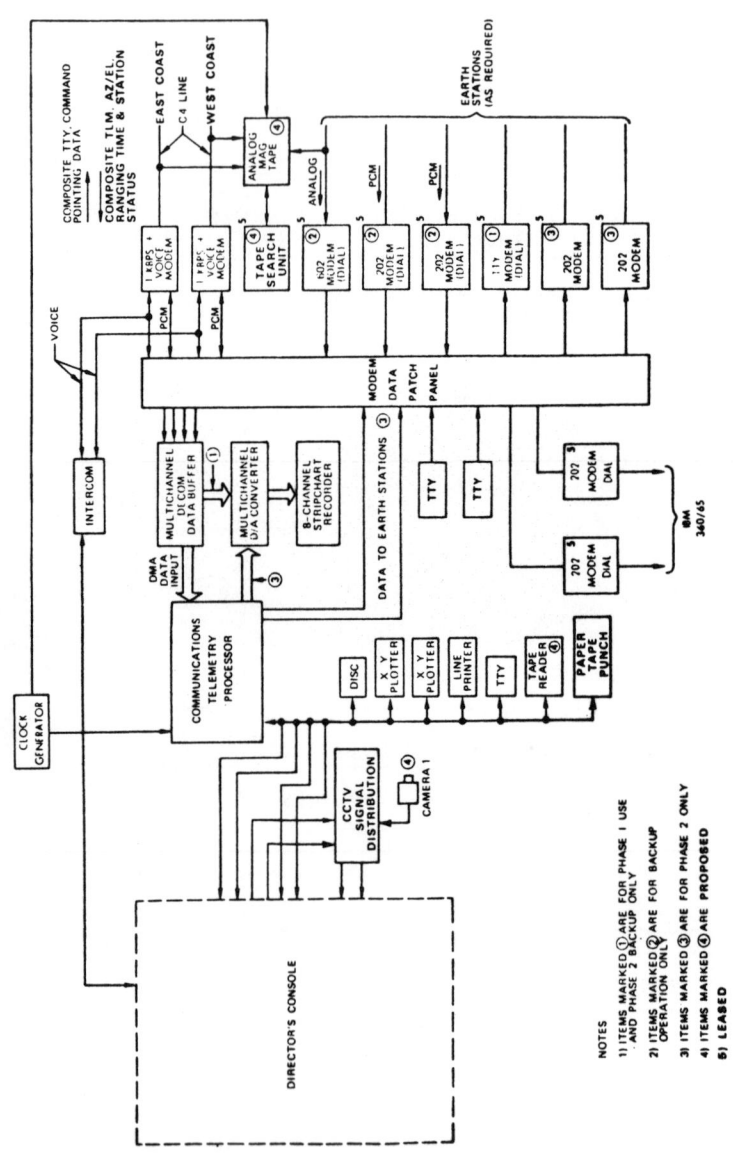

Fig. 4 System control center configuration.

the use of a synchronous controller, which can be configured remotely from the SCC. The synchronous controller derives spacecraft spin and attitude information from the telemetered data to provide synchronization information to the command generator.

C. Telemetry

The telemetry acquisition subsystem receives and demodulates spacecraft telemetry signals. A common 70-MHz IF input frequency is used for all input telemetry links to the system. The input signals are routed by a computer-controllable IF patch panel to a series of telemetry receivers for amplification and demodulation into composite video signals. The video signals, in turn, are connected via a video patch panel to an array of PSK and IRIG channels 13 and F FM demodulators, to a real-time tone detector, or to the ranging system.

D. Ranging

The slant range to the orbiting satellite in both transfer orbit and on station is determined by a multiple-tone ranging system using a group of up to four coherent audio-frequency signals. The range measurement is performed by a precision digital phase meter operating in conjunction with a multi-mode, computer-controlled range baseband processor. The range baseband processor generates four coherent high-spectral-purity audio tones (35.4 Hz, 283.4 Hz, 3968 Hz, and 27.777 kHz). The phase difference between the transmitted and received tone is measured by a precision phase meter. The measurements are normalized to eliminate spacecraft-induced phased changes, resulting in a phase shift measurement directly proportional to the distance of the spacecraft from the Earth station antenna. Each tone provides a range measurement. One cycle of the low tone measures to within 4545 km; the next two successively higher tones resolve ambiguities to within 568.6 and 37.0 km, respectively; and the high tone measures the range to within 10 m rms.

E. Data Processing and Communications

The data processors to be employed are memory-expanded Hewlett-Packard HP 2100 Microprogramable minicomputers equipped with 32 K of 16-bit core memory, two direct-memory access (DMA) channels, floating point hardware, a time-base generator, and selected peripherals.

The Earth station software structure is presented in Fig. 5. Earth station support software performs the following functions: 1) time annotation of data; 2) ranging and angle tracking data formatting; 3) antenna control (pointing information is obtained via the interfacility data link from the celestial mechanics computer and is stored for later recall for program tracking of the antenna); 4) real-time limit checking of all spacecraft data; 5) storage and transmission of time-annotated antenna tracking data (Az, El) and ranging measurements to the celestial mechanics computer; 6) local and remote monitoring and control of commanding and ranging operations from the system control center; and 7) interprocessor communications between the SCC and each Earth station.

The system control center software (Fig. 6) accomplishes the following functions:

1) Spacecraft data limit checking. Incoming S/C data value is compared with previously established limits. If a parameter is found to be beyond limits, an audible alarm is activated, and, simultaneously, a display of the out-of-limits condition is generated.

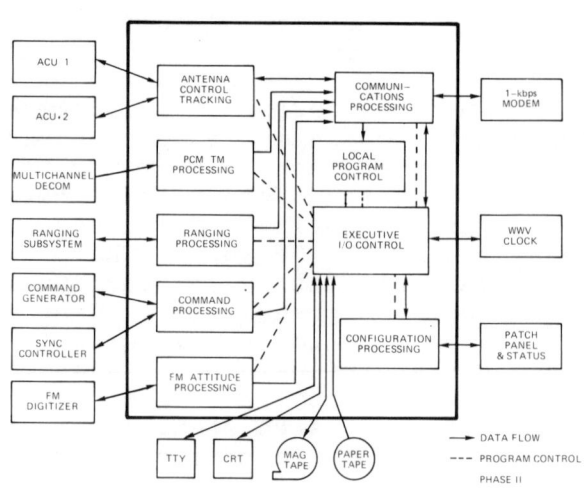

Fig. 5   Earth station software.

2) Interactive modification of data base. The capability is provided to display and/or modify,

GROUND CONTROL SYSTEM TO OPERATE SATELLITES

without interruption of data processing, all data bases, limits, calibrations, and plotting parameters.

3) Automatic system summary generation. A summary includes a) a record of all out-of-limits conditions observed; b) a listing of

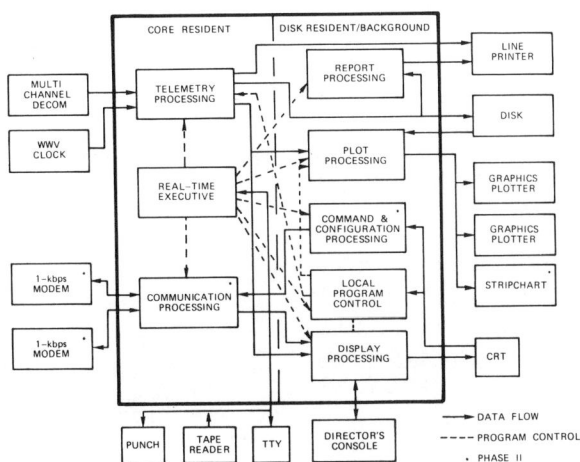

Fig. 6 System control center software.

observed values for all telemetered parameters at time intervals throughout the day; c) a statistical summary for each telemetered parameter containing daily minimum, maximum, mean, standard deviation, and number of samples; and d) a status summary showing initial status and final configuration of each spacecraft.

4) X-Y plotting. Software enables both on-line and off-line display of data on graphical plotters.

5) Command display. The software detects and displays commands as a command sequence is implemented. Spacecraft command registers, with identification and contents, time of execution, and number of execution, are displayed on a CRT; and a hard-copy record is provided on a local line printer.

6) Spacecraft status determination. Incoming data are evaluated to determine the status of spacecraft systems that are of an off/on nature, such as receivers, attitude sensors, despin motor drivers, etc. As part of the normal CRT display of spacecraft data, each pertinent status condition is presented.

7) Display of selected status and telemetry information. A summary of the present state of the

spacecraft is displayed on the CRT, including spacecraft and encoder identification, local time and GMT time, all status conditions, and all telemetered parameters with word number, description, and a value in engineering units.

8) Analog display and recording. Up to eight time-annotated parameters are displayed on a strip-chart recorder for analysis and historical record purposes.

9) Historical data access and analysis. Historical data files (on disk) are accessible via background software modules to allow plotting of data, statistical summaries, printouts of raw data at various rates, and S/C status histories.

10) Command generation. Remote command sequence implementation is provided via keyboard entry in the SCC. The communications processor automatically configures the appropriate hardware and command links to permit such command capability from the SCC.

11) Station configuration monitoring and control. The software monitors and displays the configuration and status of the Earth stations. The capability is provided to reconfigure remotely, under computer control, the applicable Earth station electronics to enable Earth station operator-free performance of routine TT&C functions.

F. Station Configuration Monitoring and Control

Highly centralized system operation from the SCC is achieved through maximum automation of routine TT&C functions. The system design goal is to achieve a one-man-per-shift normal staffing level at each Earth station. To achieve this objective, all uplink (command or range) and downlink (telemetry, video, and baseband) equipment can be switched by means of configuration switching matrices.

The switching matrices operate either in an automated mode controlled entirely by the minicomputer or manually by an operator. In either the automated or manual mode, a 16-bit word is strobed into a parallel register. The data are decoded into a preassigned address (crosspoint location) on the matrix. The decoded address sets a flip-flop that will retain the address. For example, a 5x5 matrix switch has 25 crosspoint relay addresses that incorporate independently. By addressing an appropriate crosspoint, any of the five inputs can be connected to any of the five outputs. A status matrix display is incorporated to provide the operator with visual information pertaining to the state of the input and output configuration of all switching matrices.

G. RF Subsystem

Table 2 summarizes the technical performance of the Earth stations.

VI. Conclusion

The TT&C ground system easily can be expanded to accommodate more satellites and more antennae. It combines the advantages of automatic data processing with ease of manual intervention. As operators become more familiar and confident of the system, unmanned operation at the Earth stations for night shifts, for example, becomes a possibility. This will result in further cost savings.

Table 2 Radio frequency performance of antennae

	Marisat	Full motion	Limited motion
Antenna size, m	13	13	10
G/T, db	32	32	19.6
Transmit gain, db	56.4	56.3	54.3
Polarization	Circular	Circular and linear	Linear
Transmitter power, kw	3	3	3
Low noise amplifier, °K	65	65	65
Sparing of LNA's	1 for 1	1 for 2	1 for 2
Frequency, GHz	4 and 6	4 and 6	4 and 6

References

[1] McCaskill, A. M., Neill, D. V., and Satterlee, A. A., "Launch and Orbital Injection of Intelsat IV Satellites," Comsat Technical Review, Vol. 2, No. 2, Fall 1972, p. 391.

SATCOM SYSTEM CONTROL CONCEPTS
FOR INCREASED LINK AVAILABILITY

A. N. Ince,[x] D. W. Brown,[†] and J. A. Midgley[‡]

SHAPE Technical Centre, The Hague, Netherlands

Abstract

Three different ground terminal transmit power control concepts for a Satcom system are examined. The effectiveness of constant satellite power (CSP) sharing among the carriers and adaptive satellite power (ASP) sharing is compared with constant ground terminal transmit power (CTP). It is shown that ASP offers substantial advantages over CSP in combating environmental degradations and that both can increase link availability with respect to the CTP case. The effectiveness of the ASP control technique is shown to depend on the interconnectivity of the network if multidestinational carriers are used. The measurements required for system control, as well as their accuracies, are described.

I. Introduction

The control of a satellite communication system operating in the frequency range above 7 GHz will be considered. The system to be studied comprises a number of ground terminals, some with and some without radomes, each handling many voice channels. The terminals are linked to each other by means of a geostationary satellite; the access to the satellite is by frequency division multiple access (FDMA).

Presented as Paper 74-476 at the AIAA 5th Communications Satellite Systems Conference, Los Angeles, Calif., April 22-24, 1974.
  [x]Chief, Communications Division.
  [†]Head, Techniques Branch.
  [‡]Member of the Scientific Staff.

Many satellite communication systems, particularly the military ones, are at present power-limited. In order to obtain the required link capacities and the specified performance, it is essential to set the levels of the carriers accessing the satellite very carefully. In normal system operation, the transmit and receive carrier levels and the system noise temperature are subject to variations due to external factors such as rain, cloud, wind, temperature changes, and satellite conjunctions with the sun or moon, as well as equipment aging and maintenance. To counteract the performance degradation caused by these variations, the ground terminal transmitter powers must be varied, or, in more extreme cases, the traffic capacities may have to be reduced. The interdependence of the levels of the carriers transmitted through the satellite and the multiconnectivity of the ground terminals dictate that corrective actions be taken under the control of a single authority with a full knowledge of the status of the system as a whole. Power control can increase the link availability for a given amount of traffic; alternatively, by allowing operation with reduced margins above threshold it can allow an increase in the amount of traffic which the satellite can handle at a given availability.

This paper describes and evaluates three different power control concepts that could be employed for any satellite communications network, where the signals are time-varying and the system margin is small compared with the signal variations. These control concepts are constant ground terminal transmit power (CTP), constant satellite power sharing among the carriers (CSP), and adaptive satellite power sharing (ASP). The effectiveness of these power control techniques has been investigated and their ability to increase the link availability presented.

The study originally was undertaken for the NATO Satcom system, which at present has 12 ground terminals, each transmitting one multidestinational FDMA carrier through a satellite using a hard-limiting transponder. In addition, spread spectrum multiple access (SSMA) is used in the system. The network will be augmented within a few years with many more ground terminals and a multiple transponder/multiple antenna satellite capable of operating in a quasi-linear mode.[1] Block diagrams of the two NATO satellites are given in Fig. 1.

The results presented in this paper were obtained using a computer simulation and a hypothetical network of seven ground terminals. The computer program is described in Sec. II and models the whole of the system in detail, including the satel-

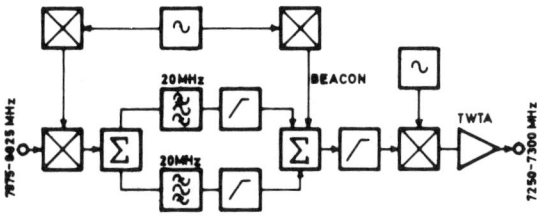

a) Hard-limited transponder

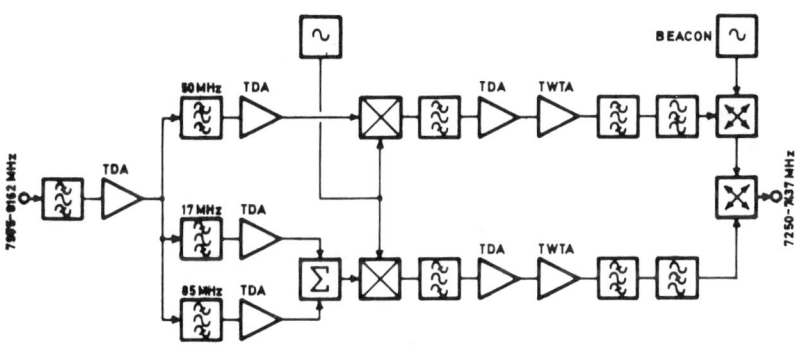

b) Quasi-linear multiple transponder

Fig. 1   Satellite transponders simulated.

lite, propagation effects, and the operation of the ground terminals. The accuracy of the computer simulation was checked against a laboratory model which was configured to simulate the NATO Satcom system. The three power control concepts are described in Sec. III and their effectiveness is compared in Sec. IV.

The implementation of system control for a network is examined in Sec. V, and an automatic data reporting system for the monitoring of the various parameters needed for system control is outlined. The accuracy with which the data can be obtained determines how well the carrier levels may be set and adjusted and, consequently, the effectiveness of the control technique. The paper examines typical measurement errors achieved in experimental and operational systems and discusses the effect of these on the control of the network.

II. Satcom System Simulation

Computer Simulation

An accurate computer model is a useful and often essential aid in the planning and operation of a Satcom system. The model allows lengthy and repetitive computations of power, noise, and interference to be carried out and conditions that rarely occur in practice to be simulated. A system simulator is, therefore, an ideal tool for the study of power control techniques and their effectiveness.

The computer simulation program produced at the SHAPE Technical Centre (STC) and used for this study is capable of simulating both hard-limiting and quasi-linear transponders and a large number of ground terminals, each handling many analog or digital channels. The measured input-output characteristic of the transponder can be used in the program if desired.

The computer program takes account of the position of the satellite, the geographical location of the terminals, their transmit powers, the gain of their antennas, and the loss introduced by the radomes covering the antennas. The up- and downlink signals propagate through a transmission medium, which is subject to variable weather conditions. The various atmospheric sources of attenuation such as oxygen, water vapor, clouds, precipitation, etc., are included in the simulation. The attenuation due to rain is calculated by integrating Eq. 1 along the slant path through the rain:

$$\gamma_r = K\, R_r(d)^\alpha \tag{1}$$

where

γ_r = absorption rate for rain, db/km

$R_r(d)$ = rainfall rate, mm/hr, as a function of distance along the path

K and α = constants for a given frequency

Based upon experimental work,[2] the STC program uses values of $K = 0.017$ and $\alpha = 1$ at a frequency of 8 GHz. The transmission of the carriers through the transponder is modeled as described in the following section and the resulting output powers and intermodulation levels computed. The signals received at the individual ground terminals are calculated and the total signal-to-noise density ratio computed for each carrier received at each terminal. The value of the signal-to-noise density ratio includes both the thermal noise at the terminal and the level of intermodulation generated in the transponder and received at the terminal.

Transponder Simulation

The most critical portion of the computer program is the transponder section. The levels of the signals and intermodulation products at the output of the hard limiters are calculated[3] from Eq. 2, where the input signal consists of N carriers with amplitude A_1 to A_N plus Gaussian noise of power P:

$$v_{\{n\}} = \int_0^\infty 1/x \; J_{n1}(A_1 x) \;\ldots\; J_{nN}(A_N x)$$
$$\cdot \exp{-[Px^2/2]} \; dx \; \cos(n_1 f_1 + \ldots + n_N f_N) 2\pi t \qquad (2)$$

$\{n\} = n_1 \ldots n_N$ denotes the signal or intermodulation component under consideration and f_1 to f_N are the frequencies of the input carriers. For example, a third-order intermodulation product of type 111 involving the input carriers 1, 3, and N is denoted $v_{10100\ldots 1}$. When the program is used to model a quasi-linear transponder the output is calculated[4,5] from Eq.3:

$$v_{\{n\}} = \int_0^\infty a v_1(a) da \int_0^\infty x J_1(ax) J_{n1}(A_1 x) \;\ldots\; J_{nN}(A_N x)$$
$$\cdot \exp{[-Px^2/2]} \; dx \; \cos(n_1 f_1 + \ldots + n_N f_N) 2\pi t \qquad (3)$$

The symbols in Eq. 3 have the same meaning as those in Eq. 2 and $v_1(a)$ is the amplitude transfer function of the nonlinear device relating the value of the output envelope to the instantaneous value of the input envelope. Amplitude-to-phase (AM-PM) conversion can be incorporated by expressing $v_1(a)$ as a complex function $v_1(a) = v_1(a) \exp j\xi(a)$, where $\xi(a)$ expresses the phase shift vs input amplitude.

When the number of carriers is small (less than 10), all of the intermodulation products up to a predetermined order are calculated on a deterministic basis and used to determine the interference levels. When the number of carriers exceeds 10, however, the program makes use of a Monte Carlo method [1]; a randomly chosen sample of the intermodulation products is used as a basis for determining interference levels. This method provides accurate results while maintaining the computer time, which depends very strongly on the number of carriers, within manageable bounds.

Laboratory Model

Although every effort was made to insure that the computer program accurately modeled the satellite transponder characteristics as specified for the in-orbit satellite, certain factors were not known accurately. There was also a limit, set by computer program complexity and execution time, on the number of types of intermodulation products which could be included in the calculations. It therefore was essential to check the accuracy of the computer simulation against measurements on a real transponder. A laboratory facility modeling the NATO Satcom system is available at STC (Fig. 2). The satellite transponder as well as the ground terminals are accurate models of those used in the actual system. Figure 3 gives a block diagram. The transponder of the laboratory simulator was accessed with several carriers according to the assigned frequency plan; accurate measurements were made both of the carrier levels and, by using a carrier cancellation technique, of the level of the intermodulation within the Carsons bandwidth associated with the carriers. This gave values for the carrier-to-interference ratios for a particular frequency plan and a particular relative setting of the carrier levels. The tests were carried out for various carrier levels and frequency plans; the computed values were always within 1.5 db of the measured results and were within 0.5 db in 50% of cases, thus validating the computer simulation.

The accuracy of the laboratory simulation of the satellite transponder was checked using the STC experimental ground terminals[6] (Fig. 4) and the actual NATO satellite. In-orbit measurements of the satellites and ground terminal measurements checked the quality of both the laboratory and computer simulation.

Fig. 2   The STC laboratory model of the NATO Satcom system.

SATCOM SYSTEM CONTROL CONCEPTS

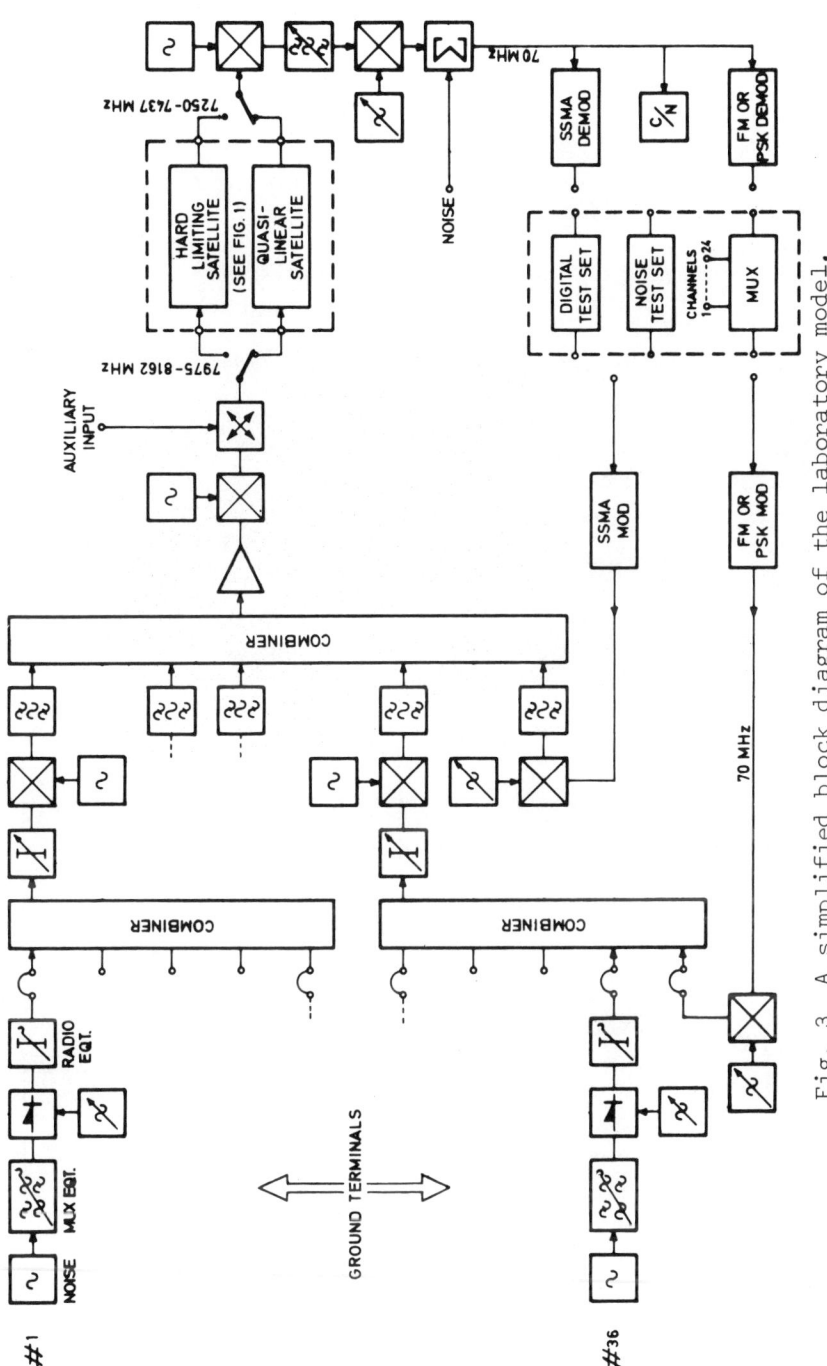

Fig. 3 A simplified block diagram of the laboratory model.

Fig. 4   The three STC experimental X-band satellite ground terminals.

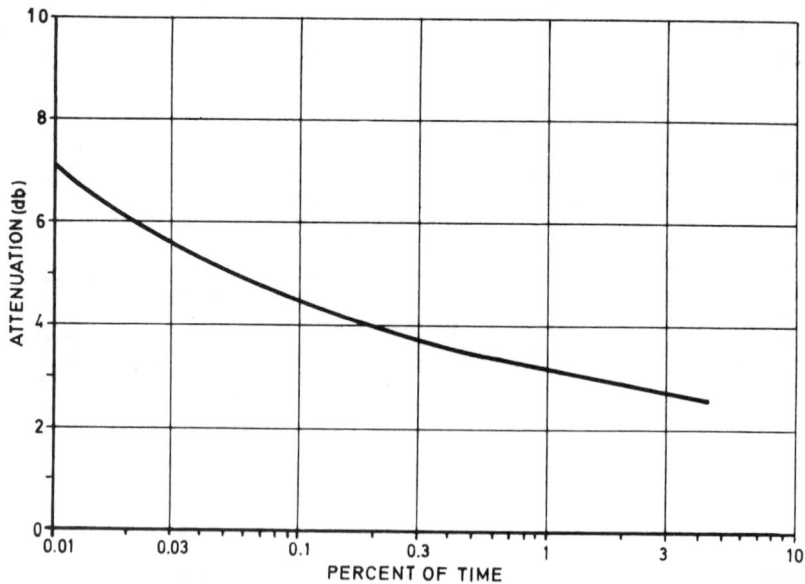

Fig. 5   Typical percentage of outage time of a ground terminal due to attenuation by all environmental effects.

SATCOM SYSTEM CONTROL CONCEPTS

III. Control Techniques

Reasons for Control

Control of the transmit levels of the ground terminals is required in order to reduce the system outage time caused by environmental effects or to offset the effects of conjunctions of the satellite with other communication satellites or with the sun or moon. The system may be controlled to maximize, say, the total number of channels or to maintain the capacity of certain selected links. Estimates have been made on the long-term statistics of the atmospheric attenuation at the NATO satellite ground terminals, and on the effects of conjunctions with the satellite. As can be seen from Fig. 5, attenuations of 4 db occur at a typical site with a radome for some 15 hr/yr. Thus with a margin of the order of 4-6 db considerable outage time can be expected unless some counteraction is taken. Using orbit determination and prediction programs developed in-house at STC[1] to provide satellite look angles for the NATO ground terminals, it has been calculated that the NATO system will be influenced by sun and moon conjunctions 0.1% and 0.03% of the time, respectively.

The elements of the Satcom control system may be divided into the following parts. First, there must be a measuring system at the terminals for determining such parameters as transmit power, receive carrier power, noise levels, and some quality measure of interest to the user such as test-tone-to-noise ratio in a voice channel. These data then must be transmitted by means of an order wire to the control center. Located at the control center is the central processor, where the computations are performed to determine the optimum ground terminal transmit powers for the control philosophy being used and the situation pertaining at that moment. Finally, orders concerning the transmit powers required must be sent to the ground terminals, where the actual transmitter power must be adjusted either manually or automatically. The practical implementation of system control is discussed in Sec. V, where measurement techniques, their accuracies, and their effect upon system outage time are also presented. The purpose of this section is to introduce the various control techniques.

Three different techniques have been studied for controlling the transmit powers in the ground terminals of the Satcom system.

Constant Ground Terminal Transmit Power (CTP)

In this control method, each ground terminal is allotted a fixed transmitter power, such that the carrier level at the satellite's output under clear-sky conditions is that necessary to give the required performance at the receiving terminals. The terminal transmitter powers are maintained at these constant values, without regard to the environmental or climatic conditions at any of the terminals. Long-term variations in equipment performance and satellite position are, however, compensated for by adjustments to the terminal transmit power assignments. CTP is a relatively weak form of control but provides a baseline for considering the other control concepts. In systems where the signal variations are small compared to the system margin, this technique may be preferred.

Constant Satellite Power Sharing (CSP)

With this technique, the transmitter power of each ground terminal is adjusted so that the output powers of the carriers from the satellite assume values that are predetermined to give the desired performance. If the environmental conditions at a ground terminal reduce a carrier's share of the satellite power, the transmitter power at the affected terminal is adjusted to make up the difference. Thus the ground terminal transmit powers are adjusted continuously so that each carrier's share of the satellite equivalent isotropically radiated power (EIRP) remains constant, irrespective of the environmental or other degradations taking place at the ground terminals.

Adaptive Satellite Power Sharing (ASP)

Under ASP control, each carrier's share of the satellite EIRP is adjusted continuously to that value which maximizes some measure of performance, such as the communication margin, throughout the network. The adjustments are effected by changing the transmit powers at the individual ground terminals. The interaction between the carriers in the transponder and the interconnectivity of the communications results in many simultaneous changes in the transmit powers for each change in the environmental conditions at a terminal. This continuous updating of the network can be achieved best using automatic control procedures.

IV. Comparison of Control Techniques

Communication Margin

In order to compare quantitatively the control techniques, it is necessary to postulate a system and to establish a criterion for the link quality. The system assumed comprises seven ground terminals accessing in FDMA the 20-MHz channel of the transponder shown in Fig. 1a. Operation with multidestination FM carriers is studied first, whereafter single destination operation is considered. The link quality criterion will be the margin in decibels above the threshold of the FM demodulator (the threshold value being defined as the value of C/η at which the departure from linearity of the demodulator characteristic is 1 db. Alternatively, a value of C/η corresponding to some channel noise could be specified as the threshold). Using the calculated values of the carrier-to-total noise densities at each receiving terminal, together with the threshold levels, the communication margins can be computed for each link in the network. This can be shown in a matrix format as in Fig. 6, which depicts a system with seven ground terminals having moderate to high interconnectivity.

The values of margins presented in Fig. 6 are those obtained using equal transmit powers from the seven terminals. As will be observed, there is a wide discrepancy in the values of the margins, the differences being accounted for by such factors as the differing angles of satellite elevation at the individual terminals, the presence or absence of radomes, the number of communication channels from particular terminals, differing intermodulation levels, etc.

Several factors should be noted concerning the communication margin matrix. An alteration in the transmit power of a particular terminal changes not only the margins with which that terminal is received but also all of the other margins in the network. This is due to the interactive effect of the carriers in the hard limiter. Increasing the transmit power of terminal 2 increases the margins with which that terminal is received, i.e., the horizontal line in the matrix, while at the same time reducing by a small amount all other margins.

The multidestinational character of the communications carriers should be emphasized; changing the transmit power of a carrier changes all of the margins with which it is received. This multidestinational feature, together with the degree of interconnectivity of the network, limits the effectiveness of power control techniques, as will become evident later.

Transmitting Terminals	Receiving Terminals						
	1	2	3	4	5	6	7
1			6.0	7.8	7.4	6.7	7.6
2			6.8	8.5	8.1	7.4	8.2
3	6.6	7.5		6.8	6.4		6.6
4	7.5	8.4	6.1		7.4		
5	7.0	7.9	5.6	7.2			
6	9.3	10.2					9.4
7	8.3	9.1	6.8		7.4		

Fig. 6 The communication margin matrix for equal transmit powers.

Transmitting Terminals	Receiving Terminals						
	1	2	3	4	5	6	7
1			2.8	4.6	4.1	3.4	4.3
2			7.2	8.9	8.5	7.8	8.7
3	3.2	7.7		7.1	6.7		6.8
4	4.2	8.6	6.4		7.6		
5	3.7	8.2	5.9	7.6			
6	6.0	10.4					9.6
7	4.9	9.3	7.1		7.7		

Fig. 7 The communication margin matrix for rain at terminal 1.

Performance Results

An understanding of the operation of the three different control concepts outlined in Sec. III may be obtained by examining a communication margin matrix where one terminal has suffered an attenuation due to adverse environmental conditions. In Fig. 7, rain at terminal 1 results in a reduction of 3.3 db in the communication margin of terminal 1. For rain at terminal 1, it is evident that both the margins for reception at that terminal, i.e., the vertical column, and the margins with which it is received, i.e., the horizontal line, have been degraded. Examining the effect of the three control philosophies on this situation reveals the following:

1) CTP would not have any effect.

2) CSP would restore the input power of terminal 1 to the satellite. Under CSP control, the proportion of satellite power allocated to each of the carriers remains constant whatever the environmental conditions prevailing at the individual terminals. The consequence of this is that all of the communications margins in the network remain constant, with the exception of those at an affected receiving terminal. Restoring terminal 1's share of the satellite power, while restoring the margins for that carrier to the values given in Fig. 6, will result in a reduction in margin at some terminals when compared with the CTP situation (Fig. 7).

3) ASP control would adjust the transmit powers at the terminals such that the output powers of the carriers from the satellite are those that maximize the minimum communication margin in the matrix.

In the previous paragraphs, we have assumed equal transmit powers from the ground terminals. In a realistic situation, it is necessary to optimize the transmit powers from the terminals such that the minimum communication margin is maximized in the clear-skies condition. This optimization gives a minimum margin of 6.4 db for the system considered.

Using the transmit powers that give this optimum condition and assuming the identical rain conditions at terminal 1 as previously postulated, CTP control gives a lowest communication margin of 3.0 db, CSP control 4.1 db, and ASP control 4.6 db. It is evident that for the particular terminal under consideration substantial gains have resulted through the use of power control.

Comparison of the Effectiveness of the Control Techniques

In order to compare the effects of the three control philosophies, the lowest communication margin in the matrix for each technique was plotted against the rain rate at each terminal. Curves for a typical terminal are presented in Fig. 8. It is not claimed that the attenuation calculated as a function of rain rate is highly accurate, since the models used for attenuation on slant paths have not been validated experimentally. The results are entirely adequate, however, for comparing the power control techniques. An outage is defined as the condition when the minimum communication margin in the matrix drops to zero. Thus, from the curves in Fig. 8, it is evident that ASP control gives better link availability than CSP, which in turn is an improvement over CTP for the particular terminal shown.

Several interesting points arose from a comparison of the diagrams similar to Fig. 8, but for the other terminals. ASP control is less effective in combating the environmental effects, the more the interconnectivity of the particular terminal; this is due to the multidestinational nature of the carriers. For most terminals, CSP gave better results than CTP control, the exception being that terminal which has the highest clear-skies link loss. For this terminal, CTP control is

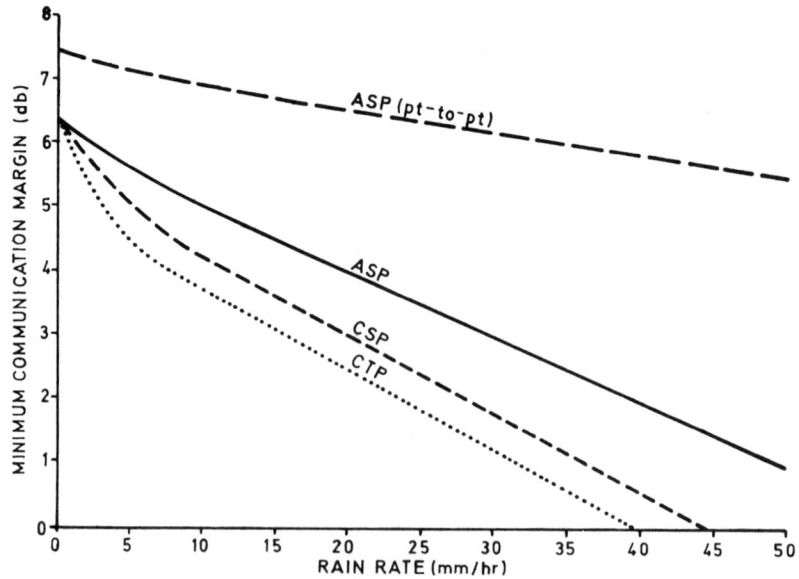

Fig. 8 A comparison of the effectiveness of the three control techniques.

preferable to CSP control. This result is to be expected from
the power-sharing among the carriers in a hard-limiting transponder explained previously.

We have just considered a system using multidestinational
carriers employing a moderate degree of interconnectivity of
the terminals. The results can be generalized by examining
two extreme cases of connectivity. If full connectivity is
present (where each terminal receives all carriers, including
its own), then for multidestinational carriers the ASP control
results will be identical with those of CSP. The other extreme to be considered is that of multiple-carrier single destination operation, as might be used in a digital Satcom system. In this condition, ASP control is highly beneficial,
since a large portion of the satellite output power is available for reallocation to the affected links, albeit at the ex-
pense of otherwise unaffected terminals. A curve for ASP control of a system carrying the same traffic but using point-to-
point links is plotted in Fig. 8 for comparative purposes. It
will be observed that in the clear-skies condition there has
been a gain of almost 1 db in the communication margin just
because all carriers no longer need to be received at the terminal having the highest link loss. The benefits of ASP control in this network employing point-to-point links are very
dramatic at high rain rates.

In order to provide an indication as to the extent of the
increased link availability provided by the various control
techniques, use was made of the rain statistics for the terminal sites. A knowledge of that rain rate at a terminal which
will cause a margin in the matrix to drop to zero, and the percentage of time that this rain rate is likely to be exceeded,
gives an estimate of the outage time due to adverse environmental effects at that terminal. For each control technique,
the sum of these outage times for all of the terminals was
computed; the results for the three techniques were compared.

For the particular network considered, CSP control procedures gave the same availability as CTP control, this being
due to the dominating influence on the availability figure of
one terminal. By coincidence, the terminal suffering the
greatest incidence of high rain rates also has the highest
link loss under clear-skies conditions. Therefore, the good
effects of CSP control at the other terminals are offset by
the actual degradation in availability due to CSP at this limiting terminal. ASP control halved the estimated outage time
over that expected for CTP and CSP control. The computations
showed that the availability value was sensitive to the posi-

tion of the satellite. A change in position toward the terminal dominating the calculated availability substantially improved the computed results for all of the power control methods considered.

Further substantial gains in availability are possible using ASP control if the interconnectivity of the multidestinational network can be reduced. If multicarrier single destination operation were to be used, then ASP could cope so effectively with the environmental degradations that no significant interruption to the communications would occur even with substantially reduced clear-skies margins.

ASP control is highly desirable, since it provides an increase in the availability for the network; this could be extremely important if the output power of the satellite were to be degraded and the available "clear-skies" margins consequently reduced. However, ASP control requires the adjustment of a large number of transmit powers in the network each time the situation changes. This is difficult to implement except on an automatic basis.

V. Implementation of System Control

Measurements

The computer simulation used to produce the results just reported did not model measurement errors, or delays and inaccuracies in the calculations and transmit power settings. The results are, therefore, for ideal implementation of the control techniques. In a practical situation the accuracy with which the system parameters can be measured and the settability of transmit levels of the carriers, as well as the intrinsic effectiveness of the control technique, determine how well the system may be controlled and, consequently, the availability of the communication links. Table 1 shows the accuracies with which the important system parameters can be measured in a carefully calibrated experimental ground terminal.[7]

Table 1 Achievable measurement accuracies

Parameter	Error, db
Transmitter power	0.3
Receiver noise	0.2
Receiver signal level above threshold	0.5

For an operational system, the measurement inaccuracies are likely to be greater than these values. Because of the measurement uncertainties in a practical situation, the effectiveness of the control techniques described in Sec. III is reduced. Since CTP control involves only the measurement of the ground terminal transmitter powers, it can be implemented more accurately than CSP control, which, in addition, involves the measurement of relative receive powers. Despite this, CSP, because of its ability to deal with multiple degradations within the network, will increase availability in a Satcom system and is therefore to be preferred over CTP.

As mentioned previously, the implementation of CSP involves the measurement of the relative receive powers of the carriers at each ground terminal. Except for the measurement errors, the relative powers as measured at a terminal are equal to the relative carrier powers at the output of the satellite. (This assumes that the difference between the highest and lowest carrier frequency is not great.) It is these relative satellite output powers which CSP seeks to maintain at values set to achieve the desired performance. The accuracy of the estimate of the satellite power-sharing can be improved considerably by averaging the measured ratios of receive powers over all of the terminals in the network. In a system employing multidestinational carriers, the ability to average across the terminals is available free, since each terminal receives a large fraction of the carriers handled by the satellite.

ASP control involves an absolute received signal level measurement or a signal-to-noise ratio measurement. It was estimated[7] that the total measurement error would be of the order of 1.2 db. Control actions would not be initiated as a result of small changes in the calculated margins; instead, step changes in the transmitter powers would be made when the indicated margins have degraded by a specified amount. A step size of 1.0 db would be appropriate for the measurement errors. It can be seen from Fig. 8 that the advantages theoretically provided by ASP in the multidestination case would be largely offset by the measurement inaccuracies. ASP control for point-to-point links is, however, so effective (Fig. 8) that substantial availability gains can be obtained even with relatively high measurement errors.

Quasi-Linear Transponders

The previous sections have discussed ground terminal power control techniques for a satellite with a hard-limiting transponder. These techniques are equally applicable to a satellite with a quasi-linear transponder. Whereas for a hard-limiting satellite only the relative carrier powers at the input of the satellite and hence only the relative transmit powers of the terminals are of importance, for a quasi-linear transponder the total incident power at the satellite input must be kept within specified limits. In other words, the backoff of the quasi-linear transponder from saturation must be controlled to a predetermined value to reduce the effects of intermodulation. An output backoff of approximately 5 db gives a carrier-to-intermodulation ratio of about 20 db.

The transmit powers required at the various ground terminals can be precalculated from a knowledge of system parameters, such as satellite transponder gain, to provide the required satellite power-sharing and the desired value of backoff under clear-skies conditions. CTP control in a quasi-linear transponder presents no new problems. Attenuation of the carriers resulting from environmental degradation gives an increase in the backoff of the satellite, with an attendant reduction in the intermodulation levels.

Long-term drift of the satellite transponder gain or variation of satellite orbital position can result in a gradual change of backoff, whichever control philosophy is adopted. It is necessary, therefore, to monitor satellite backoff and periodically adjust all of the transmit powers to restore the backoff to the desired value. This is done best by a measurement of the backoff onboard the satellite, which is then telemetered to the control station. It also can be accomplished by calculating the total satellite input power from the ground terminal transmit powers and link losses. Periodic calibrations of the satellite transponder gain then will allow backoff to be calculated.

CSP control is simple to implement for a quasi-linear transponder. Perfect implementation of the technique results in the backoff remaining constant whatever the environmental degradations at the ground terminals. The technique can be implemented as described previously for the hard-limiting transponder by measuring relative receive powers and averaging over the terminals. Alternatively, the link losses can be measured in terms of the variation in the receive level at each terminal of the satellite beacon. The measurement can be performed accurately in the ground terminal by making a relative measurement with respect to a calibration signal injected at the low-noise amplifier input. Variations of satellite beacon level either can be controlled in the satellite, or the actual beacon level telemetered to the control center. With a knowledge of link losses throughout the network, CSP control can be implemented simply by increasing transmit powers to offset the losses.

This technique can be extended to provide full ASP control. Uplink losses still would be compensated, but in addition downlink losses would be offset by an increase in transmit power of all those carriers that have to be received at the affected terminals. In a system employing multidestinational carriers and having even moderate interconnectivity, this would result in a large reduction of backoff, with an attendant increase in intermodulation interference to some carriers. Such control should not be implemented for a multidestinational network unless accurate calculations can be performed to determine the resulting redistribution and increase of intermodulation. For point-to-point operation with many ground terminals, however, ASP implemented in this way would have negligible effect on backoff. Some typical cases should be calculated, but it is to be anticipated that, if the backoff is normally such that intermodulation is considerably below thermal noise at the ground terminals, then the redistribution of intermodulation as a result of increasing a few carriers will not present a problem.

An added degree of freedom is available in a quasi-linear transponder. The backoff may be reduced to provide extra power to compensate losses at the terminals caused by environmental degradations. Intermodulation effects may, however, limit the degree to which this extra flexibility is useful.

Manual and Automatic Implementation

For CTP and CSP control, it would be possible for each individual terminal to perform its own monitor and control function. ASP control, however, necessitates a central control facility that can monitor the status over the whole network and make the necessary adjustments to optimize the performance. The network requires an automatic data reporting system, which would monitor certain of the parameters at each ground terminal and transmit this information to a system control center. A processor at the control center then would determine the optimum conditions for the network as a whole and issue orders for the required transmit power changes. The control orders could be checked and dispatched by an operator to the individual terminals, where they would be implemented manually. Alternatively, the control orders could be transmitted by the computer directly to the individual terminals, where implementation could be manual or automatic.

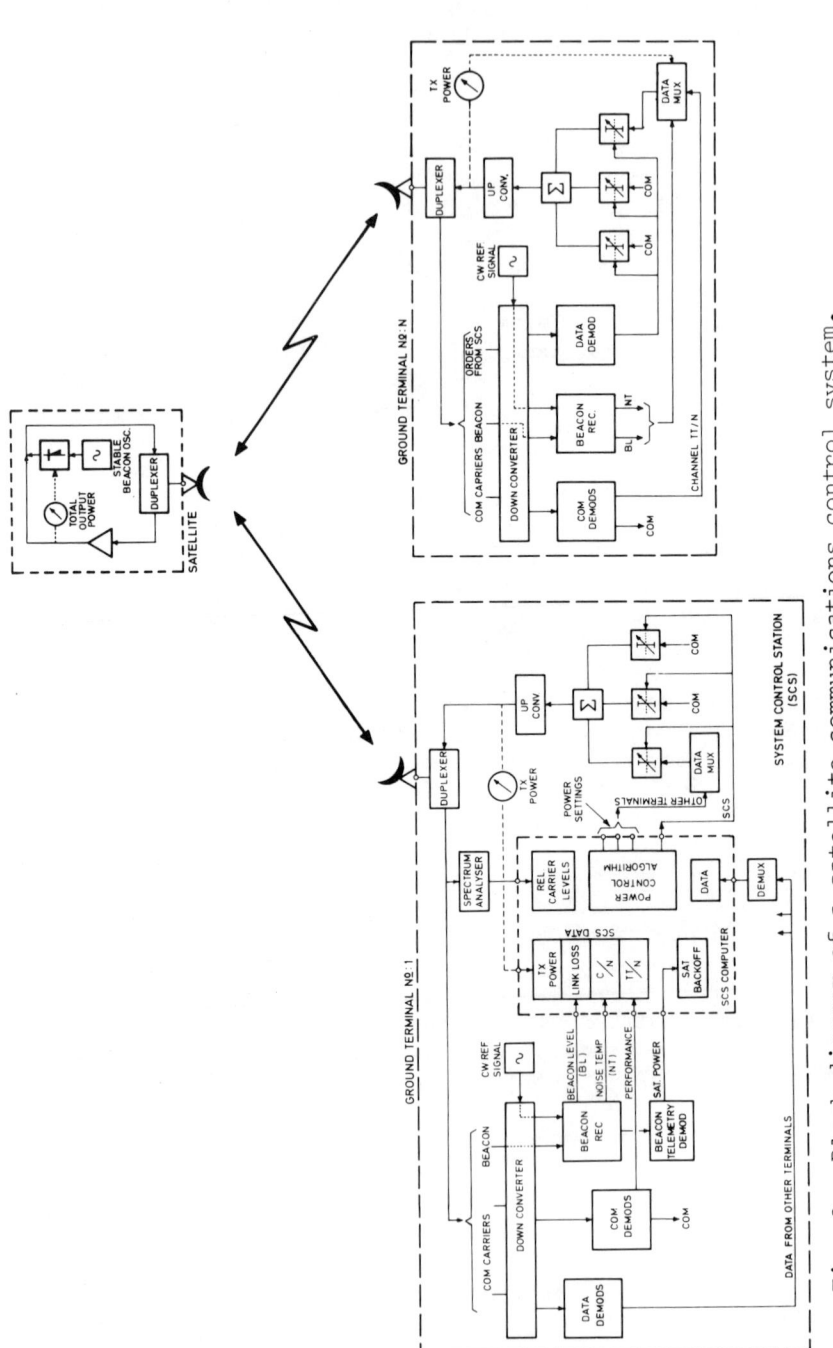

Fig. 9 Block diagram of a satellite communications control system.

SATCOM SYSTEM CONTROL CONCEPTS

Figure 9 shows a block diagram of a system for automatically implementing ASP control in a Satcom system employing a quasi-linear transponder. Many of the ideas presented above are incorporated. An accurately leveled beacon is transmitted from the satellite. A reference signal injected at the input of each ground terminal receiver is compared with the beacon level in the beacon receiver and hence an accurate measure of the beacon level at the receiver input can be obtained. The ground terminal noise temperature also can be measured in terms of the injected reference signal. Knowing accurately the path losses in the system and their clear-sky values, the noise temperature of each terminal, and the transmit levels of all the carriers, the system control computer can very simply calculate the carrier level increases required to restore the carrier-to-noise ratios at the various receiver inputs throughout the system. The total carrier power level at the satellite output is measured (Fig. 9), modulated on the beacon, and is used in the system control terminal to monitor the transponder output backoff. It is a responsibility of system control to maintain this backoff at the point which provides the optimum balance between intermodulation and satellite power; backoff would be controlled by changes in transmit power throughout the system. It can be seen (Fig. 9) that the computer has available a performance measurement for each link. This could be test-tone-to-noise ratio for an analog voice system or bit error rate for digital carriers. From these measurements the computer can monitor the effectiveness of its control actions and provide warnings of equipment malfunction. It is noteworthy that, by using measured path losses as the basis for control rather than the performance measurements, an unconditionally stable system is obtained.

Of course, control actions must be timely to be effective. There are delays inherent in the information flow, in the calculations, and to a lesser extent, if an automatic system is used, in the implementation of the control orders. Using 200-bit/s transmission, a simple algorithm based on path losses, and programmable attenuators, the total loop delay from measurement to implementation can easily be kept to 10-15 sec, which would be entirely adequate to deal with the vast majority of environmental effects.

VI. Summary and Conclusions

ASP control is theoretically superior to both CSP and CTP and, for the particular network considered, could halve the estimated outage time if perfectly implemented. For a network employing multidestinational carriers, ASP control is more beneficial, the less the interconnectivity of the ground terminals; the full benefits of ASP are realized for a network of point-to-point links, where it is very effective in increasing link availability.

The difference in availability using CSP and CTP control procedures is insignificant for the network considered. However, this is due to the dominating influence of one particular terminal in the network. In general, CSP will give a better availability result than CTP, and moreover CSP is to be preferred from the operational point of view, since it copes with simultaneous degradations at more than one terminal.

The control techniques outlined are equally applicable to satellites using hard-limiting or quasi-linear transponders. For satellites using quasi-linear transponders, an extra degree of freedom, the backoff from saturation, is available, which can under certain circumstances be varied to influence the availability of the network.

Measurement accuracy of the parameters required for system control plays an important part in determining the link availability achieved using a particular control philosophy. Enhanced accuracy would enable the availability of the network to be increased, or, alternatively for constant availability, it would allow operation with reduced clear-skies margins. Thus, increased traffic capacity could be made available or the effects of degraded satellite output power tolerated.

If the margins in a network are sufficient that outage times of 0.1% or less are experienced, then no control other than CTP would be necessary. This may be achieved with margins of 6 db or more. If, however, the system margin is small compared with the signal variations, CSP or ASP control would be advantageous and may become obligatory if the operating margin were 3 db or less. In this event, automatic implementation of the control procedure might be beneficial. Satellite output power is a valuable resource: active ground terminal transmit power control serves to make best use of this resource.

References

[1] Ince, A.N., "Design Testing and Control of an X-Band Satellite Communication System," IEEE Transactions on Communications, Vol. COM-22, No. 9, Sept. 1974, pp. 1338-1353.

[2]Blevis, B.C., Dohoo, R.M., and McCormick, K.S., "Measurements of Rainfall Attenuation at 8 and 15 GHz," IEEE Transactions on Antennas and Propagation, Vol. AP-15, No. 5, May 1967, pp. 415-423.

[3]Shaft, P.D., "Hard Limiting of Several Signals," IEEE Convention Record, 1965.

[4]Imboldi, E. and Stette, G.R., "AM-to-PM Conversion and Intermodulation in Non-Linear Devices," Proceedings of the Institute of Electrical and Electronics Engineers, Vol. 61, No. 6, June 1973, pp. 796-797.

[5]Stette, G.R., "Calculation of Intermodulation from a Single Carrier Amplitude Characteristic," IEEE Transactions on Communications, Vol. COM-22, No. 3, Mar. 1974, pp. 319-323.

[6]Ince, A.N., "Design and Calibration of X-Band Satellite Communication Ground Terminals," de Ingenieur, Vol. 84, No. 20, May 1972, pp. ET51-ET68.

[7]Ince, A.N. and Wallrabe, A., "Ground Terminal Measurement Requirements with Respect to Satellite Communications Link Availability," Conference Proceedings on "Aerospace Telecommunications Systems," AGARD-CPP-103, 1972.

CHAPTER IV—MULTIPLE ACCESS, MODULATION, AND CODING

As was pointed out in the Introduction to Chapter II, the definition of an Earth terminal excludes the technology and equipment below the intermediate frequency. This area of technology, called baseband or communications processing, encompasses modulation, multiplexing, multiple access, and coding for both the transmission path and the data source itself. The three papers in this chapter are representative of current interests in this field, which is highly responsive to the particular application of satellite communications.

A form of multiple access that has become popular of late is "single channel per carrier" (SCPC), and may be best defined as a frequency division multiple access (FDMA) method, in which each carrier is limited to one voice channel, or its data equivalent, and assigned on a *dedicated* basis. Because of its marked similarity to FDMA demand assignment methods, SCPC is sometimes used as a stepping stone into an FDMA demand assignment (DA) system, or as a *dedicated* subset of such a system. SCPC is normally applied to networks of small, lightly loaded terminals in relatively isolated circumstances or as an auxiliary capability of a large Earth station. The paper by Sanderson and Ludwig investigates the use of frequency modulation (FM) in SCPC, or its DA counterpart, for various terminal figures of merit and satellite downlink power. Whether FM or PSK is superior for this application is a major question, and this paper serves to give the FM side of the issue.

The second paper, by Kullstam, addresses a problem area of modulator-demodulator design that is becoming increasingly important as satellite communications terminals find their way into applications involving airborne and seaborne platforms. The impact of Doppler effect, once of considerable concern when medium altitude space segments were in vogue, became of comparatively little interest when geostationary satellite systems were proven practical. The current programs for aeronautical and maritime satellite communications have revitalized the studies of communications over Doppler-stressed links.

Ionospheric scintillations are the subject of a number of the papers in Chapter V in which these path degradations are discussed and attempts made to characterize or model them. The means for reducing the impact of these degradations is, however, more properly treated as part of baseband processing, and the final paper of Chapter IV treats this subject. The developing use of the lower frequency bands, such as VHF and UHF for satellite communications, coupled to the sensitivity of these bands to ionospheric scintillation, has led to investigations of various methods of combating such degradations on satellite-established links. Massey's paper provides a valuable insight into the ability of various forms of forward-acting error control to alleviate these disturbances. His comments on adaptive coding are particularly to be noted in view of the trend towards real-time adaptive control of satellite communications systems of the future.

SINGLE CHANNEL PER CARRIER VOICE
TRANSMISSION VIA COMMUNICATIONS SATELLITE

Charles C. Sanderson[*] and Lloyd G. Ludwig[+]

Hughes Aircraft Company, El Segundo, Calif.

Abstract

This paper analyzes the capacity achievable through various satellite transponder and Earth terminal links operated in a multicarrier, multiple-access mode with a single voice channel on each carrier (SCPC). Analytical techniques for maximizing this capacity in terms of the interaction of thermal and intermodulation noise in achieving specified link quality are presented. Improvements in capacity resulting from the use of voice activation and variation of modulation parameters also are described. This analysis shows that a high-capacity demand-access system can be achieved using FM modulation SCPC with voice activation.

Nomenclature

B	=	bandwidth
BW_t	=	transponder bandwidth
BW_o	=	occupied bandwidth
B_{vc}	=	voice channel rf bandwidth
N	=	total number of voice channels or carriers = BW_o/B_{vc}
N_a	=	active channels or carriers
N_s	=	number of available carrier locations in transponder (equally spaced) = BW_t/B_{vc}
A_f	=	voice activity factor

Presented as Paper 74-471 at the AIAA 5th Communications Satellite Systems Conference, Los Angeles, Calif., April 22-24, 1974.
[*]Project Engineer, Systems Laboratories.
[+]Project Manager, Commercial Systems Division.

BO_i	=	input backoff of traveling-wave tube amplifier (TWTA) relative to saturation
BO_o	=	output backoff of TWTA relative to saturation
BWR	=	bandwidth ratio = BW_t/BW_o
$(C/N)_{up}$	=	uplink carrier-to-noise ratio
$(C/N)_{dn}$	=	downlink carrier-to-noise ratio
$(C/N)_{th}$	=	carrier-to-thermal-noise ratio
(C/IM)	=	carrier-to-intermodulation ratio
sat	=	saturation of TWTA (maximum output power)
$(C/N)_L$	=	link carrier-to-noise ratio
TTNR	=	test-tone-to-noise ratio
MIF	=	modulation improvement factor
P	=	pre-emphasis advantage
W	=	noise weighting advantage
C	=	compandor advantage
Δf_{pk}	=	peak frequency deviation
f_m	=	highest baseband frequency
M	=	modulation index
F_p	=	peak factor
T	=	noise temperature of receiver system
G	=	ground station antenna gain
P_t	=	ground station transmitter power
$EIRP_{sc}$	=	ground station EIRP required for TWTA saturation with single carrier

I. Introduction

In the past 10 years, international communication by satellite has grown to reach almost every corner of Earth. Satellite technology has grown through four generations of Intelsat satellites. This communications network has utilized FM/FDMA (frequency division multiple-access) transmission, which efficiently carries the relatively high-density trunk traffic of the Intelsat system. Recently, the network has introduced the Spade system to provide more economic operations for nations having smaller volume of multidestinational, international traffic. Current expansion in the use of communication satellites involves the creation of systems for the internal communication of a single large country. Canada has pioneered in this application with its Telesat system. The Telesat system is a multipurpose system in which some satellite transponders are utilized for trunk links, some in FM/FDMA mode, some for television distribution, and some for telephonic communication to remote locations having very light traffic density.[1,2] The use of single channel per carrier (SCPC) FM modulation of a transponder in a fully variable demand-access mode for remote telephone service is the subject of this paper.

Figure 1 is a photograph of
an Earth terminal of this
type which is in service in
the Telesat network. It is
this type of telephonic service to large numbers of remote locations that is a
unique characteristic of a
national satellite system.
Because the traffic density
to each terminal is very low,
the system would not be feasible economically unless each
Earth terminal were very
small, low-cost, and suitable
for unattended operation. At
the same time, the cost-
effective use of a satellite
transponder is achieved only
if each transponder is capable of carrying a relatively
large number of concurrent

Fig. 1 Small SCPC Earth
terminal for Telesat.

telephone channels. Typical applications for this type of
service are small villages of less than 1000 population in
large underdeveloped nations, settlements in northern Canada
and Alaska, mineral development sites, offshore oil platforms,
small airports, border customs posts, remote harbors, and
pipeline relay stations.

II. Communication Link Requirements

In order to assure the achievement of a suitable Earth
terminal design for this application, a number of constraints
must be placed on Earth terminal design: 1) antenna diameters
from 10 to 30 ft, 2) fixed antenna pointing, 3) uncooled and
preferably transistorized receivers, 4) low-powered transmitters, 5) minimized baseband electronics, and 6) operation by
the subscriber.

Incorporation of this type of station into a network utilizing the current generation of satellites requires careful
planning to retain Earth terminal simplicity and low unit cost
while achieving an adequately large system (channel) capacity.
Channel utilization efficiency can be achieved only by a
demand-access system of single-channel assignment to any destination. The use of SCPC with FM baseband modulation appears
to be a method of achieving the preceding objectives.

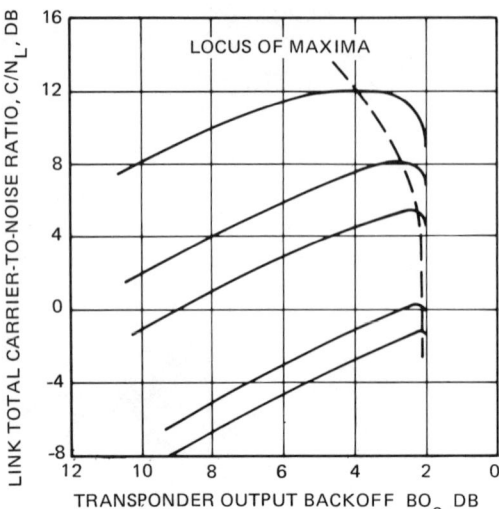

Fig. 2 Link C/N as a function of transponder backoff.

The main impediment to the use of the aforementioned system is the relatively large amount of intermodulation (IM) noise generated in the final traveling-wave tube (TWT) amplifiers of the satellite transponder when it is operated in a multicarrier mode. These IM products can be reduced by reducing the output power level (backoff) of the TWT, but this action decreases the link carrier-to-thermal-noise ratio. Figure 2 shows that there is an optimum combination of thermal and IM noise as a function of reduction of TWT output power (backoff) for each combination of satellite transmitter power (EIRP) and Earth terminal quality (G/T). Unfortunately, in the case of a large number of equally spaced FM carriers, the overall link C/N is frequently below acceptable operating levels (C/N = 9 to 13 db). The curves in the upper part of this figure represent powerful satellite links with relatively large Earth terminals. The lower curves characterize currently available satellite links with small Earth terminals. Figure 3 is a plot of the maximum C/N_L points taken from Fig. 2 for the downlink frequency bands of 4 and 2.5 GHz as a function of EIRP + G/T of the satellite-to-Earth link. Thus it can be seen that a transponder of 36-MHz bandwidth, when fully occupied by equally spaced SCPC FM carriers, cannot achieve an acceptable C/N.

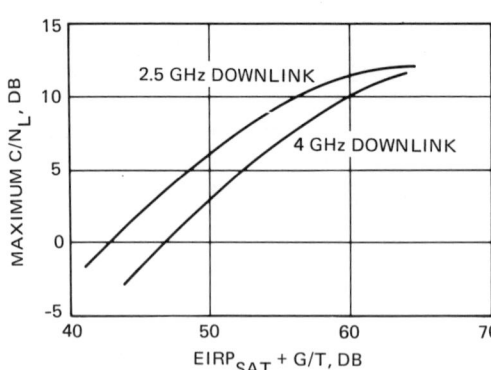

Fig. 3 Maximum link C/N as a function of ($EIRP_{sat}$ + G/T).

III. Unequally Spaced Carriers with Voice Activation†

Our studies have shown that the desired transponder capacity can be achieved with the SCPC FM, small Earth terminal, and current satellite transponders if the channels are unequally spaced and the telephone channel carriers are switched on and off by voice activation. Performance analysis results are based on the IM noise characteristics described by McClure.[3] These calculations show that the achievable capacities of the SCPC FM with voice activation are comparable to the capacity attainable by FM/FDMA and other more sophisticated baseband modulation methods.

As an example of a relatively high-capacity remote location telephone system, a transponder using a 50-rf-w TWT at 2.5 GHz was analyzed for use with a large number of Earth terminals of the type shown in Fig. 1. This link would result in a capacity of 615 concurrent voice-activated carriers at a minimum TTNR of 46 db. For low-traffic usage (0.1 Erlang per location), this system would provide service to 6150 locations. We also have shown that a significantly usable capacity can be achieved with the 6/4-GHz Telesat class of transponder.

IV. Development of Analytical Method

The standard expression for the voice channel performance is given by (in decibels)

$$\text{TTNR} = \left(\frac{C}{N}\right)_L + \text{MIF} + P + W \tag{1}$$

The TTNR is equivalent to a specific value of noise expressed in picowatts:

$$\text{TTNR} = 90 - 10 \log_{10} \text{pwp} \tag{2}$$

Thus, if the allowable noise level is 10,000 pwp (picowatts, psophometrically weighted), the corresponding TTNR is 50 db. As a point of reference, the CCIR (International Radio Consultative Committee) in Recommendation 353-2 allows for

---

†Available carrier locations N_s in transponder are equally spaced; randomness attained either by assignment or voice activation results in random assignment of carriers in these equally spaced carrier locations.

their hypothetical reference circuit, a weighted mean noise power of 10,000 pwp. Although appropriate for international or long-haul traffic, such a noise allocation extracts a large penalty in terms of power and channel capacity for low-cost, high-usage national satellite applications where such quality is not necessarily required.

Although other noise sources such as adjacent satellite interference, equipment (satellite and ground), and other fixed noise sources are inherently present in a satellite link, the TTNR or picowatt values assumed herein refer only to thermal and intermodulation noise of the space link.

V. Link C/N Optimization

Listed in Table 1 is a typical uplink power budget showing the ground station EIRP required to produce saturation (maximum output power) at the satellite with one transmitting carrier. The uplink carrier-to-noise ratio is given by

$$\left(\frac{C}{N}\right)_{up} = \frac{\text{received carrier power}}{\text{thermal noise} = KTB} \qquad (3)$$

Table 1 Uplink communication parameters

Parameters	6 GHz	2.6 GHz
Earth station EIRP, db-w		
Single carrier EIRP for saturation	82.0	68.7
Transmission loss, db		
Path loss (25° elevation angle)	-200.0	-192.7
Satellite		
Receive antenna gain, db	27.0	27.0
Receive losses, db	-1.0	-1.5
Received carrier power at		
receiver, db-w	-92.0	-98.5
Boltzmann's constant, db-w/°K-Hz	-228.6	-228.6
Satellite noise temperature		
Temperature, °K	2000	1020
Temperature, deg, db	33.0	30.1
System bandwidth		
MHz	36	35
Hz, db	75.6	75.4
Satellite noise power, db-w	-120.0	-123.1
Uplink C/N at saturation, db	28.0	24.6

VOICE TRANSMISSION VIA COMMUNICATIONS SATELLITE

Table 2 Downlink communication parameters

Parameters	4 GHz	2.5 GHz
Satellite		
Transmitter power		
Watts	5.0	50
db-w	7.0	17.0
Feed losses, db	-1.0	-1.0
Transmit antenna gain, db	27.0	27.0
Single carrier EIRP at saturation, db-w	33.0	43.0
Transmission losses, db		
Path loss (25° elevation angle)	-196.4	-192.4
Atmospheric absorption	-0.1	-0.1
Total transmission losses	-196.5	-192.5
Earth station		
Receive antenna gain, db	G	G
Received carrier power, db-w	-163.5+G	-149.5+G
Boltzmann's constant, db-w/°K-Hz	-228.6	-228.6
System noise temperature, db/°K	T	T
System bandwidth		
MHz	36	35
db-Hz	75.6	75.4
Noise power, db-w	-153+T	-153.2+T
Single carrier saturated downlink C/N, db	-10.5+(G/T)	3.7+(G/T)

Listed in Table 2 is a typical downlink (satellite to Earth) power budget. The downlink carrier-to-noise ratio for a single transmitted carrier at saturation is expressed in terms of the satellite EIRP and ground station G/T (receiving antenna gain to receive system noise temperature ratio). The resulting downlink carrier-to-noise ratio then is given by

$$\left(\frac{C}{N}\right)_{dn} = \frac{\text{received carrier power}}{\text{thermal noise} = KTB} \qquad (4)$$

The combined uplink and downlink carrier-to-thermal-noise ratio is then

$$\left(\frac{C}{N}\right)_{th} = \frac{(C/N)_{up} \times (C/N)_{dn}}{(C/N)_{up} + (C/N)_{dn}} \qquad (5)$$

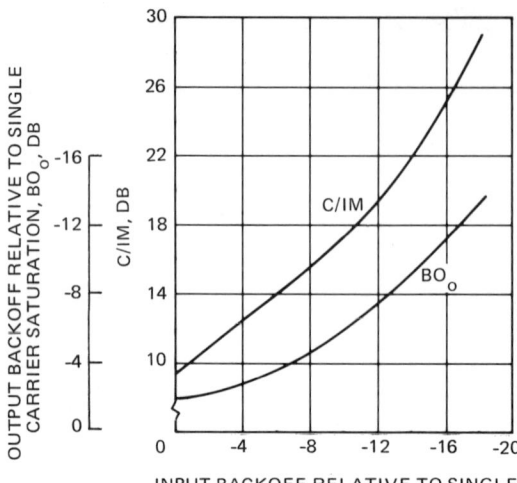

Fig. 4 Carrier-to-intermodulation ratio as function of TWT backoff.

For multicarrier transmission, the satellite repeater generates intermodulation products due to the TWT nonlinearities. The ratio of carrier power to intermodulation power, C/IM, is a function of the TWT operating point. As the total input power received at the satellite is reduced such that the TWT is operating in a more nearly linear fashion, the output power also is reduced, but the ratio of carrier to intermodulation noise power is increased. A typical TWT transfer characteristic for a large number of equally spaced carriers is shown in Fig. 4. The combination of carrier to thermal and intermodulation noise is then the link-carrier-to-noise ratio given by

$$\left(\frac{C}{N}\right)_L = \frac{(C/N)_{th} \times (C/IM)}{(C/N)_{th} + (C/IM)} \qquad (6)$$

This value of link C/N then must be above the threshold value determined by the particular FM demodulator employed. Typically, as can be seen from Fig. 2, the link C/N will not be above threshold if the full transponder bandwidth is occupied ($BW_t = BW_o$).

For the transponder bandwidth, BW_t, the carrier to thermal and intermodulation noise ratios are given by (in decibels)

$$\left(\frac{C}{N}\right)_{up} = \left(\frac{C}{N}\right)_{up-sat} + BO_i \qquad (7)$$

$$\left(\frac{C}{N}\right)_{dn} = \left(\frac{C}{N}\right)_{dn-sat} + BO_o \qquad (8)$$

$$\frac{C}{IM} = f(backoff) \qquad (9)$$

VOICE TRANSMISSION VIA COMMUNICATIONS SATELLITE 251

The link C/N then is determined from Eqs. (5) and (6). As can be seen from Fig. 2, the link C/N attains a maximum value at a specific value of backoff. Since the carrier-to-thermal-noise is a function of bandwidth (N = KTB), as is the carrier-to-intermodulation-noise,[3] with random carrier location, the occupiable bandwidth to achieve the desired link C/N may be found directly from

$$BWR = (C/N)_{req} - (C/N)_{BW_t} \qquad (10)$$

$$BW_o = BW_t/BWR \qquad (11)$$

$(C/N)_{req}$ is determined by the threshold requirement of the FM demodulator employed and the desired system margin. For "weak" (small values of satellite EIRP + ground station G/T) links, the attainable link C/N in the transponder bandwidth is small, so that the BWR required will be large, as illustrated by Fig. 3.

The preceding expressions for SCPC assume continuous transmission of the carrier. If, instead, pauses in the conversation are not transmitted, a significant improvement in link performance may be attained. With the transmit carrier switched on by a voice activator, the activity factor A_f is defined as the instantaneous number of active carriers to the total number of carriers in use. Typical talker statistics indicated that the average value of A_f is about 40%.

To maintain a desired input backoff of the TWT, the transmit power of each carrier must be increased by $1/A_f$, since only 40% of the carriers are instantaneously active. The uplink C/N for each carrier thus is increased by $1/A_f$ or 4 db when $A_f = 40\%$. Since only $A_f \times N = N_A$ of the carriers are instantaneously active, the satellite EIRP is shared by only N_A carriers, so that the downlink C/N is also 4 db higher. The C/IM at the desired operating point also is increased by 4 db; the ratio of the total occupied bandwidth to the instantaneously occupied bandwidth = $1/A_f$. Since all contributors to the link C/N are increased by 4 db, namely, $(C/N)_{up}$, $(C/N)_{dn}$, and C/IM, the link C/N is increased by 4.0 db. Thus, if voice activation is used, the link may be designed for a value 4.0 db less than that required for continuous transmission, while noting that the actual link C/N attained will be 4.0 db higher than the designed-for value due to use of voice activation. It should be noted that the improvement in C/IM attainable either by selectively spacing the carriers in the available bandwidth (improvement of $10 \log N_s/N$) randomly or by utilizing voice activation (improvement of 10 log $1/A_f$ or 10 log N/N_A), where the randomness inherent in the carrier locations is due to the talker activity, is achieved by the same process, namely, the random distribution of occupied channels throughout the transponder. The combination of both methods thus involves the selective random distribution of carriers and the instantaneous location of these carriers throughout the transponder so that the total improvement in C/IM relative to the equally spaced carrier values of Table 2 is

$$\left(\frac{C}{IM}\right)_{improvement} = 10 \log\left(\frac{N_s}{N} \times \frac{1}{A_f}\right) = 10 \log \frac{BW_t}{BW_o \times A_f} \quad (12)$$

VI. Determination of Channel Capacity and Quality

With a combined noise weighting and pre-emphasis advantage of 8.5 db for SCPC FM modulation and a desired TTNR, the remaining unknown in Eq. (1), MIF, is specified and is given by

$$MIF = 10 \log_{10}\left[\frac{3}{2} \frac{B_{vc}}{f_m} M^2\right] \quad (13)$$

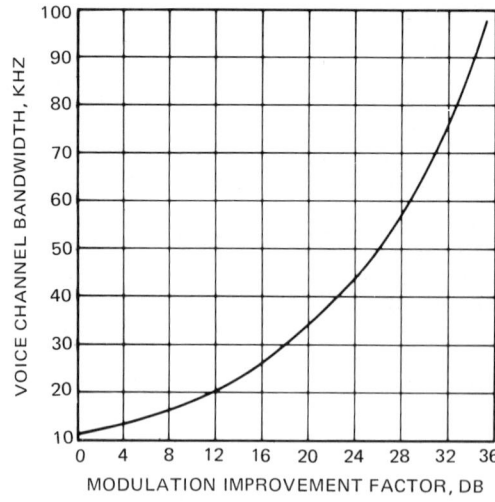

Fig. 5 Modulation improvement (MIF) vs voice channel bandwidth (B_{vc}).

$$M = \frac{\Delta f_{pk}}{f_m F_p} = \frac{(B_{vc}/2) - f_m}{f_m F_p} \quad (14)$$

Equations (13) and (14) then can be solved for B_{vc}. A plot of MIF vs B_{vc} is given in Fig. 5. The number of voice channels N attainable then is given by

$$N = BW_o/B_{vc}, BW_o \leq BW_t \quad (15)$$

If, for a given satellite EIRP, the ground station G/T is large enough, $(C/N)_L$ will be greater than the minimum acceptable value of

VOICE TRANSMISSION VIA COMMUNICATIONS SATELLITE 253

$(C/N)_L$ in the full 36-MHz transponder bandwidth. In this case, Eq. (1) still is valid except that the larger-than-required value of link C/N allows a decrease in B_{vc}, and the number of channels N is increased at a reduced MIF. Note that the link C/N for bandwidth BW_o or B_{vc} is identical, since the available carrier power is shared among the N carriers on an equal basis.

VII. Ground Transmitter Power Requirements

The ground transmitter power requirement is then given in decibels by

$$P_t = EIRP_{sc} + BO_i - 10 \log N - G + \text{margin} + [10 \log(1/A_f)] \quad (16)$$

bracket ($[\,]$) term applies only if voice activation is utilized, where margin is the required uplink margin.

VIII. Quality, Capacity, and Cost

Although it is not the intent of this paper to delve into the statistical variation of the effects of randomization or voice activation,[3] several factors are worth noting. From Fig. 2 and Eqs. (10, 11, and 15), the weaker links will have a low value of $BW_o = N \times B_{vc}$. If a high-quality circuit is required such that B_{vc} is large, the number of carriers N will be small. The statistical deviation of quality due to IM will tend to be relatively large as a result. As a practical matter, however, low values of EIRP + G/T are selected for reasons of economy for low-density, remote locations where lower channel quality may be acceptable. Also, from Eq. (16) it can be seen that larger N results in less transmit power and hence less cost. Conversely, higher-density circuits produce increased revenue, which may justify installation of a more expensive ground environment (higher G/T).

A possible compromise may be obtained by the use of voice compandors. Compandors compress the dynamic range of the input speech level before transmission (compressor) and restore the original level after reception (expandor). Low-level speech is transmitted at a higher level and the effect of channel noise is thus reduced, yielding a subjective improvement. Since the improvement is subjective, the actual companding improvement will vary with each listener. Denoting the subjective signal-to-noise (S/N) obtainable with a compandor by S/N and compandor improvement by C, Eq. (1) may be rewritten as

$$S/N = (C/N)_L + MIF + P + W + C \tag{17}$$

User evaluation tests conducted by several agencies reported companding improvements from 14 to 23 db. The required MIF and hence channel bandwidth B_{vc} are reduced, yielding a higher capacity N for a given occupiable bandwidth BW_o. The required transmitter power obtained from Eq. (16) would also be reduced since N has increased.

IX. Channel Capacity Results

The following sections and figures are provided to illustrate some typical results. The link parameters are as specified in Tables 1 and 2. Other assumptions are 1) threshold carrier-to-noise ratio, 10 db with a frequency discriminator, 7 db with threshold extension demodulator (TED); 2) system margin, 3 db (minimum) above threshold for all cases; 3) voice activity factor, 40%; 4) ground station antenna efficiency (transmit/receive), 55%; and 5) ground transmitter power margin, 3 db in all cases.

Shown in Figs. 6-8 are graphs of system capacity (measured in voice channels per transponder) for an Anik type of transponder as a function of ground station G/T and

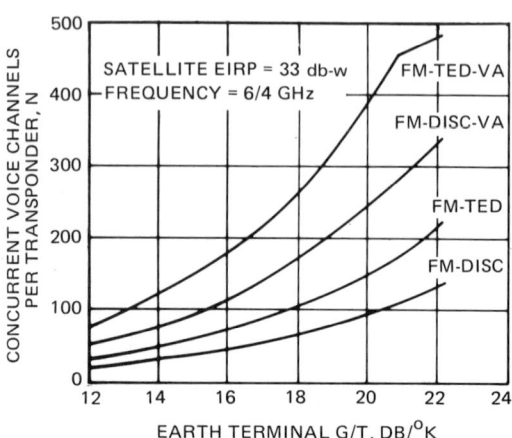

Fig. 6 Channel capacity vs G/T (7500 pwp, TTNR = 51.2 db).

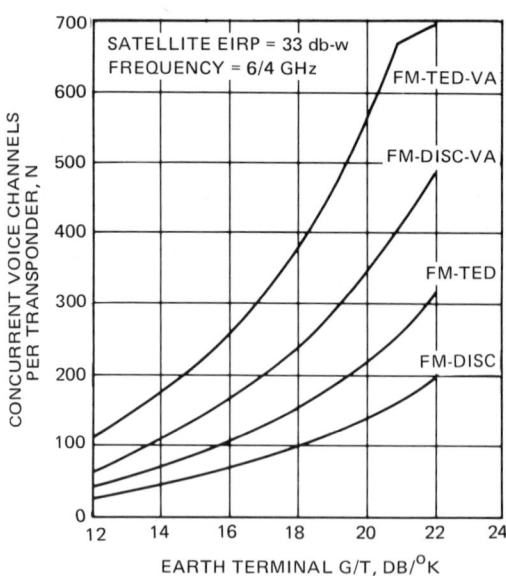

Fig. 7 Channel capacity vs G/T (25,000 pwp, TTNR = 46.0 db).

VOICE TRANSMISSION VIA COMMUNICATIONS SATELLITE

modulator/demodulator characteristics for several voice channel quality levels. The satellite EIRP is 33 db-w per transponder operating in the 6/4-GHz frequency bands.

With reference to the 7500-pwp noise allocation (Fig. 6), it can be seen that, for a specified G/T, the FM-TED-VA combination achieves the largest capacity (where VA is voice activity). Use of the threshold extension demodulator allows the link to operate at lower C/N and occupy more bandwidth BW_o. Thus, although the bandwidth required for each voice channel is larger with a TED than a frequency discriminator to achieve

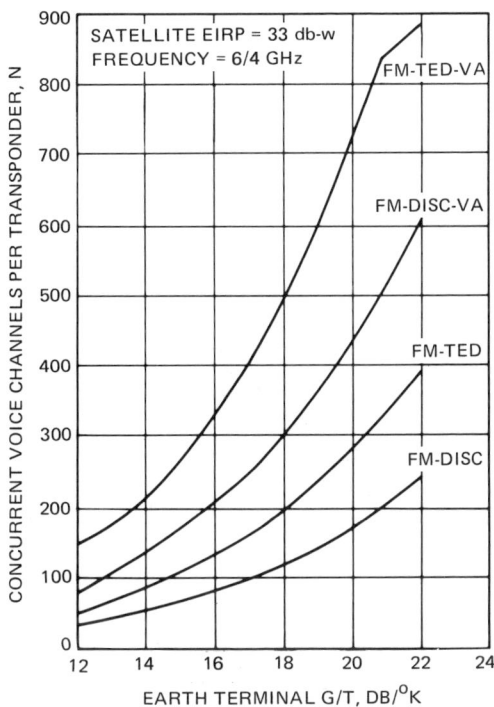

Fig. 8 Channel capacity vs G/T (57,500 pwp, TTNR = 42.4 db).

a desired TTNR [see Eq. (1)], BW_o is proportionately larger, so that N is larger. All of the links in Figs. 6-8 are power-limited for G/T 22, except for FM-TED-VA, where, at a G/T of slightly less than 21 db/°K, the occupiable bandwidth equals the transponder bandwidth. Thus, for larger values of G/T, the link is bandwidth-limited. Thus, when the FM-TED-VA is operating at values of G/T less than 21 db/°K, the capacity is $1/A_f$ times that attainable without voice activation and the same demodulator. The effects of varying the desired channel quality can be seen by comparing the capacity attainable at a specific G/T from the appropriate figures.

If compandors are utilized, the capacity attainable for any of the modulator/demodulator combinations of Figs. 6-8 is, of course, increased. A number of systems have recently been proposed based on companding, yielding a capacity of 1000 to 1500 voice channels with channel spacings of 20 to 30 kHz.

X. Transmitter Power Requirements

To illustrate a typical range of ground station transmit powers required, a transistor-preamplifier was assumed for the ground receiver, with an overall noise figure of 4.5 db at 4 GHz. Thus the receive antenna gain G is specified for a desired G/T from Figs. 6-8. The required gain then determines the required antenna diameter.

Figures 9-11 represent the transmitter power per carrier required as a function of ground antenna diameter for the corresponding values of N and quality from Figs. 6-8. The transmitter power requirement is the same for a particular demodulator regardless of whether voice activation is used, as long as the link is not bandwidth-limited, since, from Eq. (16), the increase in transmit power required with voice activation is offset by the increase in N. The choice of demodulator, however, does impact the transmit power requirement, as seen from the figures.

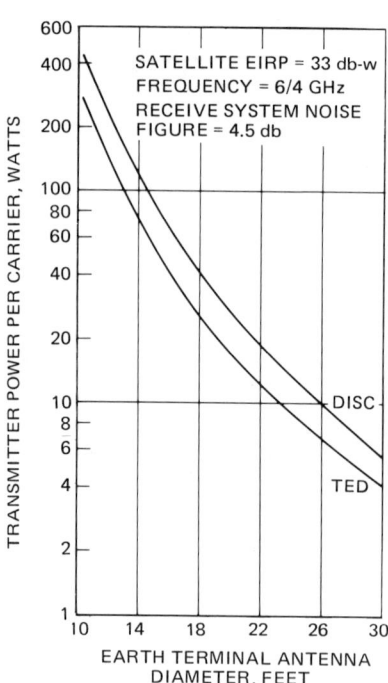

Fig. 9 Transmitter power vs antenna diameter (7500 pwp, TTNR = 51.2 db).

The preceding figures illustrate the capacities and transmit power requirements for optimal N as defined in the previous sections. If lower transmitter powers are desired to reduce the cost of the Earth terminal, reduced capacity can be traded off in two ways. The satellite repeater gain may be increased such that the required transmitter power is lessened. The uplink C/N thus is reduced accordingly. The allowable increase in gain (reduction in C/N_{up}) is determined by the amount of uplink degradation acceptable on the overall link C/N requirement, which is, in turn, a function of the type of demodulator and system margin required. If the satellite repeater gain is increased 5 db, the ground transmit power required also is reduced by 5 db or a factor of

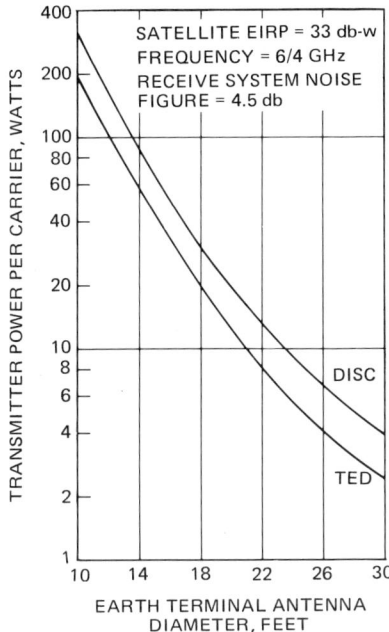

Fig. 10 Transmitter power vs antenna diameter (25,000 pwp, TTNR = 46.0 db).

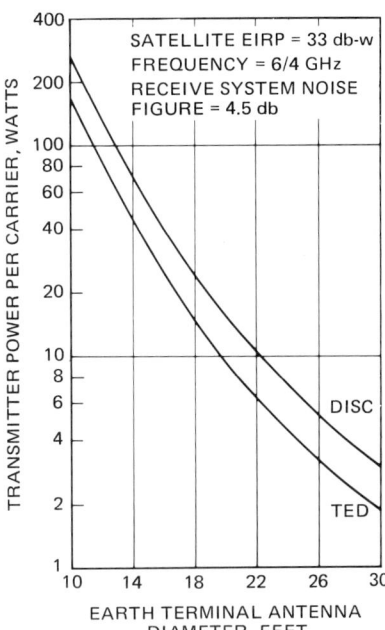

Fig. 11 Transmitter power vs antenna diameter (57,500 pwp, TTNR = 42.4 db).

3. This change results in a relatively small (less than 10%) reduction in channel capacity.

Alternatively, the transmitter power may be decreased, thereby establishing a nonoptimum backoff operating point (see Fig. 2) and reducing the channel capacity. The number of channels attainable then can be found at larger input backoff values and the resulting transmit power requirement recomputed from Eq. (16). As an example, with a 16-ft antenna, G/T = 16.4 db/°K, and a TTNR of 46.0 db, the optimal input backoff yielding the desired link C/N is about 2 db. Increasing the backoff and reducing the number of channels so that the link C/N desired remains constant, the curve of Fig. 12 may be obtained, where a large reduction in transmit power is attainable for a small decrease in N. The addition of compandors would also decrease the transmit power requirements since, as discussed in Sec. VIII, the number of voice channels N is larger. Thus, depending on the specific application, any of the above techniques add further flexibility to the overall system design.

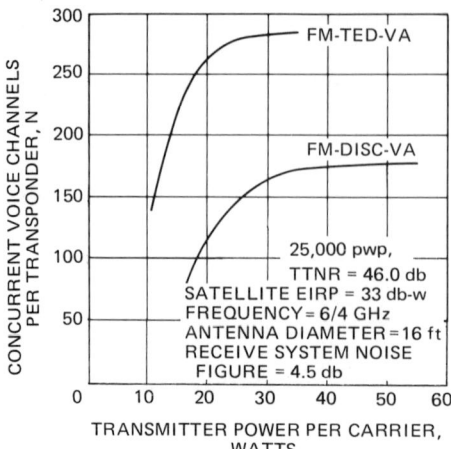

Fig. 12 Variation of capacity with nonoptimum transmitter power.

XI. High-Powered S-Band Satellite Transponder Capacity

It has been determined that a high-powered S-band-link can be added to a satellite of the Anik class while retaining the desirable features of this satellite. This additional link is achieved using a high-efficiency TWT with 50-rf-w output at saturation. Uplink frequency is 2.6 GHz, and the downlink frequency is 2.5 GHz. For applications such as Australia, Canada, and the U. S., the saturated EIRP would be about 43 db-w. As one might expect, the voice channel capacity of this link is quite large, even when very small Earth terminals are used. Figures 13 and 14 illustrate the capacities achievable with Earth terminal G/T values up to 20 db/$^{\circ}$K. For example, referring to Fig. 13, using a 10-ft-diam antenna with a transistorized preamplifier (such as that shown in Fig. 1) can produce a G/T of 9.5 db/$^{\circ}$K which has a capacity of 615 channels using an FM discriminator and voice activation. When the Earth terminal G/T is increased to 11.5 db/$^{\circ}$K (12.5-ft-diam antenna), the link is simultaneously power-limited and bandwidth-limited, and the capacity has increased to 780 channels.

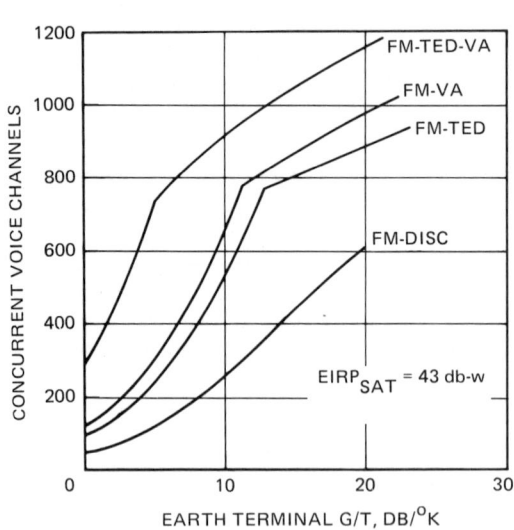

Fig. 13 Channel capacity vs G/T for 2.6/2.5-GHz link (25,000 pwp).

XII. Transmitter Power Required

The uplink transmitter power required

for each voice channel is shown in Fig. 15 as a function of Earth terminal receiving G/T. For a G/T of 9.5, the transmitter power required is 7 w for a channel noise of 25,000 pw and 3 w for a channel noise of 57,500 pw. These power levels are achievable with all solid-state transmitters, as is desirable for the proposed unattended operation.

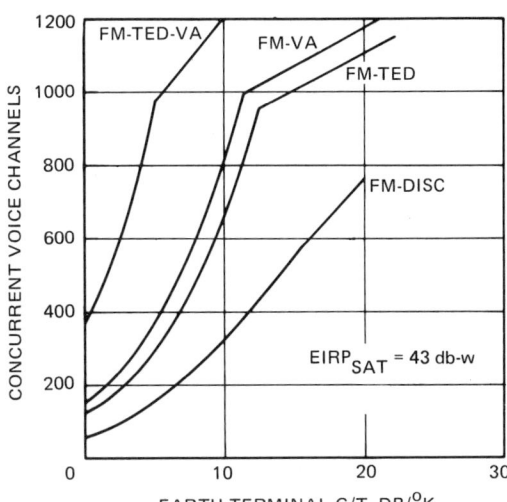

Fig. 14 Channel capacity vs G/T for 2.6/2.5-GHz link (57,500 pwp).

XIII. Summary of Results

The results of these studies show that for the S-band link the class of simple, low-cost Earth terminals having a figure of merit in the range G/T = 9 to 20 db/°K can produce practically useful channel capacity. The knee of the curves at which the link is concurrently power-limited and bandwidth-limited

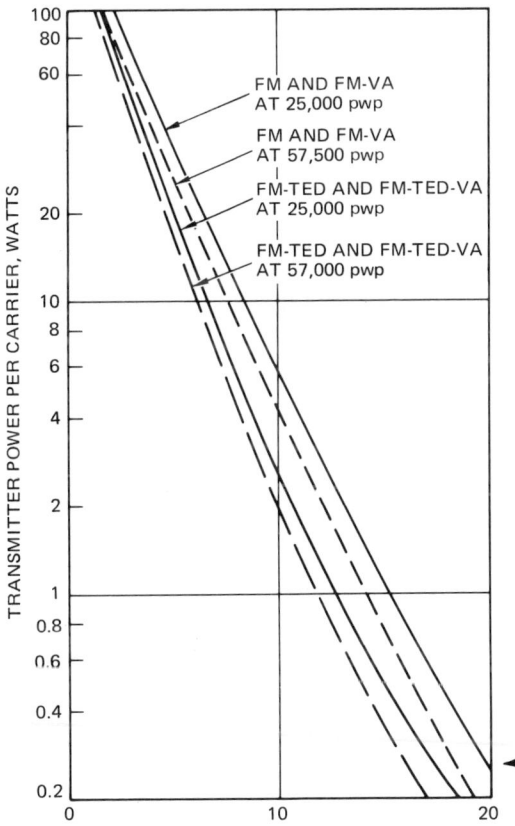

Fig. 15 Uplink transmitter power required for 2.6 link.

(35-MHz bandwidth for transponder) is in the G/T range of these small stations. For values of G/T higher than are required at this point, the increased capacity is achieved more slowly as G/T increases because of bandwidth limitations. Although capacity could be increased further by the addition of more sophisticated baseband equipment to the SCPC FM carriers, the capacity increase would not be commensurate with the added cost and maintenace requirements in the initial implementation of such a system. The initial system implementation could be augmented by the addition of such devices as delta modulators, pulse-code modulation (PCM) codecs, compandors, and/or threshold extension demodulators. These devices would provide some increased capacity but would have the main effect of improving link quality and operating margins.

The use of SCPC FM with voice activation results in a capacity that is attractive for the proposed applications and compares well with capabilities of FM/FDMA links. Uplink transmitter power for these small terminals is limited to the range of 1 to 10 w/channel with 10- to 16-ft-diam terminals. Further link optimization indicates that designs using all solid-state transmitters can be achieved.

XIV. Conclusions

This study and other investigations lead us to conclude that the current generation of communication satellites, launchable by the Thor Delta booster rocket, can provide economically justifiable multipurpose national satellite systems combining trunk telephone, TV distribution, and remote area telephone services. Telesat of Canada has demonstrated these capabilities.

The remote area telephone capability can be provided readily on an economic basis by the use of a national satellite system. The necessary small, low-cost, unattended operation Earth terminal can be achieved most readily in the initial phases of such systems through the use of modulation by SCPC FM with voice activation. The capacity achievable by this method is sufficient for the suggested applications and is competitive with modulation systems of greater equipment complexity. Installation of a high-powered 2.5-GHz transponder in a current satellite configuration would provide a service capability for several thousand service locations.

References

[1] Norman, P. M. M. and Weese, D. E., "Thin Route Satellite Communications for Northern Canada," *International Conference on Communications*, June 1971, Institute of Electrical and Electronics Engineers, pp. 11-14.

[2] Dunn, L. G., "Telephony to Remote Communities in Canada via Satellite," *International Conference on Communications*, June 1971, Institute of Electrical and Electronics Engineers, pp. 14-20.

[3] McClure, R. B., "Analysis of Intermodulation Distortion in an FDMA Satellite Communication System with a Bandwidth Constraint," *International Conference on Communications*, June 1970, Institute of Electrical and Electronics Engineers.

COMPARISON OF TWO BASIC DOPPLER COMPENSATION METHODS

P. A. Kullstam[*]

Computer Sciences Corporation, Falls Church, Va.

Abstract

High carrier frequencies being used for satellite communications in conjunction with rapidly varying range rate experienced by mobile terminals will cause a significant doppler stress on receiver functions. In particular, digital communication links employing coherent demodulation schemes often require doppler compensation to both the carrier phase and bit timing recovery estimators to secure satisfactory demodulation performance. The fundamental doppler problem is considered, and two basic compensation methods are discussed: 1) received communication or auxiliary signal tracking in the presence of doppler; and 2) separation motion information for doppler aiding. Carrier phase tracking performance using first-, second-, and third-order Butterworth-type loops is compared when subjected to a linear (step response) as well as a vibration (sinusoidal) doppler model. The results show that a higher-order loop generally will provide better doppler tracking performance. The auxiliary signal doppler compensation method has been investigated to arrive at a terminal design providing error-free doppler compensation. Finally, the motion information accuracy and delay problem associated with external source doppler compensation has been investigated, showing that accuracy of within a few percent and delay of a few milliseconds easily may be required.

Presented as Paper 74-438 at the AIAA 5th Communications Satellite Systems Conference, Los Angeles, Calif., April 22-24, 1974. This investigation was supported, in part, by the Defense Communications Agency. The author wishes to express his appreciation to D. Gan for his help in preparing the figures of this paper.

[*]Manager, Systems Analysis Section.

I. Introduction

As increasingly higher radio frequencies are being used for communication with airborne terminals, the terminal motion effects are becoming a more serious problem. This paper addresses various techniques of motion or doppler compensation for high-performance digital communication systems using (partially) coherent demodulation of phase-shift-keyed (PSK) signals, such as spread spectrum modulation, which combine low digital modulation rate with the requirement of accurate symbol (code) timing. The purpose of this paper is to provide an overall view of various doppler compensation techniques and compare their relative merits as they are applicable for systems employing digital modulation.

II. Basic Design Parameter: Range Accuracy

Relative motion between a transmitting terminal and a receiving terminal causes a shift in the received carrier frequency, rf doppler. In particular, air- and shipborne terminals will experience significant doppler effects on high-frequency signals. The relative motion generally results in a time-varying change in the length of the propagation path, the range. Basically, a shortening of the signal path will result in increased (positive doppler) received rf frequency change as well as increased digital symbol rate, whereas an increase of the path results in a decreased (negative doppler) frequency and symbol rate. Now, the instantaneous doppler shift proportional to the range rate is not what matters for coherent PSK symbol demodulation; what does matter is its effect over a given time period related to the time constants of the carrier phase and symbol time tracking loops (estimators). In other words, the requirements can be related to the range error defined by

$$r(t) = \int_{t_o}^{t} \left[v(t) - \hat{v}(t) \right] dt \qquad (1)$$

where $v(t)$ refers to the relative receive terminal velocity (range rate), and $\hat{v}(t)$ denotes (at least implicitly) the range rate estimate at the receiving terminal. Generally, the absolute range is not required, but the change in range rate caused by the relative motion is required. The benefit in viewing the doppler-effect problem as a range-accuracy (error) problem is that it focuses on the fundamental parameters that allow comparison of various doppler tracking or compensation techniques.

It is easy to translate the effect of a certain (instantaneous) range error r on the carrier phase tracking error φ and the absolute symbol timing error. Denoting the propagation velocity by c, we have the relations $r = c\tau$ and $\varphi = \omega_{rf}\tau$, which may be written as

$$\tau/T_s = r/\lambda_{symbol} \tag{2}$$

$$\varphi = 2\pi \cdot r/\lambda_{carrier} \tag{3}$$

where $\lambda_{carrier} = c2\pi/\omega_{rf}$, and $\lambda_{symbol} = cT_s$ refer to the wavelengths of the carrier and the symbol, respectively. For any given coherent PSK symbol demodulation scheme, the relative symbol timing error τ/T_s and carrier phase tracking error φ can be related to a demodulation performance loss. For example, biphase PSK modulation will experience a power performance loss because of a symbol timing error of $(1 - \tau/T_s)^2$ and a carrier phase error of $\cos^2\varphi$. Allowing, for example, a 0.5-db power loss in each case, we require

$$r \leq 0.055 \, \lambda_{symbol} \tag{4}$$

$$r \leq 0.053 \, \lambda_{carrier} \tag{5}$$

Since usually $\lambda_{carrier} \ll \lambda_{symbol}$, we can infer that the required range accuracy associated with carrier phase tracking is much greater than for the symbol timing. This does not imply, however, that it is necessarily easier to obtain adequate symbol tracking than carrier phase tracking performance, but it does imply that a carrier phase tracking loop can provide excellent doppler compensation information for doppler compensation of the symbol tracking loop.

As range and carrier phase tracking performance are related directly to each other, there are basically two approaches to the doppler problem: 1) carrier phase estimation using a feedback loop that locks on the received signal or on an auxiliary received signal, and 2) range estimation provided by an external source to the communications system such as inertial platform. First, one should note that a received signal may be either the communications signal itself or an auxiliary received signal as a satellite beacon signal; second, the doppler information obtained from a beacon signal or an inertial platform can be used only to compensate for the doppler stress on the communications system caused by the receiver terminal motion; and third, the beacon and inertial platform approach will result in open-loop compensation techniques. Therefore, it generally is advantageous to combine closed-loop carrier phase tracking on the received communications signal with either of the two compensation methods, so that only a fraction of the doppler stress needs to be accommodated by the carrier phase tracking loop.

III. Carrier Phase Tracking on Received Communications Signal

There are many carrier phase tracking loops (estimators) that may be considered. First, there are different loops dependent upon the type of loop filter being used. A first-order loop has one degree of freedom in its design, a second-order two degrees of freedom, etc. Although one may consider various kinds of higher-order loops, the most generally used may be referred to as maximum flat or Butterworth loops, specified by

$$|1 - H(j\omega)|^2 = 1/[1 + (\omega_n/\omega)^{2k}] \qquad (6)$$

where $H(s)$ is the closed-loop transfer function. $[1 - H(s)]$ then represents the input phase error transfer function. These loops all are defined by one parameter $f_n = \omega_n/2\pi$, the loop corner frequency, with k being the order of the loop. Secondly, when the received signal is PSK modulated, a modulation-free error signal is required as loop input. Either one uses a decision-feedback approach that amounts to removing the PSK modulation by "unmodulation" of the received signal using the demodulated received symbol (bit) sequence, or one uses the power loop approach, where a received PSK signal is multiplied with itself to generate a sinusoidal signal that is unaffected by the PSK modulation. Using a Costas loop approach is equivalent to the power loop approach. These methods are applicable for M-ary PSK modulation in general.

The design of a carrier phase tracking loop experiencing doppler stress represents a tradeoff between minimizing the phase error tracking variance

$$\sigma^2 = (N_o B/C)\, \eta \qquad (7)$$

in the presence of noise and the transient steady-state phase error

$$\varphi_s = 2\pi \omega_n^{-k}/\lambda \qquad (8)$$

of a kth-order Butterworth-type loop experiencing doppler caused by a range variation $r(t) = t^k/k!$. Here C/N_o is the received signal (unmodulated carrier)-to-noise density ratio, B is the loop noise bandwidth, η is a degradation factor taking into account the modulation removal (generally, $1 \leq \eta < 1.5$),

$f_n = \omega_n/2\pi$ is the loop corner frequency, and λ is the carrier wavelength. [For M-ary PSK modulation and power loop implementation,

$$\eta = M^{-2} \sum_{k=1}^{M} k! \binom{M}{k}^2 \left(\frac{N_o}{E_s}\right)^{k-1}$$

whereas for a decision feedback implementation, a first-term approximation yields

$$\eta = \left[1 + P_s(E_s/N_o) \sin^2 (2\pi/M)\right] / \left[1 + 2P_s \sin^2 (\pi/M)\right]^2$$

where E_s/N_o (= CT_s/N_o) is the M-ary PSK symbol energy-to-noise density (T_s the symbol period), and P_s is the probability of a symbol error.[2]]

The loop noise bandwidth B and the loop corner frequency $f_n = \omega_n/2\pi$ are directly proportional to each other, and so a decrease of the phase error variance σ^2 by reducing the bandwidth will increase the steady-state phase error φ_s. The proportionality factors for Butterworth-type loops are given in Table 1, obtained from Viterbi.[1] In Figs. 1-4, the normalized phase tracking error $\varphi B^m/(2\pi/\lambda)$ has been plotted for a unit step response in range (m = 0), velocity (m = 1), acceleration (m = 2), and jerk (m = 3). The following observations can be made. Generally, the higher order the loop, the more sluggish it becomes, which suggests that a loop of minimum order should give best performance. However, it is most important to realize that a first-order loop leaves a steady-state error for a step in velocity and will lose lock for a step in acceleration or jerk. (That is why first-order loop responses are omitted in Figs. 3 and 4.) Likewise, a second-order loop will completely track out a step in acceleration while leaving a residual error for a jerk step. This property of a third-order loop makes it attractive for terminals experiencing typical accelerations

Table 1 Noise bandwidth for Butterworth expressed in terms of loop corner frequency $f_n = \omega_n/2\pi$

Order of phase lock loops k	Loop noise bandwidth B	
	Exact	Approximate
1st	$(1/4)\omega_n$	$0.25\,\omega_n$
2nd	$(3/4\sqrt{2})\omega_n$	$0.53\,\omega_n$
3rd	$(5/6)/\omega_n$	$0.833\,\omega_n$

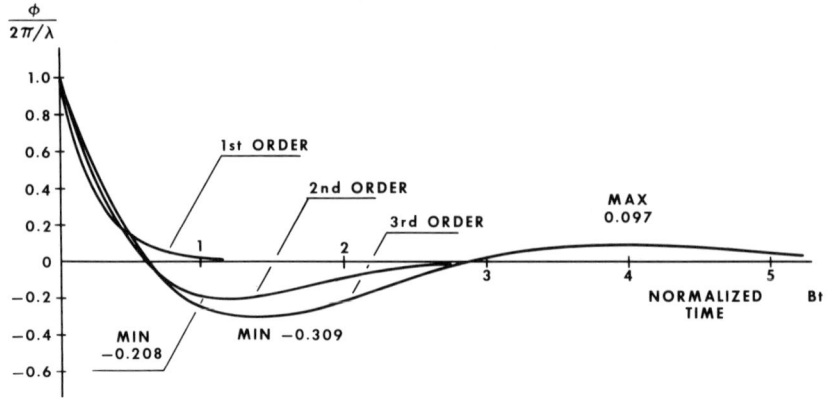

Fig. 1 Normalized phase tracking error response to a unit range step for first-, second-, and third-order Butterworth-type phase lock loops of equal noise bandwidth B.

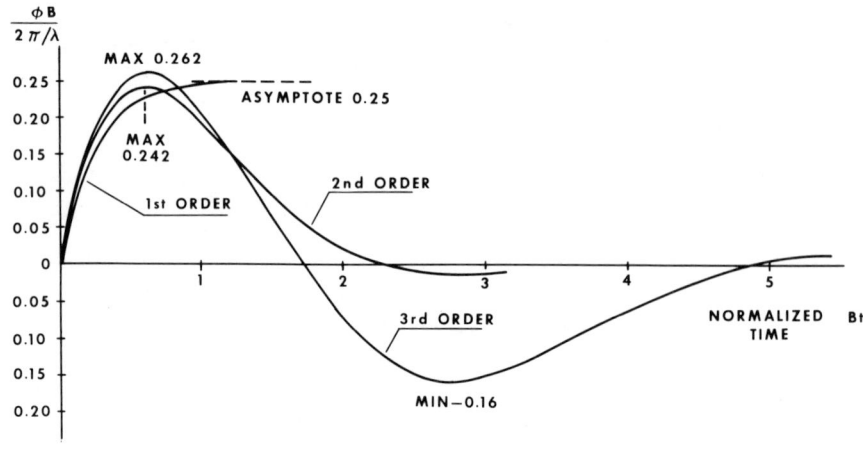

Fig. 2 Normalized phase tracking error response to a unit velocity step for first-, second-, and third-order Butterworth-type phase lock loops of equal noise bandwidth B.

such as in air- and shipborne applications. Table 2 summarizes the tracking performance of the Butterworth-type phase lock loops. Except for cases where the loop will lose lock and yield an unbounded phase error, the maximum transient phase errors are approximately the same.

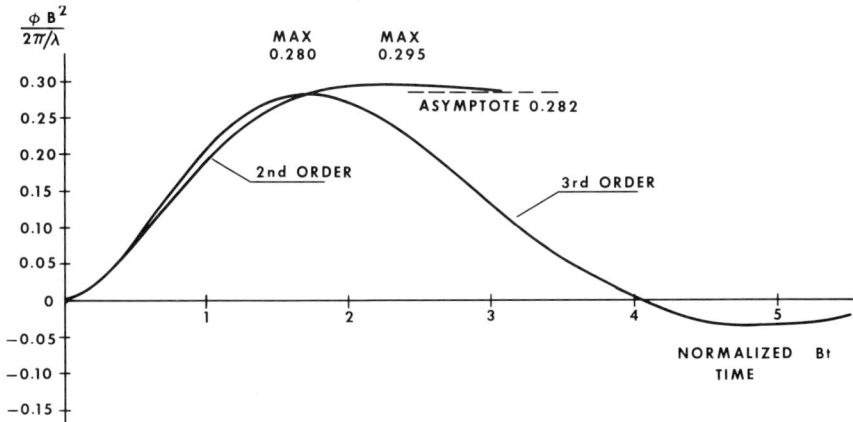

Fig. 3 Normalized phase tracking error response to a unit acceleration step for second- and third-order Butterworth-type phase lock loops of equal noise bandwidth B.

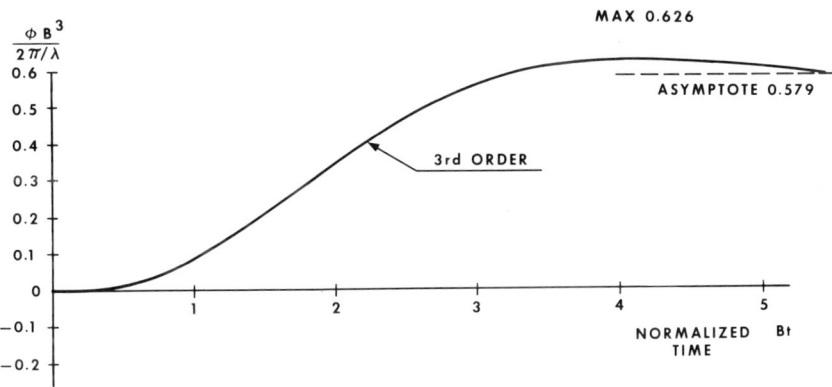

Fig. 4 Normalized phase tracking error response to a unit relative jerk step for a third-order Butterworth-type phase lock loop of noise bandwidth B.

IV. Auxiliary Signal Doppler Compensation

Tracking an auxiliary or beacon signal instead of a communications signal to obtain doppler information related to the receiver terminal can be used to provide doppler compensation of the communications signals. This amounts to the same basic

Table 2 Normalized carrier phase tracking error
$\varphi B^m/(2\pi/\lambda)$ for Butterworth-type loops

Order of loop k	Range step m = 0			Velocity step m = 1		Acceleration step m = 2		Jerk step m = 3	
	Max. value	Max. overshoot value	Steady-state value	Max. value	Steady-state value	Max. value	Steady-state value	Max. value	Steady value
1st	1	0	0	0	0.250	Unbounded	Unbounded	Unbounded	Unbounded
2nd	1	-0.208	0	0.242	0	0.293	0.282	Unbounded	Unbounded
3rd	1	-0.309	0	0.262	0	0.280	0	0.626	0.579

problem as in the previous section from a tracking point of view. Generally, however, since a beacon signal is received at a high signal-to-noise ratio, the loop noise bandwidth can be made large, and thus excellent doppler tracking capability is obtained more easily. Therefore, this section will focus on how to apply accurate doppler compensation of the communication signal using a beacon signal.

In Fig. 5, a basic terminal implementation is shown. The beacon receiver contains a carrier phase tracking loop that effectively doppler-corrects its intermediate frequency (IF) signal, commonly 70 or 700 MHz. Assuming that the received beacon signal of frequency f_b experiences a doppler shift ϵ_b, the phase-locked loop will provide a local oscillator (LO) signal of frequency $f_b - f_{IF} + \epsilon_b$ (nominally $f_b - f_{IF}$) which maintains a constant IF signal frequency $f_{IF} = f_b + \epsilon_b - (f_b - f_{IF} + \epsilon_b)$. That is, the IF is doppler-corrected. The closed phase-locked loop insures this operation, as the voltage-controlled oscillator (VCO) operating at the nominal frequency f_o (commonly 1 or 5 MHz) will show a doppler shift ϵ_o that is related to the beacon doppler:

$$\epsilon_b = M_1 M_b \epsilon_o \qquad (9)$$

where $M_1 M_b$ is the total multiplication of the reference signal up to the LO frequency of the beacon receiver. Here M_1 represents a fixed multiplication, and M_b is the multiplication provided by a programable frequency synthesizer. Clearly, the multiplication factor

$$M_1 M_b = (f_b - f_{IF})/f_o \qquad (10)$$

will provide the LO frequency of the beacon signal receiver. Now, if the beacon shows a doppler shift ϵ_b, the information-carrying received signal will experience the doppler

COMPARISON OF TWO DOPPLER COMPENSATION METHODS

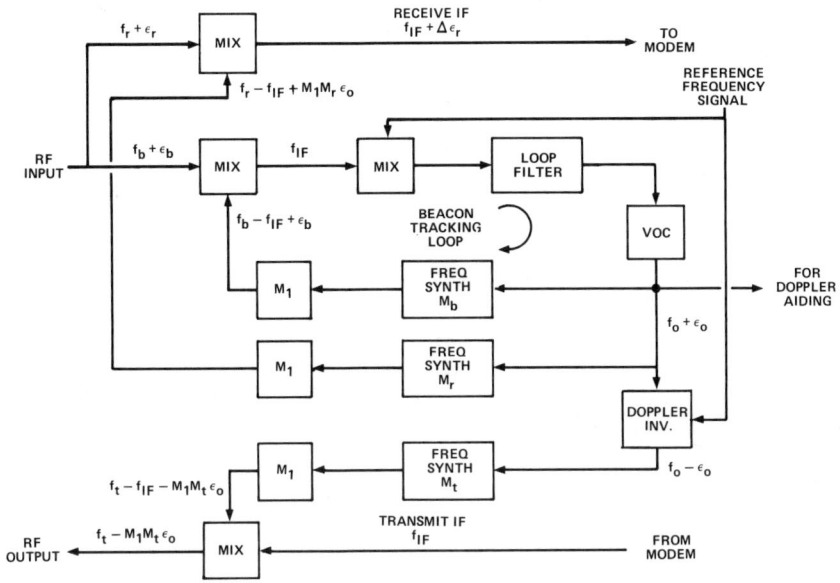

Fig. 5 Basic terminal block diagram using beacon signal doppler correction.

$$\epsilon_r = \epsilon_b f_r / f_b \tag{11}$$

Using the up-multiplied doppler-corrected reference signal produced by the beacon receiver loop to generate the LO signal for the communication signal, the resulting IF signal will show an uncompensated doppler shift

$$\Delta \epsilon_r = \epsilon_r - M_1 M_r \epsilon_o \tag{12}$$

where M_r is the appropriate multiplication factor of the frequency synthesizer to obtain the nominal IF signal frequency, i.e.,

$$M_1 M_r = (f_r - f_{IF})/f_o \tag{13}$$

From Eqs. (9-11), we then obtain

$$\Delta \epsilon_r = \epsilon_b \frac{f_{IF}(f_b - f_r)}{f_b(f_b - f_{IF})} \tag{14}$$

or

$$\Delta \epsilon_r = \epsilon_r \frac{f_{IF}(f_b - f_r)}{f_r(f_b - f_{IF})} \tag{15}$$

Similarly, the transmitter frequency also may be corrected with an inverted doppler reference signal, and the same reasoning applies again, with the result that the uncompensated doppler shift

$$\Delta\epsilon_t = \epsilon_t \frac{f_{IF}(f_b - f_t)}{f_t(f_b - f_{IF})} \tag{16}$$

where ϵ_t represents the frequency shift that the satellite-to-aircraft motion imposes on the transmitted signal. Therefore, the doppler compensation obtained by the configuration of Fig. 5 will result in compensation errors expressed by Eqs. (14) and (15). From these equations, it follows that if $f_{IF} = 0$ there would be no doppler compensation error. However, the problem is not that the IF signal frequency must vanish but that of f_{IF}, which appears in Eqs. (10) and (11). This influence can be rectified by introducing a frequency shift of $-f_{IF}/M_1$ between the frequency synthesizer output and the fixed multiplier M_1. Then $f_b - f_{IF} = M_1(M_b' f_o - f_{IF}/M_1)$, so that

$$M_1 M_b' = f_b/f_o \tag{17}$$

and likewise $M_1 M_r' = f_r/f_o$ and $M_1 M_t' = f_t/f_o$. This additional fixed frequency shift will make $\Delta\epsilon_r = \Delta\epsilon_t = 0$ as if $f_{IF} = 0$,

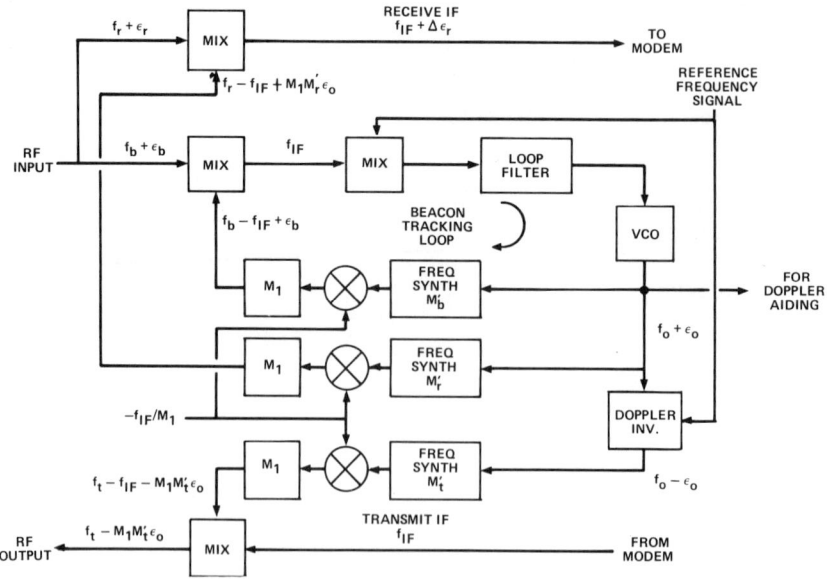

Fig. 6 Terminal block diagram with improved, "perfect" beacon doppler correction.

Table 3 Required motion compensation accuracy for
φ_{max} = 0.33 rad, λ = 3 cm (linear model)

Order of loop k	Loop noise bandwidth, Hz	Loop corner frequency, Hz	Velocity compensation		Acceleration compensation		Jerk compensation	
			δv, m/sec	Accuracy at 10 m/sec, %	δa, m/sec^2	Accuracy at 1g, %	δj, m/sec^3	Accuracy at 1g/ sec, %
1st	4	2.550	0.026	0.26	...	...	...	...
	16	10.200	0.102	1.02	...	...	...	...
2nd	4	1.200	0.026	0.26	0.087	0.89	...	...
	16	9.800	0.105	1.05	1.390	14.10	...	...
3rd	4	0.765	0.024	0.24	0.910	0.93	0.16	1.66
	16	3.06	0.098	0.98	1.460	14.80	9.35	95.40

whereas in fact f_{IF} = 70 or 700 MHz, say. Figure 6 shows the modification to the basic terminal block diagram.

The VCO output signal at nominally f_o frequency can be used for doppler aiding other tracking circuits of a receiver modem such as the symbol tracking loop. The doppler shift ϵ_o in f_o is related directly to the beacon doppler shift by Eq. (9), and, applying Eq. (10), we may write

$$\epsilon_o = \epsilon_b f_o/(f_b - f_{IF})$$
$$\epsilon_o = \epsilon_b f_o'/f_b$$
(18)

where

$$f_o' = f_o \cdot f_b/(f_b - f_{IF}) \qquad (19)$$

Equation (18) shows that the doppler on the nominal frequency f_o signal equals that of a received signal at the frequency f_o', with the difference between f_o and f_o' caused by the IF frequency shift. Again, using the compensation method of Fig. 6, this difference vanishes. This feature allows for direct conversion of the VCO signal to any frequency signal, for example, the symbol rate, providing "perfect" doppler correction.

V. External Doppler Compensation

Since doppler induced by the receiver terminal is related directly to the motion of the terminal, it is clear that motion information obtained from an accelerometer or inertial platform can be converted into range variations and be used for doppler compensation of the received communications signal. The practical feasibility of such an approach depends on the accuracy as well as the "promptness" with which one can derive the range rate. The information accuracy and delay problem includes such aspects as coordinate transformation of the inertial data into line-of-sight range rate.

In this section, we shall extend our linear motion analysis to include a sinusoidal motion model, germane for vibrational effects. The two motion models, the linear and the sinusoidal, are

$$r(t) = p\, t^m/m! \qquad (t \geq 0) \tag{20}$$

$$r(t) = p\omega^{-m} \sin \omega t \tag{21}$$

where p represents the motion parameter. For m = 0, p corresponds to a range step (linear model) or maximum range variation (sinusoidal model); for m = 1, p stands for a velocity step or maximum velocity variation, etc.

With the terminal experiencing a motion described by the parameters p and m and the external motion information source providing an estimate $\hat{p}$ of p, the resulting range error becomes

$$\delta r = \delta p\, t^m/m! \tag{22}$$

$$\delta r = \delta p\, \omega^{-m} \sin \omega t \tag{23}$$

where $\delta p = p - \hat{p}$ equals the parameter estimation error. The linear model shows that for $m \geq 1$ the range error will be unbounded with time if δp does not vanish. The sinusoidal model yields a bounded range error $|\delta r| \leq |\delta p|\omega^{-m}$, which, however, will be excessive, for low frequencies and $m \geq 1$. Therefore, if $|\delta p|$ cannot be made to vanish, one must require the terminal to have some doppler tracking capability. It is important to realize, however, that external motion compensation generally will reduce the necessary doppler tracking capability of the terminal.

In Sec. III, the phase tracking capability of Butterworth-type loops was investigated employing the linear model, and basically the nonzero steady-state, asymptotic, phase tracking error corresponds to the maximum transient phase error of higher-order loops as well. However, taking into account the maximum phase error tracking error in each case, expressed by the normalized loop tracking error $\Phi_{mk} = \varphi_{max} B^m/(2\pi/\lambda)$, the maximum residual motion parameter

$$|\delta p/\lambda| \leq \varphi_{max} B^m/(2\pi\Phi_{mk}) \tag{24}$$

Table 4 Maximum residual motion compensation error at
0.1-, 1-, and 10-Hz sinusoidal vibrations for
$\varphi_{max} = 0.33$ rad, $\lambda = 3$ cm

Order of loop k	Loop noise bandwidth, Hz	Loop corner frequency, Hz	Maximum residual velocity, m/sec			Maximum residual acceleration, m/sec^2			Maximum residual jerk, m/sec^3		
			0.1 Hz	1 Hz	10 Hz	0.1 Hz	1 Hz	10 Hz	0.1 Hz	1 Hz	10 Hz
1st	4	2.550	0.025	0.027	0.103	0.016	0.172	6.480	0.010	1.080	407.0
	16	10.200	0.102	0.102	0.143	0.064	0.643	8.970	0.080	8.050	896.0
2nd	4	1.200	0.144	0.018	0.100	0.091	0.110	6.280	0.057	0.692	395.0
	16	4.800	2.310	0.231	0.103	1.450	1.450	6.450	0.910	0.911	405.0
3rd	4	0.765	0.446	0.011	0.100	0.280	0.069	6.280	0.176	0.432	395.0
	16	3.060	28.500	0.285	0.100	17.900	1.790	6.280	11.300	11.300	395.0

can be determined for which the maximum phase error φ_{max} is limited. Being more specific, consider a 10-GHz X-band carrier ($\lambda = 3$ cm), and limit φ_{max} to 0.33 rad (corresponding to 0.5-db demodulation loss for biphase PSK). Table 3 shows the maximum residual motion parameter error that can be accommodated, as well as the compensation accuracy ($\delta p/p$) x 100 in percent which is required to accommodate 10-m/sec, 1-g (=9.81 m/sec^2), and 1-g/sec input steps, respectively. It is interesting to note that, provided the loop does not lose lock, the required compensation accuracy is essentially the same for the various loops.

Employing the sinusoidal model, and with the phase error related to the range error by the formula $\varphi = 2\pi\delta r/\lambda$, one easily can determine the maximum phase error

$$\varphi_{max} = 2\pi |1 - H(j\omega)| (|\delta r|_{max})/\lambda \qquad (25)$$

Then, for a given maximum phase error limit, the bound is

$$|\delta p/\lambda| \quad (\varphi_{max}\omega_n^m/2\pi)(\omega/\omega_n)^m \cdot \sqrt{1 + (\omega_n/\omega)^{2k}} \qquad (26)$$

for the residual motion parameter error δp, from Eqs. (6) and (23). For $m \geq k$, the order of the loop, the factor $(\omega/\omega_n)^m \sqrt{1 + (\omega_n/\omega)^{2k}}$ is steadily increasing, whereas for $m < k$ a minimum between 1 and $\sqrt{2}$ exists at $\omega/\omega_n = \sqrt[2k]{(k-m)/m}$ (see Fig. 7). Therefore, a higher-order loop (k > m) will require less compensation than an mth-order loop, assuming that the loop corner frequencies are the same. In Table 4, the maximal residual motion parameter error δp has been determined for 0.1-, 1-, and 10-Hz vibrations while limiting the maximum phase error to 0.33 rad for $\lambda = 3$ cm and taking into account that, for a given loop noise bandwidth, the loop corner frequencies are different.

Finally, we shall consider the sensitivity to a latency in the motion compensation of the terminal tracking loop. Assuming a time delay δt, the range error becomes

$$\delta r = p\, t^m/m! - p\,(t - \delta t)^m/m! \\ \simeq (p\delta t)\, t^{m-1}/(m-1)! \tag{27}$$

using the linear model and

$$\delta r = p\omega^{-m}\, \sin \omega t - \sin \omega (t - \delta t) \\ \simeq (p\delta t)\, \omega^{-(m-1)}\, \cos \omega (t - \delta t/2) \tag{28}$$

for the sinusoidal model. In both cases, $p\delta t$ and $(m-1)$ act as the new motion parameters. Therefore, we can directly apply the results obtained for required motion compensation accuracy. For $\varphi_{max} = 0.33$ rad and $\lambda = 3$ cm, Table 5 shows the result expressed in milliseconds for the maximum allowable time delay when the maximum sinusoidal acceleration and jerk are 1 g and 1 g/sec, respectively.

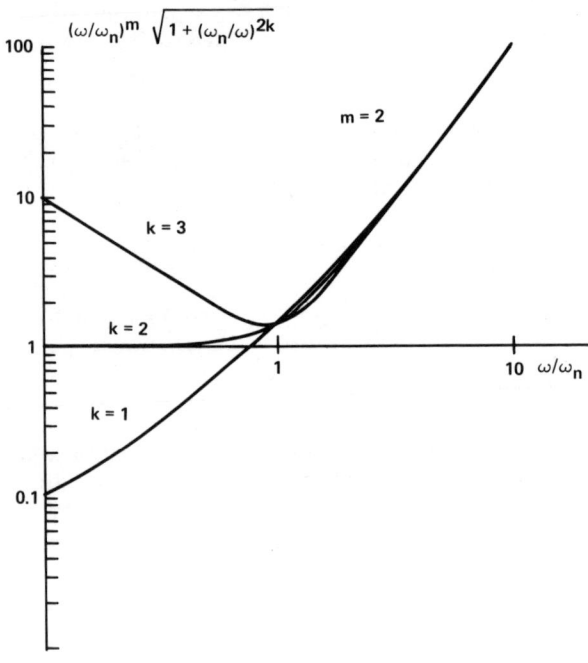

Fig. 7 Tracking factor for a second-order Butterworth-type phase lock loop.

Table 5 Maximum latency for motion compensation at 0.1-, 1-, and 10-Hz sinusoidal vibration for $\varphi_{max} = 0.33$ rad and $\lambda = 3$ cm

Order of loop k	Loop noise bandwidth, Hz	Loop corner frequency, Hz	Max. time delay at 1g acceleration, m/sec			Max. time delay at 1 g/sec jerk, m/sec		
			0.1 Hz	1 Hz	10 Hz	0.1 Hz	1 Hz	10 Hz
1st	4	2.550	2.55	2.75	10.50	1.63	1.75	660.0
	16	10.200	10.50	10.50	14.60	6.50	65.40	915.0
2nd	4	1.200	14.7	1.84	10.2	9.30	11.2	640.0
	16	4.800	235	23.50	10.5	148	657	•••
3rd	4	0.765	45.5	1.10	10.2	28.50	0.70	640.0
	16	3.060	2900	29.00	10.2	1830	183	640.0

Latency results using linear models are very close to the ones obtained for 0.1-Hz vibration of the first-order loop and 1-g acceleration and of the second-order loop and 1-g/sec jerk. Again, a higher-order loop is about equal or less sensitive to delay. The table values indicate that the maximum time delay which can be tolerated can be as low as a few milliseconds.

VI. Conclusion

This paper has addressed the doppler compensation problem of mobile terminals, which is of particular concern for coherent demodulation of PSK signals at high rf frequencies (X band), with noise bandwidth of the carrier phase tracking loop being determined by the phase variance due to receiver noise; the loop tracking capability for terminals in motion will be limited. Therefore, two other doppler compensation techniques also are investigated for their performance and use of implementation. In comparing the step and sinusoidal phase tracking responses of first-, second-, and third-order Butterworth-type loops, when normalized to equal noise loop bandwidth, we can conclude that the tracking performance improves with the order of the loop. However, if a loop of a given order can track the input variations, there is no benefit in using a higher-order loop, as the transient response of a lower-order loop will be a little bit better.

Tracking an auxiliary signal to provide doppler correction of the received as well as the transmitted communications signal was shown to yield a doppler correction error when using the conventional approach. However, a quite simple modification of the terminal design will provide accurate multiband doppler correction. It was demonstrated that very accurate motion information with little delay may be encountered easily when using an inertial platform for doppler correction. Therefore, even if such an approach is possible, great care has to be exercised in its implementation.

References

[1] Viterbi, A. J., *Principles of Coherent Communication*, McGraw-Hill, New York, 1966.

[2] Stiffler, J. J., *Theory of Synchronous Communications*, Prentice-Hall, Englewood Cliffs, N.J., 1971.

METHODS OF ALLEVIATION OF IONOSPHERIC SCINTILLATION EFFECTS ON DIGITAL COMMUNICATIONS

James L. Massey*

University of Notre Dame, Notre Dame, Ind.

Abstract

The degradation of the performance of digital communication systems because of ionospheric scintillation effects can be reduced either by diversity techniques or by coding. The effectiveness of traditional space-diversity, frequency-diversity, and time-diversity techniques is reviewed and design considerations isolated. Time-diversity signaling then is treated as an extremely simple form of coding. More advanced coding methods, such as diffuse threshold decoding and burst-trapping decoding, which appear attractive in combating scintillation effects, are discussed and design considerations noted. Finally, adaptive coding techniques, appropriate when the general state of the channel is known, are discussed.

I. Introduction

By a "scintillation channel," we mean a radio data link between an Earth station and a space satellite where the carrier frequency and Earth latitude are such that periods of significant signal variability due to ionospheric irregularities in the propagation path are experienced. The precise conditions for such "scintillations" are not yet known fully, but it has been observed in existing systems that the variability can be severe. When scintillation effects are severe,

---

Presented as Paper 74-55 at the AIAA 12th Aerospace Sciences Meeting, Jan. 30 - Feb. 1, 1974, Washington, D.C. This work was supported by NASA under Grant NGL 15-004-026 at the University of Notre Dame in liaison with the Communications and Navigation Division of the Goddard Space Flight Center.

*Freimann Professor, Department of Electrical Engineering.

one might experience peak variations of ± 10 db in signal strength relative to mean signal level. The effect is localized in stretches of perhaps several minutes interspersed between long periods of negligible variability. Within each stretch, there would be alternating intervals, roughly equal and on the order of several seconds in duration, of signal strength increase and decrease relative to the mean.

To the communications engineer, a scintillation channel is one beast in a familiar species, viz. the "fading channel." A fading channel is any channel where the path loss must be treated as a random quantity. There are three basic approaches to communications over such a channel, namely, 1) adding sufficient "margin" to a system designed for the average path loss to enable the system to ride through the periods of deepest fading; 2) using "diversity" techniques so that several separated (in time, frequency, or space) variable channels are, in effect, combined to yield the "average channel," which is nonvariable; and 3) employing "coding" techniques in which each bit of data is spread over many bauds so that again a type of "averaging" is obtained.

The "margin" approach to fading channel communications, despite its lack of elegance, is not without merit. At first blush, one might expect to pay about a 10-db margin penalty on a severe scintillation channel. This would indeed be the case if the transmitter were peak-power-limited. But if the transmitter were average-power-limited (as one might expect for the communications satellite, if not for the ground station) and if one used feedback from the other station to <u>adapt</u> the transmitter power so that the 10-db margin is applied only during the short stretches of severe scintillation, then almost no average-power margin penalty is paid. In this case, the margin approach is equally as good as, and probably simpler than, the diversity approach, which we discuss next.

The principles of time, frequency, and space diversity are alike, but, since only time-diversity appears well matched to the characteristics of a scintillation channel, we shall limit our attention to this form of diversity. In L-fold time-diversity binary data communications, each data bit controls the transmitted signal in L different bauds, these bauds being well separated in time so that the channel variability independently affects each baud. When L is sufficiently large, the "law of large numbers" insures that properly combining the received signals in these L bauds will give the same result as if the entire signal were at the average level with no variability. This is the theoretical limit to the improvement in

communications obtainable by the use of time-diversity.
(Since this theoretical limit removes all degradation due to
"fading," i.e., to signal-strength variability, one might be
tempted to look no further for approaches to the fading
channel.) Paradoxically, as we shall attempt to show in our
discussion of coding, this temptation is that of the serpent.
The key to the paradox is that there are ways to gain other
than by eliminating variability of the signal strength. For a
scintillation channel, our intuition suggests that quadruple
time-diversity (L=4) with nonuniform separations on the order
of several seconds between the L bauds controlled by each bit
probably would suffice to extract most of the advantage
possible with time-diversity, but to our knowledge no experimental studies of time-diversity on scintillation channels
have been made yet.

In the "coding" approach to communications over a fading
channel, each data bit again affects the transmitted signal in
many bauds, but, in contrast to time-diversity, more than one
bit may be controlling a given baud. Seen this way, time-
diversity is the most primitive form of coding, namely, the
so-called "repeat code" in which each data bit merely is
transmitted L times (albeit with a time lapse between each
transmission). It is well known that, for the "deep-space"
channel (the additive, white gaussian noise channel in
communication theory terminology), which is a nonfading
channel, this repeat coding gains nothing, whereas more sophisticated coding techniques presently in use are gaining about
6 db over noncoded systems at the currently acceptable error
rates of about 10^{-5} in the overall system. This suggests that
one should, in general, be able to design a coded system for a
fading channel so that it will reap the advantage of time-
diversity, i.e., remove the loss due to signal variability,
and at the same time reap the gain that coding techniques can
provide on the resultant nonfading channel. The remainder of
this paper is devoted to the description of some known coding
techniques that seem particularly suited for use with a
scintillation channel.

II. Coding for the Scintillation Channel

A. <u>Purposes of Coding</u>

It is not generally appreciated, even among communications engineers, that coding can be used for two rather distinct purposes in data communications. We shall call these
purposes "system enhancement" and "system repair." The dis-
tinction between these two purposes lies in whether coding is

an integral part of the original system design or whether
coding is considered as a design afterthought to rescue a
communications system whose error rate is intolerably high.
The term "error correcting codes" describes coding for system
repair. The term "probabilistic coding" is sometimes used to
describe coding techniques for system enhancement. From a
practical viewpoint, these two forms of coding are distinguished
by whether the demodulator makes "hard decisions" (unquali-
fied binary decisions) or "soft decisions" (decisions that include an indication of their confidence level) on the values
of the binary digits controlling the signals in the bauds
being processed. Only when hard decisions are made by the
demodulator can one speak unequivocally of "errors" made by
the demodulator. Soft decisions imply that the assigned bit
values are only tentative and anticipate that other considerations may cause some decisions to be reversed.

It is an unfortunate fact that most communications engineers at present connect coding only to its usage in system
repair, and hence they design their demodulators to produce
hard decisions. For a fading channel, this is a particularly
disastrous design choice, since the variability of the signal
level is reflected most prominently in the varying confidence
levels of the individual baud decisions. The choice of hard
decisions then effectively destroys much of the capability of
coding to remove degradation due to variable signal level.
That this choice is often made is somewhat paradoxical, since
communications engineers never use demodulator hard decisions
in a time-diversity system but always choose a "diversity
combiner," which operates on the raw demodulator output, i.e.,
on "very soft" decisions. The very vastness of the literature
on "error-correcting" codes may have biased communications
engineers to forget the value of soft decisions when the
simple repeat coding of time-diversity is replaced by a more
sophisticated and powerful code.

There are cases, of course, where "system repair" is a
perfectly sensible coding goal, viz. where the communications
system already exists but has an unacceptably high error rate.
Indeed, error correction is then the only option. In this
case, one should seek to design a coding system that gives
the desirable lower error rate without decreasing unacceptably
the data rate of the system because of the added redundancy
used for error correction. Some recent papers [1,2] provide
good illustrations of the design considerations for this type
of coding on fading channels. At heart, the "system repair"

approach does not seek to make gains from the use of coding but seeks to ameliorate losses that have resulted from earlier choices in the system design.

The considerations when coding is used for system enhancement are entirely different. One presumes at the outset that the demodulator will make soft decisions and then asks how few confidence levels can be used without significant degradation. [For instance, four different confidence levels give nearly all of the advantage possible on the (nonfading) deep-space channel, whereas a single confidence level (i.e., hard decisions) is about 2 db inferior.[3] For a fading channel, the price paid for using a single confidence level is much greater for the reasons mentioned previously.] One then seeks a coding system that can effectively use the confidence level information, is simple to implement, and provides most of the gain possible from the use of coding. This gain is at least that of manyfold time-diversity and usually will be substantially more. Moreover, the digital techniques of the coding art may well yield a coding system that is simpler to implement than the inferior diversity system that is predicated on analog techniques. In the following sections, we describe some coding techniques that appear promising for use in system enhancement on a scintillation channel.

B. Threshold Decoding of Diffuse Codes

When a demodulator decision is made, one can speak of the resultant "error digit," which is 0 if the decision is correct and 1 if incorrect. When the decision is "soft," one can specify the confidence level by the probability that the error digit is a 1. When parity-check coding is used in communications, one can always form at the receiver certain "syndrome digits," each of which is a modulo-two sum of specified error digits. The idea of "threshold decoding" is to decide whether to reverse a given demodulator decision by examining a certain set of syndrome digits, each of which includes the error digit for the decision in question as a term in its sum. The set must be "orthogonal" in the sense that no other error digit is a term in the summation for more than one of the syndrome digits in the set. These orthogonal syndrome digits then become statistically independent estimators of the error digit in question, and a threshold logic rule becomes the optimum way to estimate this error digit from this set of syndrome digits.[4] Although this coding technique is not very powerful, it has the virtue of simplicity and is, as a result, often considered in practical applications of coding.

For the use of threshold decoding on fading channels, a special type of parity-check code called a "diffuse code" has been proposed. (The reader is referred to Kohlenberg and Forney[5] for an excellent discussion of this technique in the system repair approach to coding.) In a diffuse code, the orthogonal set of syndrome digits has the further property that the error digits in any span of a given prescribed length on the channel will affect, at most, half of the syndrome digits in the set. Thus, if the code parameters are chosen correctly, one will, with high probability, obtain sufficient estimators that include only error digits from periods of high signal strength. This type of coding appears well suited to a scintillation channel but only in connection with the system enhancement approach to coding, as the use of hard decisions on a severe scintillation channel appears to be a very damaging choice, regardless of the coding technique ultimately employed.

C. Burst-Trapping Coding

The coding technique called "burst-trapping," originally suggested by Tong[6] for use on the telephone channel, also appears to have some attractive features for use on a scintillation channel. In burst-trapping, a short block code (of the type ordinarily employed for correction of randomly located errors in the system repair approach to coding) is modified for use in correcting occasional "bursts" on the channel, i.e., short intervals when a large number of errors are made by the demodulator. The basic idea is quite simple. Instead of using all of the redundancy of the code for error correction, only part of the redundancy is used for this purpose, with the remainder used to detect when uncorrectable error patterns are present. With proper design, one can be virtually certain to detect such error patterns, even when they contain a very large number of errors, as would be expected if the block in question falls within a burst. The basic block code is modified slightly to the extent that the block codeword also is added modulo-two to a subsequent block at such a distance from the first that it is unlikely that both blocks will lie simultaneously within a burst. When the first block is decoded in the error correction mode, this decoded block is stored to be subtracted modulo-two from the latter block, so that the latter block then can be decoded in the same fashion as the former. But when an uncorrectable error is sensed in the former block, one postpones decoding of this block until the latter block arrives. It then is supposed that the latter block lies in a "good" stretch on the channel, i.e., the demodulator made no errors in estimating the bits in this block, and the redundancy of the code then is used to solve from the codeword in the latter block for the codeword from the first block, which has been added modulo-two to the second, as well as the codeword in the second block. This requires that the redundancy be at least 50% in the original block code. Less redundancy can be used, but this entails adding each codeword to several later codewords; e.g., 33 1/3% redundancy suffices if each block is added to two later blocks.

Burst-trapping, although a form of system repair coding since the demodulator is allowed to make hard decisions, is a reasonable approach to a fading channel such as a scintillation channel in which the channel makes, at definite intervals, sharp transitions from weak signal-level conditions to strong signal-level conditions. The time delay between the two (or more in the case of less than 50% redundancy) blocks used together in decoding must be chosen carefully to match the nominal transition time between weak and strong signal-level conditions. For a scintillation channel, this appears quite feasible.

The strongest objection to the use of burst-trapping on a scintillation channel is that the redundancy of the code is wasted during the long periods between scintillation events, so that the channel repair coding actually degrades the system in these "quiet" times on the channel. For this reason, only low-redundancy burst-trapping appears attractive. The special, almost-periodic oscillation of a scintillation channel during scintillation events suggests further that the several blocks to which a given block is added modulo-two be closely spaced mutually, unlike the scheme proposed by Tong for the telephone channel, in which the spacing between these blocks is the same as that between the first of these blocks and the block that is being added to each.

The burst-trapping scheme can be modified so as to become a system enhancement method to some degree. One could use apparent signal strength of the digits in a received block as the adaptation signal to change decoding modes rather than use the detected error capability of the code. The only use of the code redundancy in this modification would be for the purpose of solving for the codeword in a block where weak signal levels were detected from the latter blocks to which this codeword had been added modulo-two. One definitely would wish to use a low-redundancy code in this modified form of burst-trapping.

It should be mentioned that burst-trapping is a block coding version of Gallager's "time-diversity coding." Gallager's scheme, a good description of which, in the channel repair mode, is given by Kohlenberg and Forney,[5] uses a convolutional code rather than a block code. However, the main features of our discussion of burst-trapping apply to Gallager's scheme as well, and the convolutional codes used in this scheme may, in fact, offer some implementation advantages over the block codes used in burst-trapping.

D. Other Coding Techniques

In the two previous subsections, we have discussed how two well-known coding techniques can be modified for system enhancement on a scintillation channel. Conceptually, the modifications that permit the use of soft decisions can be thought of as ways to adapt the coding system to the actual state of the channel rather than to treat impartially all of the decisions made by the demodulator. It seems to us that this will be the essential feature of an effective coding system on a scintillation channel.

One also could consider system enhancement coding techniques more powerful than threshold decoding for use on a scintillation channel. The two most obvious candidates are sequential decoding[3] and Viterbi decoding.[7] Both of these schemes utilize convolutional codes rather than block codes. The codes to be used with either technique should be "interleaved," which is a coding technique for spreading out the encoded digits so that only a small number of the digits affecting the decoding at any given time can come from a "burst" on the channel.

Finally, we should point out that we have considered only "forward-acting" coding schemes to this point, but consideration also should be given to the use of "automatic repeat request" (ARQ) systems on scintillation channels. ARQ is an attractive possibility if one can tolerate complete loss of communications during the several minutes of a scintillation event, provided that the data intended originally for this interval are postponed rather than "erased." ARQ coding systems are of the channel repair variety, but they use little redundancy and hence do not degrade an uncoded system substantially between scintillation events. On the other hand, ARQ systems provide none of the gain possible to attain with coding of the system enhancement type.

III. Remarks

We have attempted in the foregoing to place the considerations for choosing a coding system for a scintillation channel into the proper perspective. We have shied away from specifying a particular coding system or even the parameters of a possibly attractive coding system, since the scintillation channel is not yet well-enough understood to make this possible and sensible. Rather than this lack of present knowledge being a disadvantage for designing future communications systems for scintillation channels, we see it as a positive advantage, since as a consequence the demodulator design for scintillation channels has not yet been "frozen" into the hard decision type. The scintillation channel thus offers a unique opportunity to consider coding of the system enhancement type, which should provide substantial gains over the system repair type of coding which too often is the only form of coding that is, or can be, considered in data communications.

References

[1] Goldberg, B., Moyes, E. D., and Quigley, J. E., "Interleaved Viterbi Decoding Applied to Troposcatter Channels," Conference Record, IEEE National Telecommunications Conference, Nov. 26-28, 1973, Atlanta, Ga., Vol. II, pp. 21C-1 to 21C-6.

[2] Tsai, S., "Evaluation of Burst Error Correcting Codes Using a Simple Partitioned Markov Chain Model," Conference Record, IEEE National Telecommunications Conference, Nov. 26-28, 1973, Atlanta, Ga., Vol. II, pp. 21E-1 to 21E-4.

[3] Jacobs, I., "Sequential Decoding for Efficient Communication from Deep Space," IEEE Transactions on Communications Technology, Vol. COM-15, No. 4, Aug. 1967, pp. 492-501.

[4] Massey, J. L., Threshold Decoding, M.I.T. Press, Cambridge, Mass., 1963.

[5] Kohlenberg, A. and Forney, G. D., Jr., "Convolutional Coding for Channels with Memory," IEEE Transactions on Information Theory, Vol. IT-16, No. 5, Sept. 1968, pp. 618-626.

[6] Tong, S. Y., "Burst-Trapping Techniques for a Compound Channel," IEEE Transactions on Information Theory, Vol. IT-15, No. 6, Nov. 1969, pp. 710-715.

[7] Viterbi, A. J., "Convolutional Codes and Their Performance in Communication Systems," IEEE Transactions on Communications Technology, Vol. COM-19, No. 5, Oct. 1971, pp. 751-772.

CHAPTER V—TRANSMISSION PATH EFFECTS

One of the trends of satellite communications systems has been the greater use of portions of the frequency spectrum other than the SHF region that spawned most of the technical and operational growth of the satcom technology. The UHF and EHF segments are of particular interest, the former for its promise of low-cost mobile terminals and the latter for very wide spectral allocations that minimize interference with terrestrial communications. Each of these regions also has unique transmission path characteristics. The EHF band and the upper portion of the SHF band are impacted to a significant degree, proportional to the frequency of interest, by rain attenuation. The UHF band and the lower part of the SHF band are influenced by ionospheric scintillations. Because of the importance of information on these phenomena to satellite communications systems engineers, five papers on these phenomena have been selected for inclusion in this Chapter.

The first paper focuses on the rain attenuation measurement program of Bell Telephone Laboratories as directed to the possible use of the 18/30-GHz allocation for future generations of AT&T satellites. The trade-offs of system margin, Earth station power, path availability, etc., will provide interesting reading, and will be noted by those concerned with the planning and design of systems in the 11/14-GHz allocation, which may precede use of the 18/30-GHz band.

The next four papers discuss various aspects of scintillation as observed in the VHF, UHF, and SHF bands. Crane's paper reviews the current information on measurements and shows that ionospheric scintillations will influence communications systems operating in the auroral and polar regions, as well as nighttime reception near the geomagnetic equation. The implications for polar-orbit, highly elliptical orbit, and equatorial-orbit satellite systems is obvious. The paper by Crane serves well as a tutorial introduction to scintillation phenomena. The subsequent paper by McClure concentrates on the nocturnal equatorial irregularities, while Fremouw concentrates on the modeling and prediction aspects and proposes a scintillation model.

The final paper by Blank and Golden is directed at defining error models for data transmission via transionospheric communications. It is concluded that the classic burst-error channel behavior is a characteristic of the VHF equatorial ionospheric channel.

IMPACT OF RAIN ATTENUATION
ON 18/30-GHZ SATELLITE SYSTEMS

D. Jarett[*] and L. D. Spilman[+]

Bell Telephone Laboratories, Inc., Holmdel, N. J.

Abstract

Communication satellite systems at 18/30 GHz must have a far larger margin against rain attenuation than systems using the 4/6-GHz bands. This has such a serious effect on system design and viability that considerable study and exercise of ingenuity are warranted. This paper reviews propagation data now available, discusses the areas in which information is deficient, justifies the importance of measurement programs now being undertaken, and demonstrates how this information will be used in system design.

1. Introduction

Commercial communication satellite service has begun in the 4- and 6-GHz frequency bands, where rain attenuation is relatively small. There are already indications (for instance, in the difficulty in finding orbital positions for all of the initial applicants for U.S. domestic service) that it will be necessary to look to higher frequency bands to provide the

---

Presented as Paper 74-496 at the AIAA 5th Communications Satellite Systems Conference, Los Angeles, Calif., April 22-24, 1974. The 18/30-GHz system study was done by the Satellite Systems Engineering Department, and the rain attenuation measurements for New Jersey were supplied by D. C. Hogg of the Antenna and Propagation Research Department. The writers acknowledge fruitful discussions with members of Hughes Aircraft Company on the interaction of new satellite subsystems with satellite design.

[*]Supervisor, Future Satellite Systems Group; presently Manager, Advanced Communications Satellite Systems, TRW Systems Group, Redondo Beach, Calif.

[+]Member of Technical Staff, Satellite Systems Engineering Department.

amount of service that may be required. The new and presently unused bands at 18 and 30 GHz[1] are particularly inviting, because their large 2500-MHz bandwidth accommodates digital transmission more comfortably. This, and the greater antenna directivity possible within dimensional constraints on satellites, permits consideration of reuse of the spectrum several times by a single satellite. All of this does not come free, however. Rain attenuation and the thermal noise it causes at the Earth station receiver are significant at 18 and 30 GHz, and this makes for quite a different system design than at 4 and 6 GHz.

Early work at Bell Laboratories demonstrated the potential of the 18/30-GHz frequencies for high-capacity satellite communications,[2] and recent work has shown that significant benefits also can be derived for smaller, more limited-capacity satellites operating at these frequencies.[3] The key to realizing the potential of these frequencies lies in the validity of the assumptions underlying the system design. One of those assumptions, a major one, is that excessive rain attenuation which exceeds the system margin can be limited to a relatively small percentage of the time (just a few minutes per year) at each Earth station location.

Recognizing the importance of rain attenuation on system design, a substantial effort to understand the rain environment was begun some years ago at Bell Laboratories in New Jersey. This effort, on terrestrial paths as well as Earth-space paths, has been very fruitful, and confidence in Earth-space path rain margins necessary for the New Jersey area is increasing steadily. Although attenuation is only one of the properties of rain being investigated, it is the controlling one for the system designs presently envisioned. Other considerations, such as cross-polarization coupling due to nonspherical raindrops, become very important for systems using orthogonal polarizations to double the capacity. This is to be done on the American Telephone and Telegraph (AT&T) 4/6-GHz satellite system,[4] and possibly the larger 18/30-GHz satellites. Coherence across 2500 MHz is also of interest for systems using broadband signals to fill the entire bandwidth. However, the model system considered here divides the 2500-MHz bandwidth into from 5 to 7 channels/beam, each accommodating a 274-Mbit/sec PSK (phase shift keyed) digital signal.

This work on Earth-space path rain attenuation in New Jersey has advanced to the point of knowing what questions to ask about the rain environment in other parts of the country, and how and with what equipment to get the answers. The two most important questions relate to the expected attenuation distribution, with and without diversity Earth stations, and the dependence of diversity improvement on separation. These questions and more are being addressed at two sites in different climatic regions: Denver, Colo. and Atlanta, Ga. A separate paper[5] describes those experiments, and another paper[6] reports some preliminary results, compares them with extrapolations from New Jersey measurements, and points out how expected fade margins impact on transmission reliability.

This paper relates the measurement program discussed in Refs. 5 and 6 to 18/30-GHz communication satellite system design, first by reviewing briefly the presently available New Jersey data, and second by describing the effects of rain on the design of a particular 18/30-GHz system.

A satellite sized for the Atlas-Centaur launch vehicle is assumed. This might be called a "medium"-capacity 18/30-GHz satellite ("small"-capacity satellite could be launched by a Delta, and "large"-capacity by a Titan or larger vehicle). This size satellite would be aimed at an application requiring high-capacity point-to-point communication links between only a few large Earth stations (5 to 10). This application derives from a previous 18/30-GHz satellite system study.[7,8]

Some satellite parameters are determined by the constraints of the launch vehicle, and the system configuration usually is determined by the assumed operational time frame and the application. The system configuration depends on the operational time frame because of the long lead time for development of satellite hardware, and because the terrestrial network that a satellite must complement is changing with time.

For the model 18/30-GHz system considered, an Earth-station-to-Earth-station reliability objective of 99.98% of the time has been assumed for consistency with long-haul terrestrial radio relay systems. Thus, propagation considerations should meet at least a reliability of 99.99%. This 99.99% must be divided between the up and down paths of a given link between two Earth stations.

Section 2 reviews rain attenuation measurements in New Jersey which are presently used in system design, and discusses briefly the future measurement program. Section 3 then considers the impact of an assumed rain fade margin on the design of a specific 18/30-GHz system.

2. Rain Attenuation

Attenuation by rain at 18 and 30 GHz and its implications for radio systems have been recognized and dealt with for some time.[9] In 1965,[10] the theoretical treatment of rain attenuation was used to correlate point rainfall rates with measured path attenuation with limited success. Calculations similar to those in 1965 have extended the predictions to higher frequencies.[11] These calculations of attenuation are summarized in Fig. 1, and the relative positions of the 18- and 30-GHz frequency bands are indicated. This figure shows that, even for moderate path lengths (e.g., 10 km) and rain rates (e.g., 25 mm/hr), the attenuation on the down path would exceed 20 db, and on the up path it would exceed 50 db.

Fortunately, the Earth-space path through rain is limited in extent, first by the high elevation angle of the path required for synchronous satellites, and second by the gross structure of rainfall.[12] But even these limitations in path length are not sufficient at all locations to limit the expected attenuation to values that can be accommodated by wideband (274 Mbit/sec) satellite communication systems.

The first extensive measurements of rain attenuation at these frequencies for an Earth-space path in New Jersey[13] showed that an excessive fade margin would be required to meet the path availability objective of 99.99%. To provide these large down-path margins would be very costly in satellite prime power and/or satellite capacity.

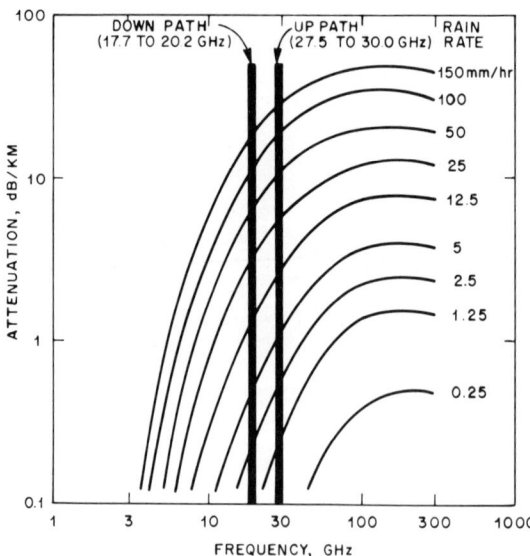

Fig. 1 Calculated attenuation due to rain.

These measurements, together with investigations of the temporal and spatial distribution of heavy rainfall,[14] made it clear that, for a viable satellite system to be constructed at these frequencies, some form of ground

IMPACT OF RAIN ATTENUATION ON SATELLITE SYSTEMS

station space diversity would be required. The actual form (i.e., separation and base line orientation) had to be determined.

Rain Attenuation with Diversity Terminals

It has been established that radiometers can be used reliably to infer attenuation due to rain on an Earth-space path, and this has made these relatively simple, fixed-pointing instruments a natural choice for multiple site measurements.[13,15] The accuracy of the inferred attenuation is dependent primarily on the absorber temperature assumed for the rain environment.[13,17,19] Comparison of radiometer measurements with direct measurements using the sun as a source has established a representative absorber temperature[13] for 16 GHz of between 270° and 280°K for attenuation measurements less than 10 db. At 16 GHz, 10 db is about the maximum attenuation that can be measured reliably due to uncertainty in the apparent absorber temperature. But for Earth-space paths with space diversity, this value is sufficient to provide a measure of the diversity improvement.

Presently, three path diversity experiments using radiometers have been completed at Bell Laboratories in New Jersey,[16-18] and others are now in progress. The relative location of the diversity terminals for the New Jersey experiments is summarized in Fig. 2. In each case, the antenna beam of the radiometers is directed toward the geostationary orbit at an elevation angle of 32° and an azimuth angle of 226°.

The first experiment,[16] covering four months of 1969, showed that separations of 11.2 and 14.4 km provided a substantial reduction in the total time a given level of attenuation was exceeded (a factor of 50 at 10 db attenuation), but the separation of 3.2 km showed little reduction. For the second experiment,[17] the 3.2-km separation was extended to 19.2 km along the baseline already established by the other two sites.

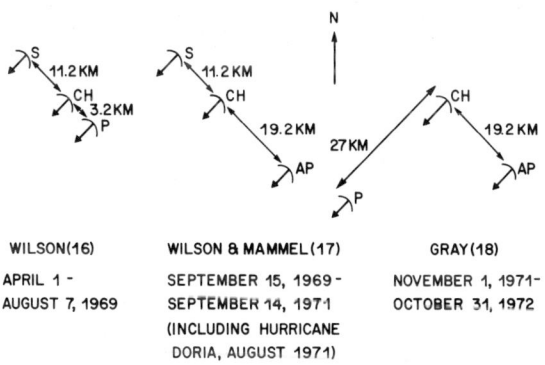

Fig. 2 Path diversity experiments.

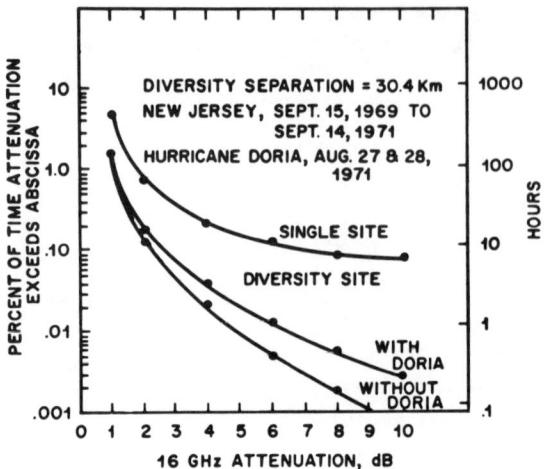

Fig. 3 Diversity statistics.[17]

This configuration, shown in Fig. 2, operated for 2 yr, giving the first long-term diversity statistics. The length of the measurement period is very important, since it is the tails of the distribution in which there is most interest, and it is just these tails that fluctuate greatly if, as is normally the case, they are derived from relatively few rain events.

Figure 3 shows the averaged distribution of attenuation for the single sites, and the distribution for two diversity sites separated by 30.4 km. One very interesting aspect is the effect on the diversity distribution of the two days of hurricane Doria (1971). There is some question as to which distribution to use in system design: the one including Doria, the one not including Doria, or somewhere in between. This is discussed in Ref. 6.

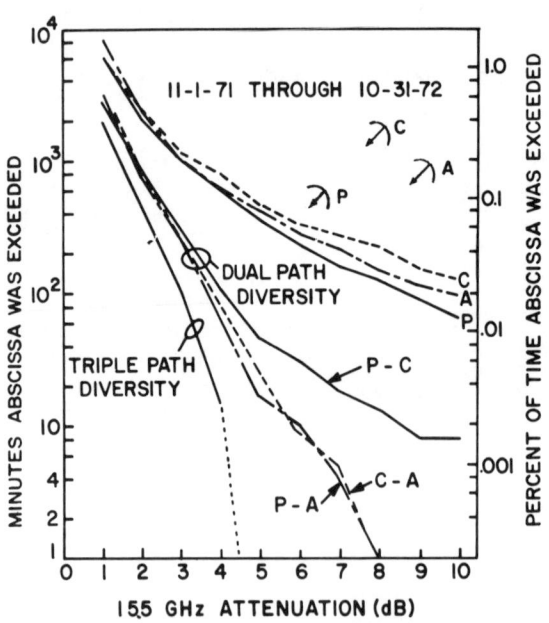

Fig. 4 Diversity statistics.[18]

The third experiment,[18] which operated for 1 yr, changed the baseline configuration to that shown in Fig. 2. Figure 4 shows the distributions for all four diversity site combinations (the fourth combination is three-site diversity).

IMPACT OF RAIN ATTENUATION ON SATELLITE SYSTEMS

Note that the distribution for the 19-km separation shows more improvement than the distribution for the larger 27-km separation, and similar improvement to the diagonal baseline of 33 km. Although it is tempting to draw the conclusion that baseline orientation is just as important as separation, the experimenters caution that relatively few events constitute the tails of the distribution, and thus more data are needed before such conclusions can be reached.

Reference 6 shows how the preceding measurements at 16 GHz are extrapolated to the 18- and 30-GHz frequencies of interest here. Recent work[22] combined the foregoing results with Earth-space path-diversity results obtained in Ohio. It is then suggested[22] that little more diversity gain (reduction in attenuation for a given percent time) should be expected for terminal separations greater than 30 km. Since a high-capacity interconnecting link must be provided between terminals (if an existing link is not available for such use), excess separation with little or no additional improvement would be costly.

Number and Length of Fades

The distributions just discussed give the total cumulative time (or percent time) a level of attenuation is exceeded. The time during which the attenuation exceeds the designed fade margin is an outage, and the number and duration of these outages is of interest, especially for very high-capacity systems. A rain margin determined solely by its ability to meet a given cumulative yearly outage objective may not be sufficient if limitations also are placed on the maximum outage duration or the maximum number of outages. It is not presently clear what such limitations should be, if any.

Figure 5 is a histogram of the rain events that make up the

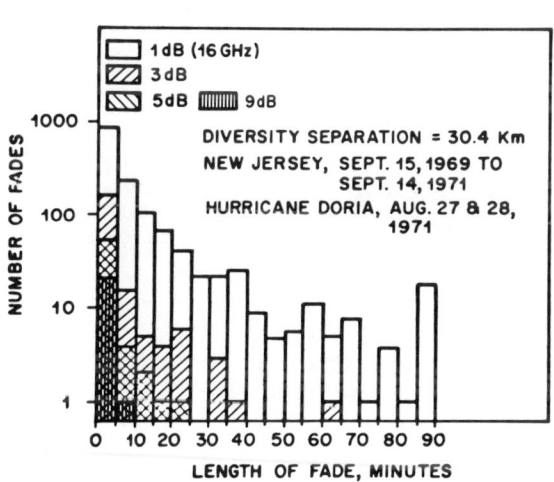

Fig. 5 Number of times attenuation exceeds 1-, 3-, 5-, 9-db level vs length of fade.[17]

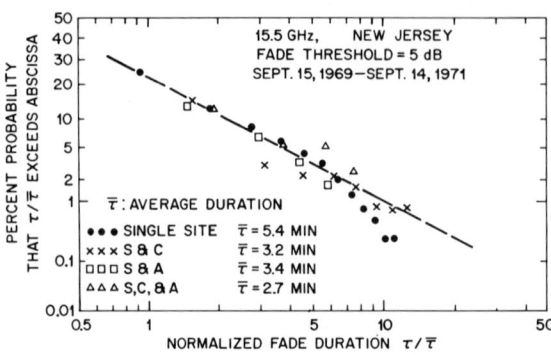

Fig. 6 Lognormal distribution of fade duration.

distribution of Fig. 3. This histogram shows, first, the rapid decline in the number of rain fade events as the threshold attenuation is increased, and, second, the fade duration distribution as a function of the threshold attenuation (1, 3, 5, or 9 db). The distribution of fade duration described by the histogram has been shown to be approximately lognormal,[23] as shown in Fig. 6.

Considerations of rain fade duration become very important for systems employing down-path power control.[3] This particular technique eliminates large d.c. power requirements on the satellite by sizing the satellite transmitters for clear-weather operation and increasing the transmit power only during a rain fade that exceeds a preset threshold attenuation. For the time that the fade is in excess of the threshold, batteries supply the increased power demand. The ability of the batteries to supply the required demand depends strongly on rain fade duration. This power-control technique will be discussed further in Sec. 3.

Cross-Polarization Coupling Caused by Rain

Interference from the depolarization of two orthogonal polarizations due to nonspherical raindrops is a matter of much interest for systems using both polarizations to double the capacity,[4] or to provide additional interference isolation as between adjacent beams. Depolarization can be expressed as a function of attenuation (linear horizontal, linear vertical, or circular), frequency, and angle of incidence with the drops of rain (canting angle and elevation angle). Recent theoretical predictions[20] have shown very good agreement with terrestrial path measurements of depolarization.[21,33] The results indicate that, for the frequencies of interest (18 and 30 GHz) and for the expected rain margins (approximately 10 and 20 db, respectively), the cross-polarization discrimination ratio due to depolarization by rain will be approximately -30 db for two linear polarizations and an assumed raindrop canting angle

IMPACT OF RAIN ATTENUATION ON SATELLITE SYSTEMS

appropriate for terrestrial paths. For the same canting angle, depolarization on an Earth-space path is expected to be less than on a terrestrial path because of the more circular projections of the raindrops apparent on an elevated Earth-space path.

For Earth-space paths, the angle between the incident polarization and the local vertical at the Earth station can vary significantly, depending on the relative location of the satellite and Earth stations. For angles approaching 45°, the cross-polarization discrimination can be an order of magnitude worse than the -30 db stated previously.[20,21] For satellites with area coverage antenna beams, the polarization angle cannot be controlled for all Earth station locations, and thus depolarization will vary at the different Earth stations. For multibeam satellite systems, such as considered here, the polarization of each beam can be rotated individually to match the local vertical for each Earth station location. This will help to make the depolarization on an Earth-space path better, or at least no worse, than that predicted for terrestrial paths.

Future Measurements Outside of New Jersey

For some time now, studies of 18/30-GHz systems have made various assumptions about expected rain margins at locations outside of New Jersey. Some have provided the same margin as derived in New Jersey to all locations, recognizing that it may be excessive for some but shy for others. Other studies have extrapolated New Jersey data to other locations by comparison of rainfall rates. In one such study,[24] isohyets of equal probability of hourly rainfall of more than 1 in. have been drawn for the contiguous United States and normalized to New Jersey (Fig. 7). Although other methods of extrapolation are used,[6] they show a high degree of

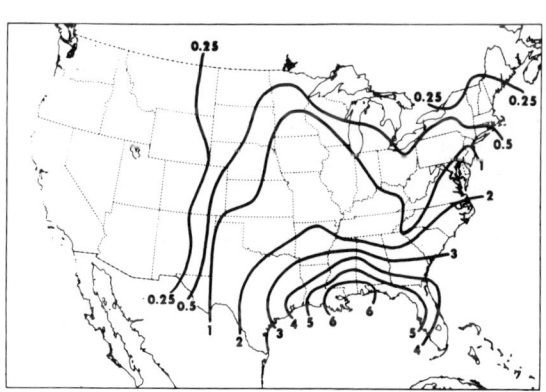

Fig. 7 Isohyets of equal probability of hourly rainfall of more than 1 in.

similarity. Once this is done, there remains the question of extrapolating diversity results the same way, since it is fairly certain that diversity terminals will be needed at any eastern and southeastern location (see Fig. 7). In the absence of measurements, only estimates based on extrapolations are available. But recently a program was initiated to answer the basic questions about rain attenuation at two sites in different climate regions. One site near Atlanta, Ga. consists of three terminals to measure diversity improvement vs separation. A similar arrangement exists for the site near Denver, Colo.

Present plans call for AT&T domestic communication satellites to be in service in early 1976. The communications channels of these satellites will operate in the 4- and 6-GHz frequency bands. Each of these satellites will include beacons operating in the vicinity of 18 and 30 GHz for experimentation on Earth-space path propagation, in addition to the experiments just described. As described in the application by AT&T to the Federal Communications Commission (FCC),[4] this 18/30-GHz beacon experiment will permit measurements of depolarization by rain and investigation of bandwidth limitations imposed by propagation considerations, in addition to measurements of rain attenuation (horizontal and vertical polarization) and diversity advantage.[25] The beacons will be directed toward the 48 contiguous states, and measurements will be made at selected sites throughout this service area. A principal experimental facility will be established at Bell Laboratories in Holmdel, N. J., with capability to make more detailed measurements pertinent to the foregoing interests. The orbital parameters of the satellite will be available to steer the Earth antennas over the small excursions of the satellite (±0.1° geocentric angle). Other entities could participate, and indeed all those who might be interested in making measurements from their own terminals are encouraged to participate. The overall technical characteristics of the beacons are described in Ref. 26.

3. System Impact

The previous section described some properties of rain which must be accommodated in system design. In the final analysis, almost every aspect of a 18/30-GHz system is influenced by the excess transmission margin that must be provided. The following paragraphs point out some considerations that surfaced in recent investigations of small-(Delta launch) to medium-(Atlas-Centaur launch) size 18/30-GHz multibeam communication satellites. Larger 18/30-GHz satellites (Titan launch and above) may present different limitations. These considerations show that the technical characteristics of the

IMPACT OF RAIN ATTENUATION ON SATELLITE SYSTEMS   301

Earth stations and satellites could vary greatly, depending on the assumed Earth-space path rain margins.

One of the benefits of a multibeam 18/30-GHz satellite is the possibility of reusing the entire 2500-MHz bandwidth in each beam. To do this, adequate isolation between the beams must be provided for the desired mode of transmission. For digital PCM-coherent phase shift keying, the effect on error rate of the carrier-to-thermal-noise ratio (C/N) and the carrier-to-interference ratio (C/I) is well understood for the case of an ideal channel,[27] but the effects on error rate due to channel distortions are more difficult to predict, especially when multiple contributions exist in the complete end-to-end link. One approach in allowing for these distortions is to retain the ideal channel relation between error rate, C/N and C/I, by simply adding a fixed margin to each of these quantities, i.e., an effective thermal noise margin and an effective interference noise margin. Although it can be argued that these margins also should be a function of error rate, in lieu of measurements a fixed margin was assumed here. By doing this, the total end-to-end requirements on C/N and C/I are determined for the desired error rate.

Since transmission requirements are set for each one-way way link between two Earth stations, analysis for the case of rain in the up path can be separated from rain in the down path, assuming that it does not rain simultaneously at the two Earth stations. For uncompensated up-path rain attenuation, $(C/I)_{up}$ and $(C/N)_{up}$ degrade linearly with rain attenuation, and at some up-path attenuation the degradation causes the minimum error rate to be exceeded. This value of attenuation

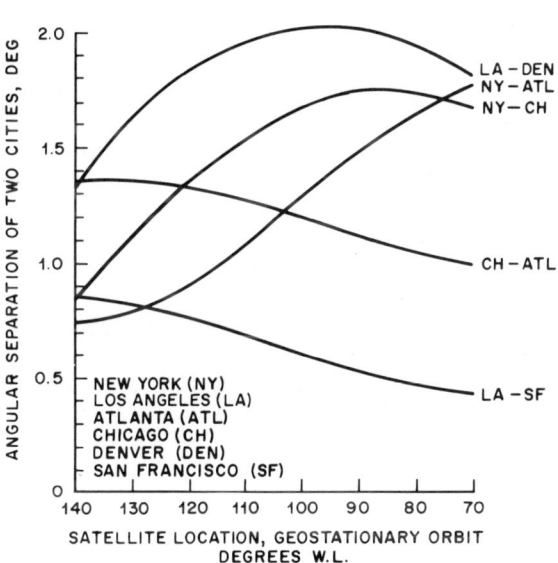

Fig. 8 Angular separation of a few major cities as seen from geostationary orbit.

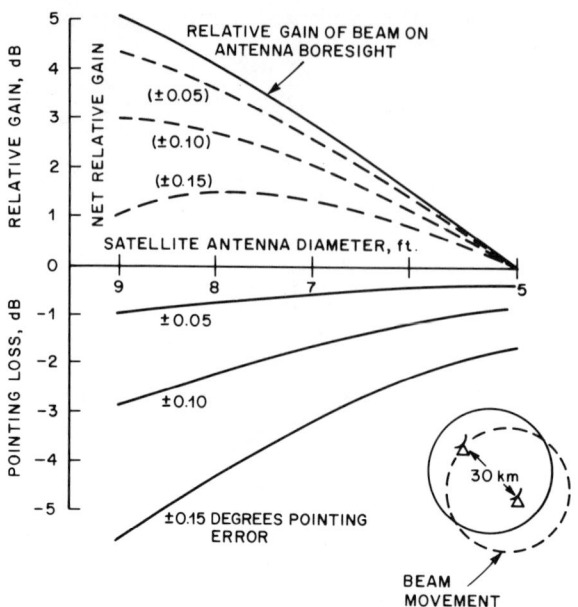

Fig. 9 Net relative antenna gain allowing for satellite pointing errors.

is dependent on the clear weather operating point, $(C/I)_{total,clear}$ and $(C/N)_{total,clear}$. For the particular design considered here, the single-conversion 30- to 18-GHz satellite transponder and the standard Bell System bit rate of 274 Mbit/sec determined the up-path C/N during clear weather.

The clear weather C/I was dominated by inter-beam interference, the magnitude of which depended on satellite and Earth station location. Figure 8 shows the angular separation of a few major cities as seen from geostationary orbit. The satellite antenna directivity was limited, first, by the maximum aperture allowable in the launch vehicle shroud when integrated with a particular spacecraft, and, second, by the pointing error assumed for the satellite. Figure 9 shows the net satellite antenna gain achievable after pointing losses are subtracted, relative to the net gain of an antenna with a diameter of 5 ft. Since deployable apertures were not considered, the range of antenna diameters accommodated by both Delta and Atlas-Centaur launch vehicles is as shown in Fig. 9. These considerations (satellite location, Earth station locations, satellite antenna configuration) set the clear weather C/I for the up and down paths.

With the initial operating point set as just described $(C/N_{total,clear}, C/I_{total,clear})$, uncompensated up-path rain attenuation of 6 db was found to degrade $(C/I)_{up}$, $(C/N)_{up}$, and therefore the error rate to the minimum value acceptable. Thus, for up-path rain attenuation in excess of 6 db, Earth station power increase was required to keep the error rate from degrading further:

$$\begin{Bmatrix} \text{Earth station} \\ \text{power increase} \end{Bmatrix} = \begin{Bmatrix} \text{maximum rain} \\ \text{attenuation, db} \end{Bmatrix} - \begin{Bmatrix} \text{allowable satellite} \\ \text{receiver input} \\ \text{backoff, db} \end{Bmatrix}$$

For multiple rf channels transmitted from a single Earth station feed, the maximum transmit power can be limited by hardware considerations. For the particular satellite design just discussed with 6-db allowable up-path reduction, Fig. 10 shows how the Earth station diameter must change with rain margin for various assumed limitations on maximum power permitted per rf channel. For example, this particular design would call for 50-ft-diam antennas if the maximum allowable transmit power is 200 w per rf channel at a geographic location where the rain attenuation at 30 GHz is 20 db. These numbers will change for other system designs, particularly for ones using larger launch vehicles (Titan or above) where more directive antennas and less limiting transponders could be implemented due to relaxed weight constraints.

Rain in the down path can be compensated for in a number of ways,[28,3] depending on the specific application. Two methods will be discussed here:
1) "full-power" systems, which simply overdesign transmission margin by the expected rain attenuation;
2) a method of down-path power

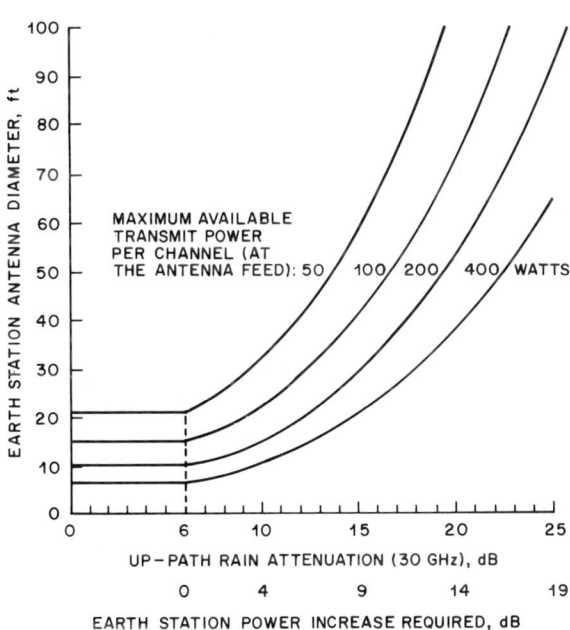

Fig. 10 Minimum Earth station antenna diameter based on up-path rain attenuation and transmit power.

control which nominally operates at clear weather conditions, and increases power only during excessive rain; the increased power required during rain is supplied by the batteries provided for eclipse operation.[3] These two methods of providing down-path rain margins have been described in a previous paper,[3] where it was shown that implementation of the power control technique permitted the capacity of a particular satellite to increase by a factor of 3 to 4 over the capacity possible with continuous "full-power" down-path transmission. This large-capacity increase resulted from balancing the total solar power requirement of the transmitters with the number of transmitters allowable by weight constraints. Since the total available solar power was fixed, the balancing had to be done with transmit power. For other satellite designs where the solar power can be increased, the balancing can be done without requiring power-controlled transmitters by increasing the solar power (e.g., by larger sun-oriented paddles) and/or by increasing the Earth station antenna diameter. Thus, one way of comparing these two approaches is to weigh the impact of power-controlled transmitters with the impact of larger solar paddles and/or larger Earth station antennas. The more costly of the two, in terms of reliability or capacity, will depend on the specific satellite and system configuration. Studies to date for small-(Delta) to medium-(Atlas-Centaur) size 18/30-GHz satellites have shown that some form of power-controlled transmitters is preferable. But the implementation of such a system, including telemetry and control techniques, still needs further work. Recent work by Hughes Aircraft Company, under contract to Bell Laboratories, on some of the hardware involved, including a multipower TWT, shows good results.

The measurements to be made in Colorado and Georgia could influence the foregoing tradeoff. If the rain attenuation statistics show that only a few large rain fades are experienced per year, the infrequent use of power control would be more desirable than the continuous operation of an oversized transmitter and satellite. On the other hand, numerous fades could make the reverse situation true.

In evaluating other uses of power-controlled transmitters, it was suggested[29] that they also could be used to eliminate down-path outages due to the sun which could occur for a few days twice each year when the sun passes directly behind the satellite.[30] For a down-path frequency of 20 GHz, the effective blackbody temperature of the sun can be estimated conservatively to be 12,000°K.[31] The increase in noise, or reduction in C/N_{down}, depends on the initial operating temperature of the Earth station. For the worst case of a cooled receiver and a high antenna elevation angle, the clear weather system noise temperature could approach 150° at the input to the preamplifier. When the sun is directly in the main beam of the Earth station antenna, the antenna noise temperature could increase to about 5300°K, for an elevation angle of 30° corresponding to an atmospheric loss of 0.6 db:

$$T_{b,sun} \cong 12,000(10^{-.6/10}) + 300(1-10^{-.6/10}) \cong 10,700°K$$

$$T_{a,sun} \cong (A_e/\lambda^2)(T_{b,sun}) \left[\iint_{sun} p_n d\Omega_s \right] \cong 5300°K$$

A_e is the effective area of the antenna aperture, and Ω_s is the solid angle subtended by the radio sun. A gaussian approximation to the antenna beam has been assumed for p_n, and the A_e/λ^2 term is derived from a 50-ft-diam antenna of 50% efficiency at 20 GHz.

Referring the antenna temperature to the input of the Earth station preamplifier (1 db of rf loss) results in an overall system temperature of about 4200°K. The increase in noise or reduction in down-link C/N is then approximately 10 log (4200/150) = 14.5 db. If a 5-db reduction in down-link C/N is permitted before the satellite transmitters are increased,[3] then about 9 db of transmitter power increase will be required. This value, 9 db, is the power increase available from the presently envisioned satellite transmitter configuration. The battery support required for the approximately 5-min duration would be much less than needed for the worst rain condition (6-db power increase for 1 hr) previously considered.

Countering sun noise with increased transmitter power would allow 18/30-GHz satellites to be spaced as close together in the geostationary orbit as interference considerations would permit. This spacing then would allow use of multiple-beam, single-reflector, Earth station antennas[32] to reduce the number of antenna apertures required per Earth station. The savings would be even greater for Earth stations requiring diversity terminals.

Down-path rain not only attenuates the signal but also increases the effective noise temperature of the Earth station antenna. For very heavy rain, the antenna temperature approaches 270°K for down-path frequencies near 18 GHz.[13,15,17,19] With an antenna noise temperature this large, there is a question of whether cooled preamplifiers would be beneficial, and therefore desirable, or whether simpler and less costly uncooled preamplifiers would be more appropriate. The effect of the Earth station receiver noise temperature on down-path transmission during rain can be evaluated by writing the system temperature as $T_{s,rain} = T_{rec} + 290(1-\ell) + T_{a,rain}(\ell)$, where $T_{a,rain} = T_{a,clear} + 270(1-t)$. The parameter t is the transmission coefficient of rain, and ℓ is the coefficient for rf component losses (ℓ = 0.8 for the 1 db of loss assumed). The expression for $T_{a,rain}$ is conservative, but for rain attenuation less than 10 db it is adequate. Figure 11 shows the sensitivity of the system temperature to rain attenuation for various receiver temperatures, all relative to a receiver temperature of 50°K. The ordinate also could be interpreted as the relative change in Earth station G/T. This figure shows that, for the same satellite transmit power, an uncooled receiver of 250°K would reduce the clear weather down-path C/N by 3.7 db, but during rain it is only 2.0 db worse than a cooled 50°K receiver. For "full-power" satellites, which transmit a high power continuously, the 2.0 db is of more interest, since this type of system sizes the satellite transmitters for worst rain conditions. But for power-controlled systems that size the satellite transmitters for clear weather operation, the 3.7-db improvement becomes equally important. Other receiver temperatures would give different results, but for power-controlled satellites a cooled Earth station receiver looks beneficial, whereas for a full-power satellite the cost, complexity, reliability, and maintenance of a cooled vs uncooled receiver must be considered before a decision is made.

This section has summarized briefly some considerations that surfaced in recent investigations of an 18/30-GHz system. The results

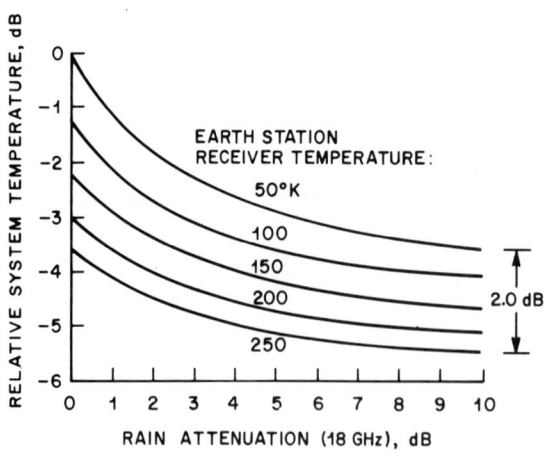

Fig. 11 Sensitivity of system temperature to rain attenuation and receiver temperature.

discussed here stem from a few basic assumptions: first, the launch vehicle was restricted to the Delta or Atlas-Centaur; second, the transmit power from the Earth station was limited to about 150 w per rf channel at the antenna feed; third, an expected satellite pointing error of ±0.1° was assumed; and fourth, a single-conversion 30- to 18-GHz satellite transponder was chosen. If any of these assumptions change, the results would change. Work since the time of these investigations has suggested that future studies would be more optimistic about Earth station transmit power and satellite pointing errors, and double-conversion transponders would be considered seriously so as to allow greater flexibility in intercity connections.

4. Conclusions

It has been recognized for many years that good answers to the basic questions about Earth-space propagation are necessary for optimal design of an 18/30-GHz satellite system. The comprehensive Earth-space propagation program in New Jersey already has begun to answer many of these questions, particularly the ones related to diversity Earth terminals. A complementary program now is needed to answer the same questions for other possible Earth station locations. Two companion papers by Bergman[5] and Lin[6] of the Radio Engineering Department of Bell Laboratories discuss a program now underway which is planned to accomplish this task. Beyond this program, in early 1976 the 18- and 30-GHz satellite beacon experiment previously discussed will provide an additional opportunity to obtain information on rain attenuation as well as data on depolarization by rain and bandwidth coherence.

References

[1] "Final Acts of the World Administrative Radio Conference," Report and Order in Docket 19547, July 1971, Amendments to the Table of Frequency Allocation (Part 2 of the Rules), effective March 1, 1973, Federal Communications Commission.

[2] Tillotson, L. C., unpublished work, Jan. 3, 1966; also "A Model of a Domestic Satellite System," Bell System Technical Journal, Vol. 47, No. 10, Dec. 1968, pp. 2111-2137.

[3] Bell Laboratories, unpublished work, 1972; also Lundgren, C. W. and Spilman, L. D., "A Method of Providing Rain Margins for 18/30 GHz Communications Satellites Without Increasing the Solar Power Requirement," International Conference on Communications, 1973, Seattle, Wash.

[4]"Application for a Domestic Communications Satellite System," original application filed Oct. 19, 1970, latest amended application dated March 29, 1973, American Telephone and Telegraph Co.

[5]Bergmann, H. J., "Site Diversity Experiment at 18 GHz," AIAA Paper 74-497, April 1974, Los Angeles, Calif.

[6]Lin, S. H., "Outage Estimation for Earth-Satellite Radio Links (at 18 and 30 GHz)," AIAA Paper 74-498, April 1974, Los Angeles, Calif.

[7]"An Integrated Space/Earth Communications System to Serve the U.S.," FCC Docket 16495, Dec. 15, 1966, Bell System proposal.

[8]Evans, H. W., "Technical Background, AT&T Domestic Satellite Proposal," AIAA Paper 68-411, April 1968, San Francisco, Calif.

[9]Hogg, D. C., "Path Diversity in Propagation of Millimeter Wave Through Rain," AP-15, No. 3, May 1967, Institute of Electrical and Electronics Engineers; also "Millimeter-Wave Communication Through the Atmosphere," Science, Vol. 159, No. 3810, Jan. 1968, pp. 39-46; also "Statistics on Attenuation of Microwaves by Intense Rain," Bell System Technical Journal, Vol. 48, No. 9, Nov. 1969, pp. 2949-2962.

[10]Medhurst, R. G., "Rainfall Attenuation of Centimeter Waves; Comparison of Theory and Measurement," AP-13, No. 4, July 1965, Institute of Electrical and Electronics Engineers.

[11]Setzer, E. D., "Computed Transmission Through Rain at Microwave and Visible Frequencies," Bell System Technical Journal, Vol. 49, No. 5, Oct. 1970, pp. 1873-1892.

[12]Hogg, D. C., "Intensity and Extent of Rain on Earth-Space Paths," Nature, Vol. 243, No. 5406, June 8, 1973, pp. 337-338.

[13]Wilson, R. W., "Sun Tracker Measurements of Attenuation by Rain at 16 and 30 GHz," Bell System Technical Journal, Vol. 48, No. 5, May-June 1969, pp. 1383-1404.

[14] Semplak, R. A. and Keller, N. E., "A Dense Network for Rapid Measurement of Rainfall Rate," Bell System Technical Journal, Vol. 48, No. 6, July-Aug. 1969, pp. 1745-1756.

[15] Penzias, A. A., "First Result from 15.3 GHz Earth-Space Propagation Study," Bell System Technical Journal, Vol. 49, No. 6, July-Aug. 1970, pp. 1242-1245; also Ippolito, L. J., "Millimeter-Wave Propagation Experiments Utilizing the ATS-5 Satellite," Fall 1970 URSI Meeting, Columbus, Ohio.

[16] Wilson, R. W., "A Three-Radiometer Path-Diversity Experiment," Bell System Technical Journal, Vol. 49, No. 6, July-Aug. 1970, pp. 1239-1242.

[17] Wilson, R. W. and Mammel, W. L., "Results from a Three-Radiometer Path-Diversity Experiment," Conference on the Propagation of Radio Waves at Frequencies Above 10 GHz, April 1973, London, Institute of Electrical Engineers; also authors' supplementary notes to the report.

[18] Gray, D. A., "Earth Space Path Diversity: Dependence on Base Line Orientation," G-AP International Symposium, Aug. 22-24, 1972, Boulder, Colo.

[19] Gray, D. A., "The Effect of Scatter on Radiometric Observations of Rainfall," 1972 Spring Meeting USNC/USRI, April 13-15, 1972, Washington, D.C.

[20] Chu, T. S., "Rain-Induced Cross-Polarization at Centimeter and Millimeter Wavelengths," Bell System Technical Journal, Vol. 53, No. 8, Oct. 1974, pp. 1557-1579.

[21] Semplak, R. A., "Simultaneous Measurements of Depolarization by Rain Using Linear and Circular Polarizations at 18 GHz," Bell System Technical Journal, Vol. 53, No. 2, Feb. 1974, pp. 400-404.

[22] Hodge, D. B., "Space Diversity for Reception of Satellite Signals," Inter-Union Commission on Radiometeorology (I.U.C.R.M.) Colloquium, Oct. 1973, Nice, France.

[23] Lin, S. H., "Statistical Behavior of Rain Attenuation," Bell System Technical Journal, Vol. 52, No. 4, April 1972, pp. 557-581.

[24] Evans, H. W., "Attenuation on Earth-Space Paths at Frequencies up to 30 GHz," *International Conference on Communications*, 1971, Montreal, Canada.

[25] Cox, D. C., "Design of the Bell Laboratories 19 and 28 GHz Satellite Beacon Propagation Experiment," *International Conference on Communications*, 1974, Minneapolis, Minn.

[26] Briskman, R. D., Latter, R. F., and Muller, E. E., "Call for Help," *Institute of Electrical and Electronics Engineers, Spectrum*, Vol. 11, No. 10, Oct. 1974, pp. 35-36.

[27] Rosenbaum, A. S., "PSK Error Performance with Gaussian Noise and Interference," *Bell System Technical Journal*, Vol. 48, No. 2, Feb. 1969, pp. 418-442.

[28] Robbins, W. P., "Techniques for Dealing with the Effects of Bad Weather in a Communication Satellite System," *International Conference on Communications*, June 1972, Philadelphia, Pa.

[29] Brady, D. M., private communication, Oct. 1973.

[30] Lundgren, C. W., "A Satellite System for Avoiding Serial Sun-Transit Outages and Eclipses," *Bell System Technical Journal*, Vol. 49, No. 8, Oct. 1970, pp. 1943-1972.

[31] Wrixon, G. T. and Hogg, D. C., "Absolute Measurements of Solar Flux at 16 and 30 GHz," *Astronomy and Astrophysics*, Vol. 10, No. 2, Feb. 1971, pp. 193-197.

[32] Ohm, E. A., "A Proposed Multiple-Beam Microwave Antenna for Earth Stations and Satellites," *Bell System Technical Journal*, Vol. 53, No. 8, Sept. 1974, pp. 1657-1665.

[33] Barnett, W. T., "Some Experimental Results on 18 GHz Propagation," *Conference Record, National Telecommunications Conference*, Dec. 1972, Houston, Texas; also Lin, S. H., unpublished work, July 1973.

MORPHOLOGY OF IONOSPHERIC SCINTILLATION

R. K. Crane[*]

Massachusetts Institute of Technology, Lexington, Mass.

Abstract

Small-scale ionospheric irregularities in the F region can cause fluctuations in the amplitude, phase, and angle of arrival of vhf, uhf, and shf signals traversing the ionosphere. Under some conditions, the power level fluctuations or scintillations at vhf and uhf may become severe, with 12-db signal level increases and fades in excess of 30 db being observed. Current information about the probabilities of occurrence of severe fades is derived from a number of experiments using either radio star or satellite-borne sources. The measurements are generally of signal level only and have been used to calculate scintillation indices to characterize scintillation intensity. An examination of the global distribution of scintillation indices shows that scintillations are of importance to communication system performance primarily in the auroral and polar regions and at night near the geomagnetic equator.

Introduction

This paper is directed toward providing information to communication systems designers first about scintillation as observed in a single experiment, second about the adequacy of the existing models used to interpret scintillation data, and finally about the variation of scintillation with geophysical

---

Presented as Paper 74-52 at the AIAA 12th Aerospace Sciences Meeting, Jan. 30-Feb. 1, 1974, Washington, D.C. This work was sponsored by the Department of the Navy. The author acknowledges the help and support of J. V. Evans and R. H. Wand of the Millstone Hill radar site in making their data available for analysis and of T. M. Turbett for the data processing. The percentage occurrence data used in preparing Figs. 17-19 were provided by R. H. Wand.

[*]Staff Member, Lincoln Laboratory.

parameters. Scintillation due to electron density fluctuations has been observed on line-of-sight paths through the ionosphere at frequencies ranging from 20 MHz to 6 GHz. Frequencies between 100 and 400 MHz are emphasized in this paper because of their immediate concern in system design. An exhaustive literature exists on the subject of ionospheric scintillation. The references cited in this paper are not complete and are intended only to be illustrative. When possible, the references will be to measurements made at frequencies in the 100- to 400-MHz range.

Early radioastronomical observations of sources of small angular extent exhibited significant intensity fluctuations. Spaced receiver measurements made at the Jodrell Bank Experiment Station[1] at 81.5 MHz showed that the intensity fluctuations originating in the F region of the ionosphere were correlated with the occurrence of spread-F and formed a random intensity pattern on the ground, with a correlation distance of approximately 4 km.

Current knowledge of received signal fluctuations or scintillation caused by the ionosphere has been derived from a large number of observations of amplitude fluctuations made at meter, decimeter, and centimeter wavelengths. The observations have been made over the past two decades at a number of locations using radio stars or satellite-borne sources. The data generally have been recorded on strip charts. Various scintillation indices have been used to characterize the appearance of the recorded data on the charts. The scintillation indices often were estimated subjectively or, in recent years, calculated using the extreme values observed on short sections of the recording.[2] Information about the dependence of scintillation on geophysical parameters such as invariant latitude, magnetic activity index, and sunspot number has been published from summaries of the behavior of the qualitative scintillation index values compiled from available experimental data.[3-8]

Prediction of fading statistics for the design of communication systems requires more than a cataloging of data from available observations. Most of the experimental data are from observations of limited duration for path geometries, frequencies, and locations different from the system to be designed. To obtain fading statistics for system design, either additional experiments must be made using precisely the frequency and path geometry of the proposed system, or one of the diffraction or scintillation models must be used to interpret available data. The recent discovery of scintillation at 4 and 6 GHz in the equatorial region[9] was a surprise because it was not predicted using available data and the thin-phase-screen, gaussian correlation function model.

Booker et al.[10] proposed that diffraction by fluctuations of electron density in the ionosphere could cause the observed scintillation and that the effect of the electron density fluctuation could be modeled by a thin-phase changing diffraction screen. Currently, scintillation phenomena usually are modeled as being caused by a thin screen, using the refinements to the original analysis made by Mercier[11] and by Briggs and Parkin.[12] The refinements included the introduction of a gaussian spatial correlation function to describe the anisotropic fluctuations in electron density at and above the screen. Measurements of the spatial and temporal correlation properties of the fluctuations observed at the ground often are characterized by the scale size or correlation distance for the ionospheric diffraction screen using the gaussian correlation function assumption.[13,14]

Recent observations have shown that the region of the ionosphere causing the fluctuations is often quite thick,[15] and the thin-screen model may not be adequate. The effects of random fluctuations of refractive index (or electron content) in a thick region may be analyzed using the Born (single-scattering) approximately to the wave equation[16] or the Rytov method (method of smooth perturbations)[17] when the scintillation index is not too large. In the limit of weak scintillation, all three approximate methods are identical. For strong scintillation, multiple scattering must be taken into account, and none of the models are adequate.

Millstone Observations

Experiment Description

Observations of scintillation at 150 and 400 MHz were made during the period January 1971 to March 1973 using U.S. Navy Navigation System satellites and receivers at the Millstone Hill Radar Facility.[18] The satellites transmitted phase-coherent signals, which were recorded simultaneously at the receiver site. The uhf (400-MHz) receiver system included the 84-ft Millstone Hill antenna equipped for elevation and azimuth tracking and simultaneous observations using right- and left-hand circular polarization; a phase-lock tracking receiver; and analog-to-digital conversion of the principal polarization channel automatic gain control (AGC) voltage, error channel signals, and the in-phase and quadrature orthogonal polarization channel voltages (referenced to the principal polarization signal). The data were samples at 15 times/sec, together with time, antenna pointing angle, and the vhf data, and they were recorded on digital magnetic tape for post-measurement analysis. The vhf receiver system provided in-phase and quadrature voltages for the 150-MHz signal referenced to the phase of the uhf signal divided down by the ratio of the frequencies. The vhf antenna was an 11-element yagi mounted on one of the feed struts of the Millstone antenna.

The satellites were in circumpolar orbit at an approximate altitude of 1000 km and were tracked from horizon to horizon. For each pass, the average signal levels at each frequency varied by about 10 db. The vhf receiver system had a predetection bandwidth of approximately 250 Hz. The uhf predetection bandwidth was approximately 10 kHz, and the AGC system had an effective bandwidth (closed loop) of 250 Hz. The signal-to-noise ratio for optimum observing conditions was approximately 35 db at uhf and 25 db at vhf. The receiver system was calibrated prior to each satellite pass.

Amplitude Fluctuations

Sample observations of received signal level at both frequencies are shown in Fig. 1. The data are for a pass during the most severe magnetic disturbance that occurred during the experiment. (The planetary 3-hr magnetic activity index, Kp, equaled 8^+ at the time of the pass.) The quiet conditions were observed to the south, the disturbed conditions through the auroral region to the north. Each 1/15-sec sample is displayed. Under quiet conditions, some fluctuations are observed at each frequency. Some of the variation at 150 MHz is also due to receiver noise.

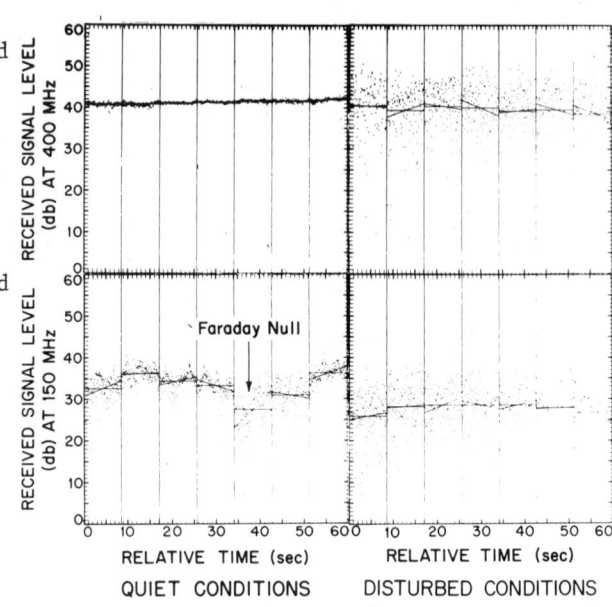

Fig. 1 Uhf and vhf amplitude measurements obtained from a pass of object 3133 on Aug. 4, 1972.

Faraday fading of the linearly polarized vhf signal also is evident, as noted in the figure. Under disturbed conditions, peak-to-peak level changes of 45 db at 400 MHz and 40 db at 150 MHz are evident.

The data displayed in Fig. 1 are for roughly the same elevation angles; the midpoint elevation angle for the 1-min quiet period was 7.6°; the midpoint elevation for the disturbed period was 8.1°. For lines of sight to the satellite at these elevation angles, the undisturbed signal levels should be identical for observations to the north and to the south of the receiver site. The detailed Faraday null structure of the 150-MHz signal will change, however, because of differences in the mean properties of the ionosphere. The sampled data were analyzed in overlapping 8.5-sec intervals. The mean and the linear least-square fit line for the logarithm of the signal amplitude are displayed between the vertical lines for alternate analysis intervals. The short analysis intervals were chosen to provide the best straight-line fits to the variation in signal levels caused by the satellite, receiver geometry, and Faraday fading at 150 MHz. The rms variation of the observed values about the least-square lines was computed to characterize the intensity of the fluctuations. For each analysis interval and frequency, the rms variation of received power about a least-square straight-line fit to the observed power values also was calculated. The latter rms value, when normalized by the mean value of received power for the analysis interval, is the S_4 index proposed by Briggs and Parkin.[12]

The values of S_4 for each analysis interval (with midpoints spaced by 4.3 sec) are displayed in Fig. 2 for the same satellite pass as for Fig. 1. The satellite rose in the south, and the relatively quiet conditions are evident for the first 5 min of the pass. Data for elevation angles below 2° are contaminated by tropospheric scintillation and surface multipath and are not displayed. The analysis of Briggs and Parkin indicates that $S_4 = 1$ is a limiting value for strong scintillations, although values as high as 1.5 are possible for the right combination of scale size and distance from the thin screen. Although strong scintillation requires a consideration of multiple scattering, no analytical multiple scattering model is available. Experimental data both for ionospheric scintillation and tropospheric scintillation at optical frequencies[19] show that, for strong scintillation or multiple scattering, a limiting value is reached.

An obvious limiting value is shown in Fig. 3. This figure is a plot of the rms variation of the logarithm of the received signal, σ_χ, vs time for the same pass. In Fig. 3, the 95% confidence limits for the estimated error in calculating σ_χ also are depicted. The confidence limits are based upon the number of times different electron density irregularities are observed within the first Fresnel zone due to satellite motion. The data show that σ_χ reaches a limiting value of approximately 5.6 db at each frequency. The limiting values depicted in Figs. 2 and 3 are calculated values for a Rayleigh received signal amplitude distribution. The spread of S_4 values about the limiting value of 1 at 150 MHz in Fig. 2 may be due either to sampling error or to a different signal amplitude distribution.

The empirical signal amplitude distributions for the 2 min depicted in Fig. 1 are shown in Fig. 4. The data show nearly identical distributions for the two frequencies and disturbed conditions. The uhf

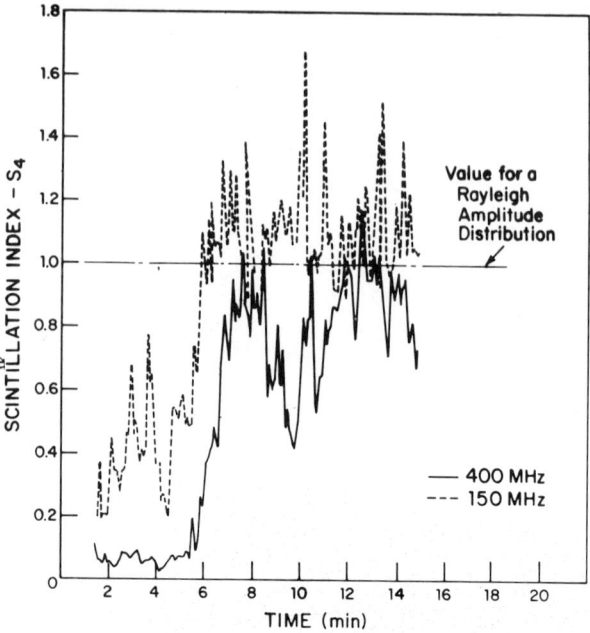

Fig. 2 S_4 vs time for pass of object 3133 rising at 0411 GMT on Aug. 4, 1972 (Kp = 8+).

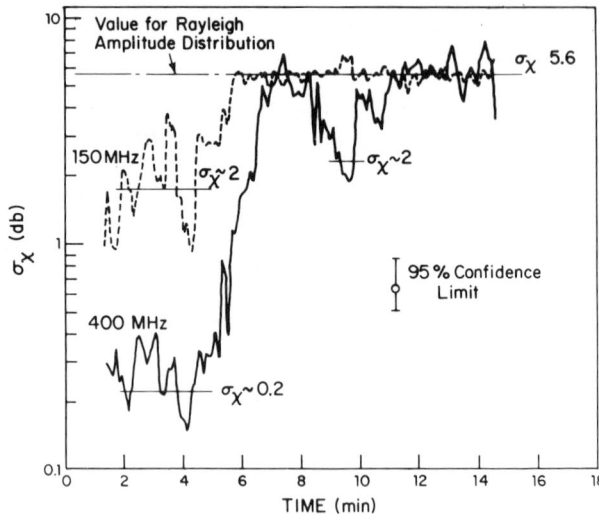

Fig. 3 σ_χ vs time for pass of object 3133 rising at 0411 GMT on Aug. 4, 1972 (Kp = 8+).

empirical distribution function appears to be log-normal for quiet conditions. Tropospheric scintillation at optical frequencies also appears to have a log-normal distribution in the limit of strong scintillation.[20] The distribution functions, are, however, definitely not log-normal for ionospheric scintillation under disturbed conditions. Bischoff and Chytil[21] proposed the use of the Nakagami-m distribution as an approximation to the empirical distributions, with $m = 1/S_4^2$. Although the Nakagami-m distribution is not theoretically correct for the thin-screen diffraction problem,[22] it may provide a reasonable approximate distribution and has been used for the construction of long-term amplitude distribution functions.[23] The Nakagami-m distribution

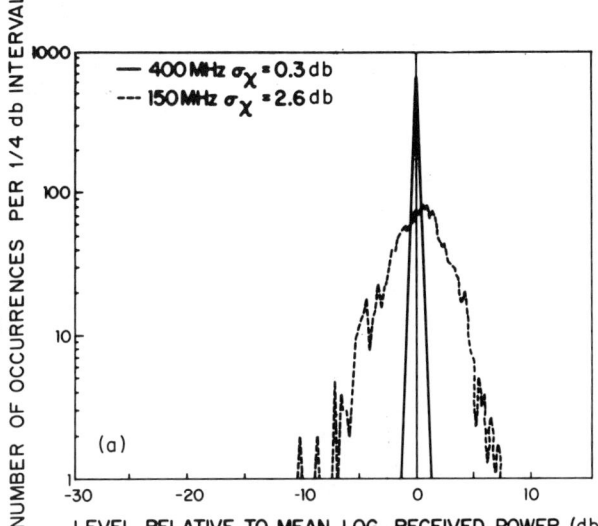

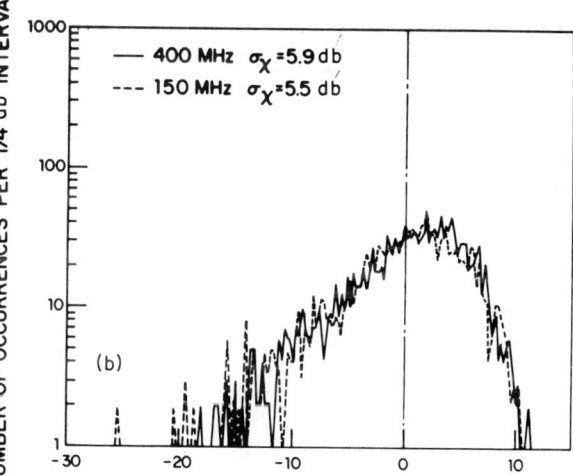

Fig. 4 Empirical amplitude distribution functions for data shown in Fig. 1.

reduces to the Rayleigh distribution for $S_4 = 1$ and approaches the log-normal distribution for σ_χ less than 1 db.[24] The empirical distribution functions depicted in Fig. 4 were tested against both the log-normal and Nakagami-m distributions for the calculated σ_χ values. Using the Pearson χ^2 distribution test and a 0.05 significance level, the distributions depicted in Fig. 4 were neither log-normal nor Nakagami-m. The vhf distribution for disturbed conditions had an m-parameter of 1.0 and tested to be Nakagami-m (Rayleigh) at a 0.01 significance level. The vhf distribution function for the minute preceding the disturbed minute (between 11 and 12 in Fig. 3) also tested as being Rayleigh with a 0.05 significance level. The uhf distribution for disturbed condition had an m-value of 0.92 and tested to be different from a Nakagami-m distribution at reasonable significance levels.

The Nakagami-m distribution, although not identical to the observed distributions, does provide a useful approximation for relating the various forms of scintillation index used in the reduction of experimental data. Calculated values of S_4 and σ_χ for the satellite pass depicted in Figs. 2 and 3 are shown in Fig. 5. The relationship between S_4 and σ_χ calculated using both the Nakagami-m and log-normal distributions together with a weak scatter approximation also are shown. Of the distribution functions shown, the Nakagami-m provides the best estimate of the S_4-σ_χ relationship for the entire range of observed values. The usefulness of the Nakagami-m distribution for approximately relating the various signal variance values proposed by Briggs and Parkin[12] and for relating the extreme value indices to other measures of scintillation is documented by Bischoff and Chytil.[21]

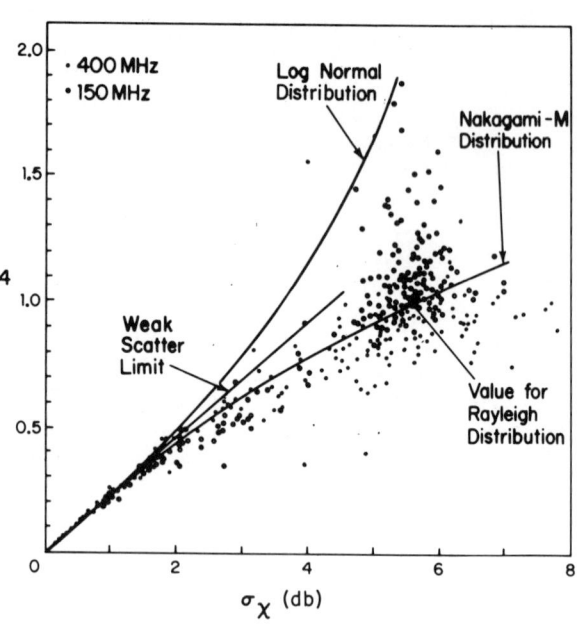

Fig. 5 S_4 vs σ_χ for pass of object 3133 rising at 0411 GMT on Aug. 4, 1972 (Kp = 8+).

MORPHOLOGY OF IONOSPHERIC SCINTILLATION

The vhf amplitude fluctuations depicted in Fig. 1 appear to be more rapid for disturbed than for quiet conditions. The temporal behavior of scintillation can be depicted quantitatively by computing distributions of the time durations the signal is below or above preset thresholds. Empirical distribution functions of duration below and above a level 3 db below the mean log received power for each of the analysis intervals in the 2 min of data shown in Fig. 1 are presented in Fig. 6. The data show that the duration distribution functions are approximately exponential, with different slopes for times above and below the -3-db threshold. The average fade duration is given by the value of time duration required to reduce the number of observations by 1/e. For the 3-db threshold and disturbed conditions, the average fade duration was 0.08 sec at 400 MHz and 0.05 sec at 150 MHz. The fade rate is the reciprocal of the average fade duration for a 0-db threshold. For the quiet conditions depicted in Figs. 1 and 4, the fade rates were 3.8 Hz at uhf and 6.2 Hz at vhf. For disturbed conditions, the fade rates were higher, being 7.2 Hz at uhf and 9.2 Hz at vhf.

The temporal behavior of scintillation may also be characterized by empirical correlation functions or power spectra. Power spectra for selected time periods from the pass depicted in Fig. 3 are depicted in Figs. 7 and 8. The power spectra were calculated using detrended log received power data from each analysis interval. The data for each interval were weighted parabolically prior to calculating the Fourier transform, and the resultant power spectra were averaged over 13 analysis intervals within 1 min. The resultant confidence limits for the power spectra estimates are shown in the figures. The spectra represent signal plus noise. The receiver noise levels also are depicted on each figure. The receiver bandwidths prior to sampling are approximately 250 Hz, and, with a sampling rate of only 15/sec (Nyquist frequency of 7.5 Hz), considerable aliasing is possible in the reported spectra. The data reported here were obtained from a satellite tracking analysis program which required the relatively slow sampling rate.

The uhf spectra for the quiet and disturbed times are identified on the figure. The dashed spectra are for minute intervals between 2 and 5 min, as shown in Fig. 3. The horizontal lines with σ_χ values in Fig. 3 represent the duration spanned by the spectra displayed in Fig. 7. The dashed lines are for quiet or weak scintillation conditions. The quiet data are barely above receiver noise and are not useful in describing scintillation phenomena. The dot-dashed spectrum, for minute 9-10, is for a period when the strong scintillation limit was not reached. The spectra approximate a power law for frequencies greater than 1 Hz. Power law power spectra have been reported by Rufenach,[25] and the power law form is evident in the data reported by Elkin and Papagiannis.[26] A line with a slope of -3 is drawn on the figure, but the best fit slope for the power spectra may lie between -2 and -3. The fluctuations of the power spectra with frequency are due to the limited statistical accuracy of the reported values. Sample sizes of 1 min were chosen because the process obviously is not stationary over longer time intervals (except perhaps when the

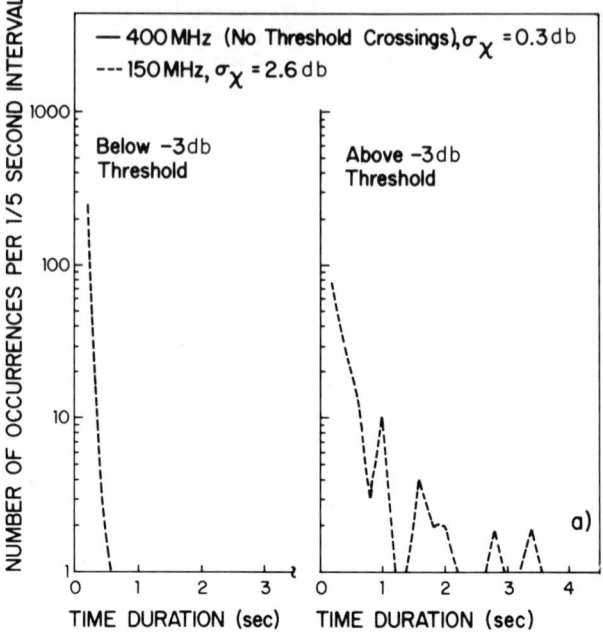

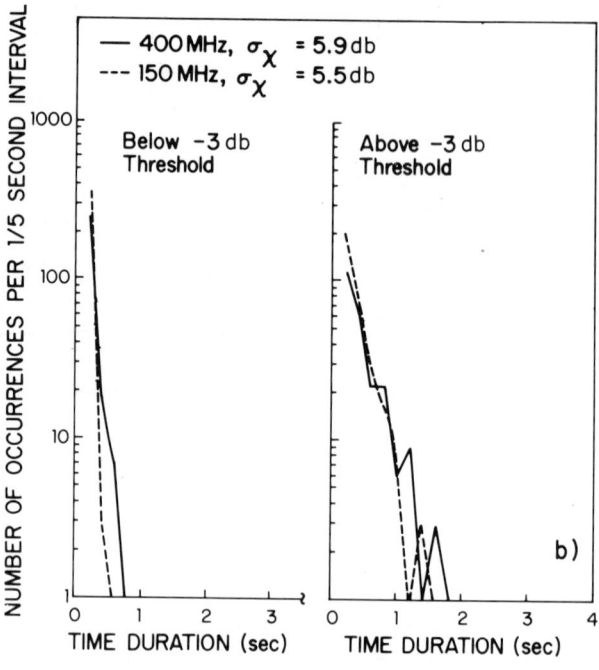

Fig. 6 Empirical distribution functions for durations below and above -3 db threshold relative to the mean log received power for the data shown in Fig. 1.

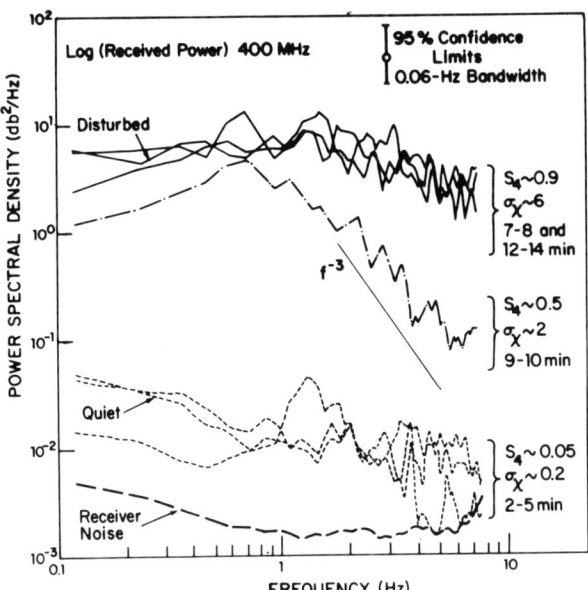

Fig. 7 Uhf log amplitude power spectra for selected minutes from the pass of object 3133 rising at 0411 GMT on Aug. 4, 1972 (Kp = 8+).

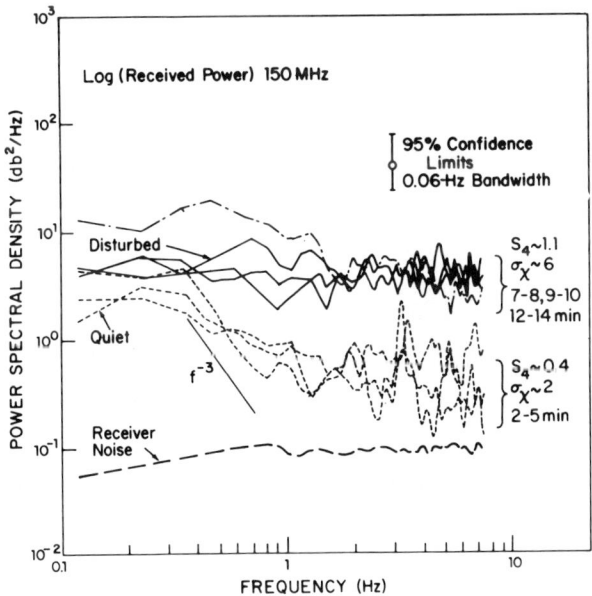

Fig. 8 Vhf log amplitude power spectra for selected minutes from the pass of object 3133 rising at 0411 GMT on Aug. 4, 1972 (Kp = 8+).

strong scintillation limit is reached, as shown in Fig. 3) and may well not be stationary even over 1 min. Shorter sample lengths were not chosen, because the statistical error would become significantly larger.

The spectra for disturbed or strong scintillation limit fluctuations are represented by solid lines. These data show little change of level with frequency, implying that severe aliasing is present in the data and the sampling rate was not high enough to represent the strong scattering case adequately. The data were obtained as a part of a general propagation study, and higher sampling rates would have compromised the other elements of the program. What is evident in the data is relatively little change in the low-frequency variance energy and significant increases at higher frequencies. The correlation time therefore decreases and the spectrum spreads (fade rate increases) as the strong scintillation limit is reached.

The vhf spectra for the disturbed period are also flat spectra. For quiet conditions, the σ_χ value is approximately the same as for the dot-dashed curve in Fig. 7. As in Fig. 7, the spectra at lower frequencies have almost the same levels as for the strong scintillation limit. For quiet conditions and frequencies above 1 Hz, the signal-variance-to-noise ratios are not large enough to provide an adequate measure of the shape of each spectrum.

Phase Fluctuations

The phase of the signals from the satellite fluctuates when scintillation occurs. The differential phase path length was measured using the vhf in-phase and quadrature voltage values. The phase reference for the vhf signal was the phase of the uhf signal divided down by the ratio of the two frequencies. The differential phase path length values, reported in terms of phase change at 150 MHz reference the initially reported phase value, are shown in Fig. 9. The differential phase value at a sample instant is computed from the reported in-phase and quadrature voltage values and can be determined only modulo 2π. In processing the data, the assumption is made that the phase cannot change by more than π rad between successive samples. The assumption is adequate only if the data are samples at a sufficiently high rate. The in-phase and quadrature channel bandwidths were 250 Hz prior to sampling, and it is possible that phase shifts greater than 2π can occur between sampling times.

The differential doppler values for quite conditions (same observation times as for Fig. 1) show a relatively

narrow spread of
values about the
origin, suggest-
ing that no phase
ambiguity prob-
lems occurred.
The differential
phase values
showed a smoothly
changing trend
caused by the
change in inte-
grated electron
content along the
path (see Gar-
riott and da-
Rosa[27]). The
data show an
increase in phase
fluctuations at
the Faraday null
caused by a

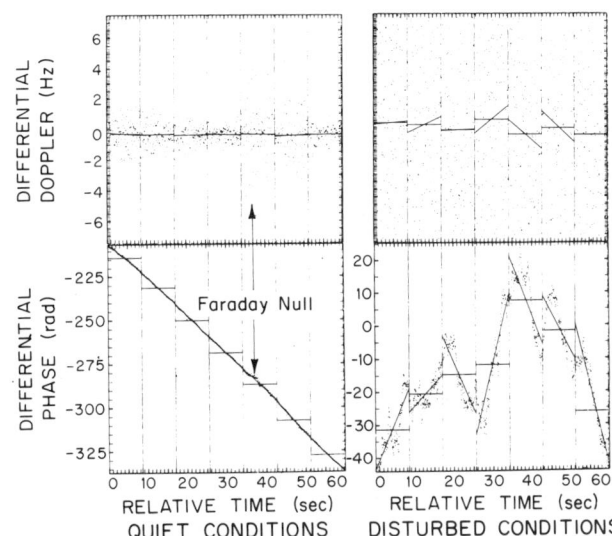

Fig. 9 Differential phase and differential doppler measurements obtained from a pass of object 3133 on Aug. 4, 1972

decrease in signal-to-noise ratio. The differential doppler data for disturbed conditions show a nearly uniform distribution of points between $\pm$ 7.5 Hz ($\pm \pi$ change between successive samples). For these data, occasional phase ambiguities are quite likely which introduce errors into the differential phase values. The low-frequency differential phase fluctuations apparent in Fig. 9 for disturbed conditions may be due to actual changes in total electron content or to random inclusions of 2π rad phase ambiguities.

The rms variations in differential doppler and phase are depicted in Fig. 10 for the entire pass. The differential doppler values appear to show a strong scintillation limit at 4.3 Hz, corresponding to the $\pi/\sqrt{3}$ rms value for phase change between successive observations of a Rayleigh process. The rms variations in differential phase show more uncertainty for strong scintillation because of possible phase ambiguities.

Power spectra for the differential phase fluctuation observations are shown in Fig. 11. The dashed curves correspond to weak scintillation and a sampling rate adequate to measure differential phase unambiguously. These power spectra show a region of generally linear (power law) decrease until the data are contaminated by receiver noise. For frequencies below 0.3 Hz, the spectra are increasing with decreasing frequency, although at a rate lower than for frequencies from 0.3 - 0.6 Hz. The data at the low frequencies are, however, contaminated by the curve-fitting, detrending procedure used to prepare the data for transform analysis. Phase difference power spectra observations reported by Porcello and Hughes[28] show reasonably convincing power law power spectra over a range from 0.1-10 Hz for satellites in orbits similar to

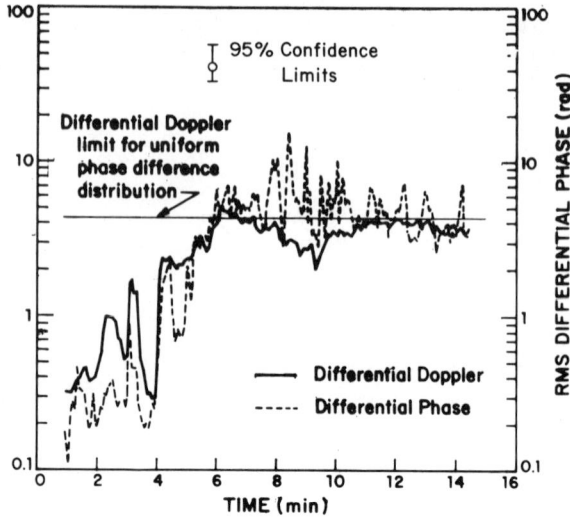

Fig. 10 Rms variations in differential phase and differential doppler for pass of object 3133 rising at 0411 GMT on Aug. 4, 1972 (Kp = 8+).

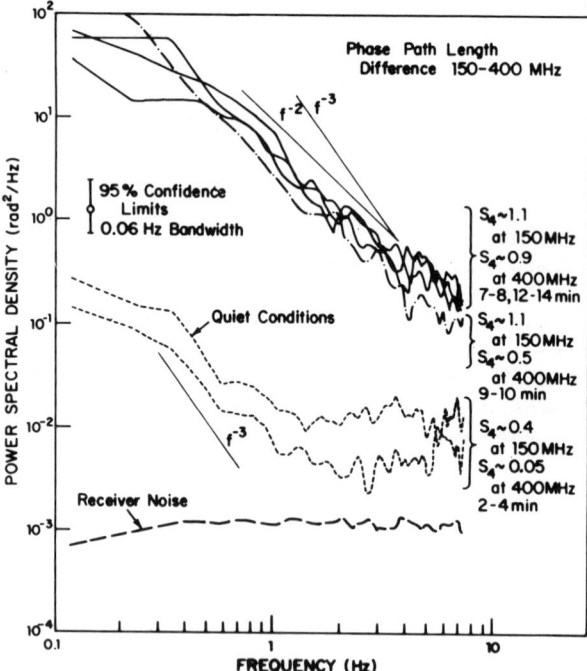

Fig. 11 Differential phase power spectra for selected minutes from the pass of object 3133 rising at 0411 GMT on Aug. 4, 1972 (Kp = 8+).

those used for the Millstone measurements. The slopes of the power spectra observed by Porcello and Hughes[28] ranged from -2.8 to -3.0. The strong scintillation data also show a power law behavior caused by the low-frequency fluctuations evident in Fig. 9.

Depolarization

Simultaneous observations were made on both left- and right-hand circular polarizations at uhf. The transmissions were nominally right-hand circular but in practice were elliptically polarized. The polarization state changed slowly with changes in satellite receiver station geometry. The orthogonal polarization receiver was gain-controlled by the primary polarization AGC signal. The AGC control system was effective in removing fluctuations of limited dynamic range which occurred simultaneously on both channels at frequencies up to 250 Hz. For strong fluctuations with peak-to-peak spreads of more than 20 db, the AGC system did not remove all of the simultaneous fluctuations from the orthogonal channel output, and the residual fluctuations were detected. For weak scintillation, only fluctuations on the orthogonal channel which were not correlated with the principal polarization fluctuations would be detected. For strong scintillation, the residual fluctuations had to be correlated with the principal polarization fluctuations to detect the presence or absence of uncorrelated fluctuations in the orthogonal polarization channel. The signal-to-noise ratio for uncorrelated fluctuations was in excess of 20 db for typical satellite receiver geometries.

The rms variation of the log of the orthogonal channel amplitude and the correlation coefficient between the log of the orthogonal channel output and the log of the principal channel output is shown in Fig. 12 for the entire pass (see Fig. 3 for principal polarization variation). For weak scintillation, the output is near receiver noise, and no correlation is evident. After 7 min, the scintillation is much stronger, and a low-level variation is evident in the orthogonal channel data. This residual output is, however, highly correlated with the scintillation in the principal polarization channel. The data therefore show no uncorrelated fluctuations.

These observations show that the fluctuations on both polarizations are correlated, and polarization diversity systems will not be useful in combating ionospheric scintillation at uhf. If significant uncorrelated orthogonal polarization fluctuations were present, they would have a σ_χ value near those in Fig. 3 which obviously were not observed in the data presented in Fig. 12, even for the strongest scintillation levels. Similar conclusions have been drawn by Whitney and Ring[29] from observations made at 137 MHz for scintillation levels below the strong scintillation limit and by Koster[13] for strong scintillation at 137 MHz in the equatorial region.

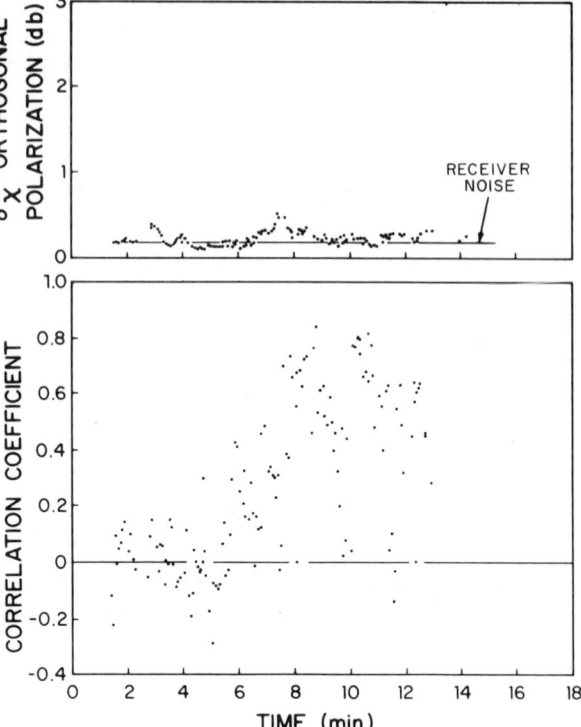

Fig. 12 Rms variation in the output from the orthogonally polarized channel and correlation coefficients between the principal and orthogonal channel outputs for pass of object 3133 rising at 0411 GMT on Aug. 4, 1972 ($K_p = 8^+$).

Observations at frequencies lower than 54 MHz show that orthogonal circular polarization channels may fade independently,[30] and diversity is possible. At frequencies below about 50 MHz, the ordinary and extraordinary rays may be separated by more than the radius of the first Fresnel zone, and the electron density fluctuations causing scintillation will not be correlated for two orthogonal polarizations.[31] Sufficient separation between ordinary and extraordinary ray paths for the fluctuations to become independent is not possible at frequencies above 100 MHz.

Carrier Frequency Dependence of the
uhf and vhf Observations

Frequency Dependence of the Scintillation Index

The data just presented are provided to illustrate the characteristics of scintillation as observed at uhf and vhf.

These data can be used to provide information about scintillation at other locations, frequencies, and path geometries only if a model is available for their interpretation. In the limit of weak scintillation, the available models discussed all relate the power spectrum of amplitude (or log amplitude) as observed on the ground to the power spectrum of the electron density fluctuations modified by the effects of the scattering process (Fresnel filtering). The power spectrum of temporal changes observed on a line-of-sight path to a low-orbiting satellite may be related to the power spectrum of spatial fluctuations of electron density by assuming that the electron density fluctuations do not change during the time the line of sight sweeps through the disturbed region of the ionosphere.

Early models of scintillation assumed that the spatial correlation function for electron density fluctuations was approximately gaussian, with different scale sizes along and perpendicular to the magnetic field lines.[12] The gaussian model implies a gaussian power spectrum, with power spectral densities decreasing rapidly for spatial frequencies above the reciprocal of the scale size. In the limit of high frequencies corresponding to spatial frequencies larger than the reciprocal of the Fresnel zone size, the spectrum observed on the ground should be identical to the two-dimensional spectrum of electron density fluctuations observed in a plane normal to the direction of the propagation path. For a gaussian model, the high-frequency limit should have a parabolic shape with increasing negative slopes for increasing frequency when observed for weak scintillation. The dot-dashed curve in Fig. 7 has sufficient signal-variation-to-noise-variance ratio to show the slopes of the spectrum to be nearly linear in shape for high frequencies. The shape is indicative of a power law power spectrum rather than a gaussian power spectrum.[25]

Power spectra at both uhf and vhf with a reasonably high signal-variance-to-noise-variance ratio and for weak scintillation are shown in Fig. 13. These spectra are for the same 1-min observation periods and for a time period when the individual rms log amplitude values for each analysis interval changed little at both frequencies. (The process was nearly stationary.) Both spectra show power law high-frequency regions. The scintillation models predict that the power spectral density values should increase by the square of the ratio of the carrier wavelength. The best-fit straight lines to both observed spectra having a wavelength-squared separation are shown on the figure. The slope of these lines is approximately -3. This corresponds to a power law dependence for the three-deminsional spatial power spectrum of electron density fluctuations with an index of 4 ($S \propto k^{-p}$, S = power spectral density, k = wave number, p = index) and a one-dimensional spectrum with an index of 2. The data presented in Fig. 13 are for observations to the northwest at an elevation angle of 18°. For the satellite receiver station geometry, the Fresnel zone size at a height of 300 km was 0.7 km at 400 MHz and 1.1 km at 150 MHz. The ray moved at approximately 1.0 km/sec through the ionosphere (velocity perpendicular to the line of sight at 300 km). The frequency at which each of the spectra flatten (1.5 Hz at 400 MHz; 0.9 Hz at 150 MHz) is approximately the ratio of the ray motion to Fresnel zone size, and higher frequencies correspond to scale sizes smaller than the Fresnel zone size.

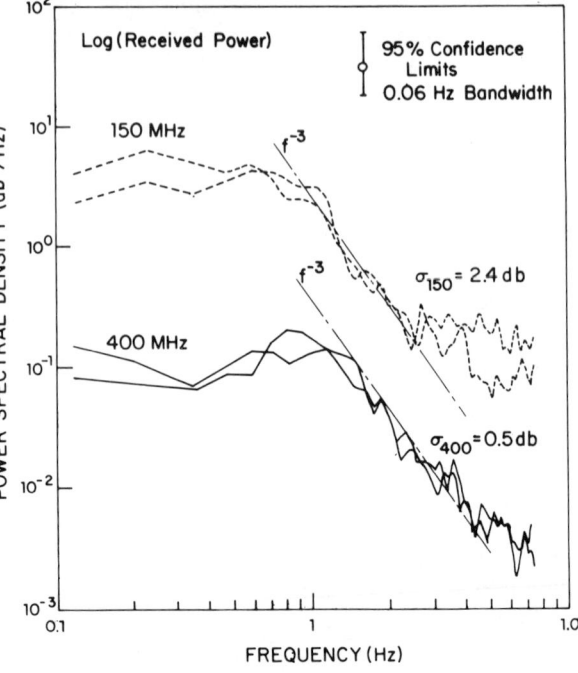

Fig. 13 Power spectra for log received power fluctuations at both uhf and vhf for the same 1-min observation period of the pass of object 3133, Aug. 5, 1972 at 0334 GMT (Kp = 8+).

The weak scintillation theory for a power law spectrum predicts a scintillation index (S_4 or σ_χ) frequency dependence given by[32]

$$(\sigma_{\chi_1}/\sigma_{\chi_2}) \propto (\lambda_1/\lambda_2)^{(p+2)/4} = (\lambda_1/\lambda_2)^\eta$$

where λ is wavelength, the subscripts refer to the carrier frequencies, and η is the spectral index. For a three-dimensional power law index of 4, the spectral index is 1.5. Using this spectral index, σ_χ at 150 MHz should be 4.4 times σ_χ at 400 MHz. For the data displayed in Fig. 13, σ_χ at 400 MHz was 0.5 db, σ_χ at 150 MHz was 2.4 db, and the predicted value using

$\eta = 1.5$ is 2.3 db, well within the measurement error of the observed value. The relationship between σ_χ at 150 MHz and σ_χ at 400 MHz for a pass with reasonably high signal-variance-to-noise-variance ratios is shown on Fig. 14. The weak scintillation limit curve corresponding to a spectral index of 1.5 is shown, together with the strong scintillation limit. The data appear to lie along the weak scintillation estimate curve until the strong scintillation limit is reached, and then σ_χ remains at the latter value.

The frequency dependence of ionospheric scintillation has received considerable attention in the literature. The thin-phase-screen, gaussian correlation function model predicted a spectral index of two

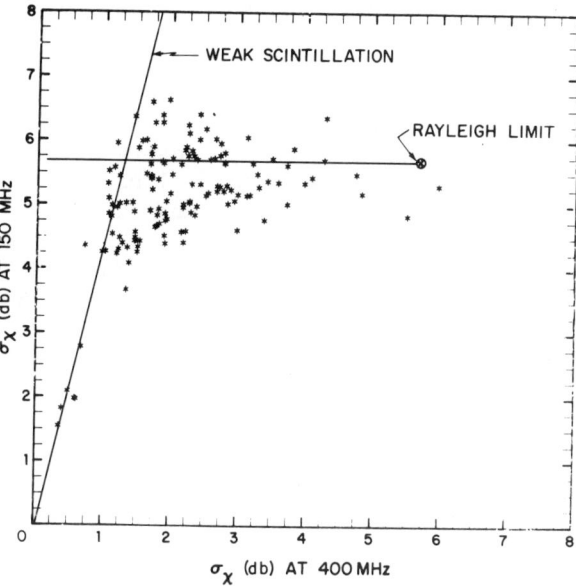

Fig. 14 Rms variation of log received power at vhf vs simultaneously observed value at uhf from pass of object 3133 rising at 2342 GMT on Aug. 4, 1972 ($Kp = 8^+$).

when the scale size was larger than the first Fresnel zone radius (near-field limit) and of one when the scale size was smaller (far-field limit).[12] The power law power spectra model predicts a single value for all cases (provided the power law holds over all scale sizes). Experimental verification of the frequency dependence predictions is difficult because the scintillation index must be less than the saturation limit value at both the high and low frequencies. For strong scintillation at both frequencies, the empirically determined spectral index would be zero. For a random selection of observations, the empirical spectral index should lie between 0 and 1.5. Signal-to-noise and measurement dynamic range problems inherent in many of the early measurements could further degrade the estimates of spectral index. From Fig. 3 for the period between 2 and 5 min, the ratio of σ_χ values is approximately 10, implying a spectral index of 2.3.

As shown in Fig. 8, the lower frequency was contaminated by receiver noise, causing a fictitiously high spectral index estimate.

Simultaneous observations of radio star scintillation reported by Basu et al.[33] showed a spectral index value of approximately 1.5 for frequencies at 112 and 224 MHz and a lower value for the 63-, 112-MHz frequency pair. The lower value for the lower frequency presumably is caused by strong scintillation data. These observations were made at the Sagamore Hill Radio Observatory, less than 100 km from Millstone. Aarons[34] reported observations at frequencies between 22 and 39 MHz from Arecibo, Puerto Rico under weak scintillation conditions and found a median spectral index of 1.6. Amplitude scintillations observed by Lawrence et al.[4] at 53 and 108 MHz show, for S_1 values at 53 MHz below 0.01 ($\sim S_4 = 0.7$), a median spectral index value of 1.5. Early observations reported by Chivers[35] made at Jodrell Bank Experimental Station at several frequencies between 36 and 408 MHz using radio stars had mean spectral indices ranging from 1.9 to 2.1. Observations also made at Jodrell Bank at 79 and 1390 MHz and reported by Chivers and Davies[36] showed, in the limit of weak scatter at 79 MHz and just detectable fluctuations at 1390 MHz, a value of η near 1.5.

Aarons et al.[37] and Allen[38] argue, for observations made above 63 MHz from the Sagamore Hill Radio Observatory, that a spectral index of 2.0 best fits their data in the limit of weak scatter, although the plots of scintillation index vs η curve have a median value of 1.6 as long as the scintillation index at 113 MHz is between 5 and 25% (S_4 at 63 MHz less than 0.6 and measurable scintillation at 113 MHz). Recently reported observations from the same observatory at frequencies of 137 and 412 MHz by Whitney et al.[23] show, for similar weak but measurable scintillation conditions, a distribution of η values between 1.2 and 1.8, with a mean value of 1.49 ± 0.05.

Only a limited number of spectral index estimations have been made for equatorial regions. Blank and Golden[39] report $\eta \sim 0.2$ to 0.3 for frequencies between 137 and 400 MHz. They comment that the results pertain, in part, to strong scatter. Taur[40] reported a value of $\eta \sim 2$ for equatorial measurements at 4 and 6 GHz. The latter observations were for weak scintillation and may indicate either the existence of an inner scale where the power law region ends and a different power spectrum shape occurs or the problems of detecting small fluctuations in noise.

Correlation of Amplitude Fluctuations at Two Frequencies

The rms amplitude fluctuations decrease with increasing frequency for weak scatter. They are also correlated over a wide frequency range. Calculations of the expected correlation coefficient for power law indices ranging from 3 to 4 are shown in Fig. 15. Similar calculations for gaussian spectra have been made by Budden.[41] Since the available power spectra and frequency dependence data support the power law model, the gaussian model predictions will not be presented. Measured correlation coefficients for the Millstone data and from available papers also are

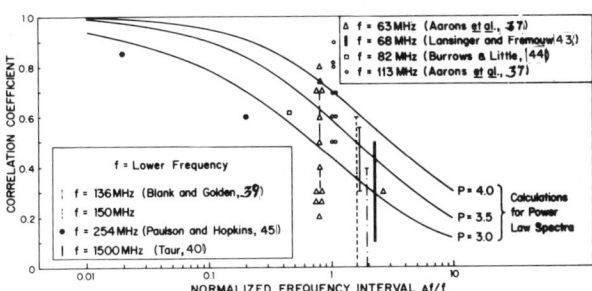

Fig. 15 Two frequency correlation functions for amplitude fluctuations.

presented. The p = 4 curve provides the best estimate based upon the preceding analysis and should represent the upper bound for measurements for each Δf value. The lower frequencies used in the two frequency observations are listed in the figure. For strong scintillation, the correlation coefficient should be lower than the calculated values. For low-frequency observations, the first Fresnel zones at each frequency may not overlap because of different ionospheric refraction at each frequency. For an elevation angle of 10°, rays at 150 and 400 MHz are separated by more than 5 km at a height of 300 km. The Millstone observations depicted in Fig. 1 show no correlation between the two frequencies. Since the elevation angles are less than 10° for the data depicted in Fig. 1 and the first Fresnel zone radii are less than 1.3 km, no correlation was expected.

The data displayed on Fig. 15 show reasonable agreement with the calculations. The data reported by Taur[40] are the only data for weak scintillation only. These data, with a lower frequency at 1550 MHz for an equatorial station, show better agreement for p between 3 and 3.5 than at 4. This again may indicate a change in the power spectrum of electron density fluctuations at small scale sizes in the equatorial region. The observations just reported for auroral and mid-latitude sites are all in reasonable agreement with a power law power spectrum model with an index of 4 for the three-dimensional fluctuations of electron density. The situation for equatorial regions is not as convincing, and more data are required. It is, however, evident that the correlation coefficients are reasonably high over a wide frequency range. This implies that extremely wide frequency separations ($\sim$ 2 to 1) are required to provide adequate frequency diversity operation, and frequency diversity is not useful in combating the effects of scintillation.

Dependence of Scintillation on Geophysical Parameters

Morphological studies of the dependence of scintillation on geophysical parameters have shown that scintillation is most severe and prevelent in and north of the auroral zone and near the geomagnetic equator (equatorial region within $\pm 15°$ to $20°$ of the geomagnetic equator). These studies of available scintillation observations show that the severity of scintillation also depends upon the elevation angle of the line of sight to the satellite, time of day, season of the year, the degree of magnetic disturbance, and sunspot number. In most of the observations, the exact nature of the dependence could not be determined because of the limited duration of the observations and possible correlations between some of the geophysical and geographical parameters. For observations of a radio star, the elevation angle and time of day are correlated for each season, and the diurnal elevation angle dependences cannot be separated. For observations of satellites in low circumpolar orbits, the effects of elevation angle and geomagnetic latitude cannot be separated without making simultaneous observations from a number of locations.[42]

Scintillations may be caused by electron density fluctuations anywhere in the ionosphere. Early studies showed strong correlation with the occurrence of spread-F and very weak correlation with sporadic-E and spread-E. This implies that the majority of scintillation occurrences are caused by F-region irregularities. Studies of the heights of regions causing scintillation deduced from the rates of change of fluctuations observed using low orbiting satellites or from simultaneous observations at more than one station on the ground show that the electron density irregularities may occur over a wide range of heights, 200-600 km, both in the auroral and equatorial regions.[46-48] The radar data of Pomalaza et al.[15] show that the irregularity region is usually several hundred kilometers thick.

Using the Rytov method to analyze weak scintillation due to a thick irregularity region, the variance of the log amplitude (σ_χ^2) is proportional to the integral of the product of the variance of the electron density fluctuations (intensity) and distance from their regularities to the receiver over the thickness of the irregularity region for plane waves incident on the ionosphere and a three-dimenional power law power spectrum with an index p = 4. The scintillation index, therefore, depends both upon the square root of the extent of the irregularity region along the line-of-sight path and upon the square root of the distance from the receiver to the irregularity region. Both the distance to the irregularity region and the extent of the region along the ray vary with elevation angle. If the size intensity and location of the irregularity region are known, the scintillation index may be computed from the model for any combination of frequency and path geometry for σ_χ < 2-3 db (weak scintillation). It is difficult, however, to deduct the intensity of the irregularities from observations using low orbiting satellites because the scintillation index value changes may be due to a change in size, a change of intensity, or a change in elevation angle.

The Millstone data for the satellite pass reported in Figs. 3 and 10 were replotted vs the invariant latitude of the line of sight to the satellite at a height of 300 km and are presented in Fig. 16. These data show a relatively rapid

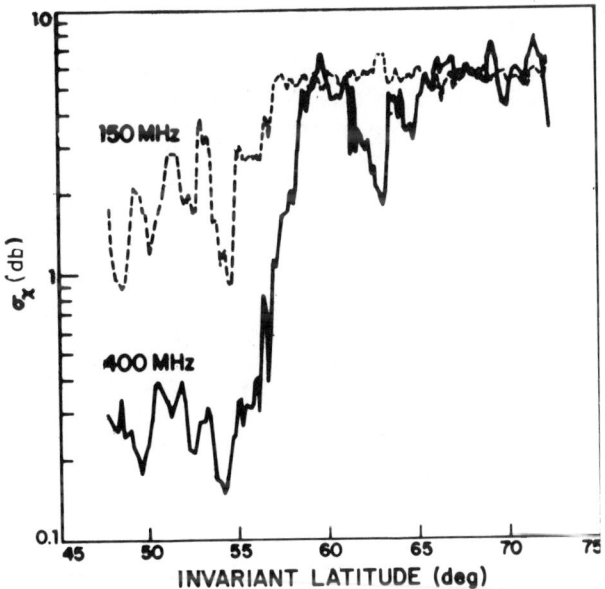

Fig. 16 Latitude dependence of scintillation for pass of object 3133 rising at 0411 GMT on Aug. 4, 1972 (Kp = 8+).

change from weak to strong scintillation as the variant latitude Λ changes from 56° to 59°. This sudden change may be described as a scintillation boundary in analogy with the boundary observed by Aarons et al.[3] The scintillation boundary considered by Aarons et al.[3] is defined as the latitude where their extreme value scintillation index SI exceeds 50% at 40 MHz. This translates to an S_4 value of 0.02 (σ_χ = 0.08 db) at 400 MHz, using spectral index of 1.5 and the relationship between SI and S_4 reported by Bischoff and Chytil.[21] The 400-MHz observations made at Millstone do not have sufficient sensitivity to observe the boundary defined at 40 MHz, although an apparent boundary sometimes is evident in the data. The abrupt changes in σ_χ and the apparent differences in σ_χ values to the north and south of Millstone (Λ = 56°) shown in Fig. 16 indicate changes in either the intensity or extent of electron density fluctuations with invariant latitude, the elevation angles being approximately the same for latitudes equispaced above and below 57° for the pass depicted in the figure. The data reported in Figs. 1 and 9 correspond to 51° and 64° for quiet and disturbed conditions to the north and south of the boundary, respectively.

Auroral and midlatitude scintillation regions are separated by the scintillation boundary. The position of the boundary changes with time of day, magnetic activity, and possibly sunspot number. Aarons[49] reported that the position of the boundary defined using the mean SI values for a number of observations ranged from 54° to 76°, depending upon time of day and magnetic activity. The diurnal variations of S_4 at 400 MHz tabulated from Millstone data for detrended 3-sec observation periods[50] are shown in Fig. 17. Two invariant latitude bands were analyzed, 44°-46°, corresponding to positions always within the midlatitude region (south of the boundary), and 64°-66°, corresponding to locations within the nighttime auroral region. Data from a total of 2471 passes were analyzed. For the two invariant latitude bands used, the elevation angles ranged from 3° to 14°. The elevation angle values for each pass, however, generally were different in each band. Over the range of elevation angles, the S_4 or σ_χ value may change by a factor of 2. The elevation angles should have nearly the same occurrence distribution for each band, and no elevation angle bias is expected.

The occurrence percentages of S_4 values above 0.2, 0.4, 0.6, and 0.8 (σ_χ values of 0.9, 1.8, 2.8, and 4.2 db, respectively) were chosen to characterize the observations. For $S_4 > 0.8$, the 64°-66° data show a morning minimum and an afternoon maximum. Data reported by Aarons[49] for Narssarssuaq hav-

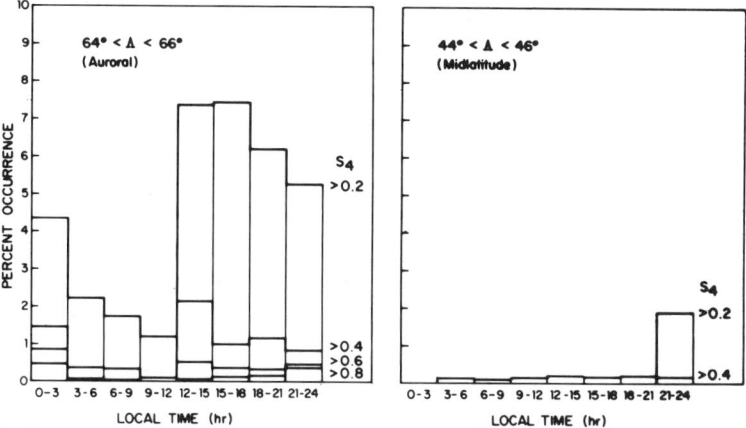

Fig. 17 Diurnal variation of scintillation at 400 MHz for auroral and midlatitudes for 2471 satellite passes observed between January 1971 and March 1973.

ing an invariant latitude of 63° and an elevation angle of 18° show, at 136 MHz, a morning minimum in mean SI and a nighttime maximum occurring between 2100 and 0500 hr. The Millstone afternoon maximum is not evident in the 136-MHz data. Aarons et al.[3] reported occurrence percentages for SI > 60% at 136 MHz (equivalent to $S_4 > 0.15$ at 400 MHz) which displayed a morning minimum and a nighttime maximum. Although the Narssarssuaq data are for nearly the same invariant latitudes as the Millstone data and the elevation angles differ slightly, the latter data show an afternoon maximum not present in the former. The Millstone data are for observations to the north, the Narssarssuaq data for observations toward the south, and differences in propagation direction relative to the field lines might be important. The 113-MHz data reported by Aarons[49] for radio star observations toward the north at an invariant latitude of 66° at the same elevation angles as for Millstone show relatively higher mean SI values in the afternoon than in the morning, displaying the same general trend as the Millstone data. The 113-MHz data also showed a lower mean SI value averaged over all times of day than the 136-MHz data. The lower SI values are presumably due to differences in sunspot number; the 113-MHz data were taken at the minimum of the sunspot cycle and the 136-MHz and Millstone data near the maximum. The shape of the diurnal variation curves may be affected by differences in the intensity and extent of the irregularities along and normal to the field lines.

The midlatitude region observations show significantly smaller percentages of occurrence for each of the 3-hr intervals. The number of satellite passes associated with the observed occurrences with $S_4 > 0.2$ ranges between 0 and 2 for all intervals except 2100-2400 hr, implying that sufficient data were available to deduce a diurnal trend. Midlatitude observations by Aarons et al.[51] at 54 MHz show mean SI values to have relative maxima at noon and midnight. Equatorial region observations show scintillation to be a nighttime phenomenon, with occurrences rare between sunrise and sunset. Observations reported by Aarons et al.[4] for Huancayo, Peru (equatorial) show that at 136 MHz the scintillations were most prevalent at 2200 hr local time, with a 60% occurrence for SI > 60. This translates to a 60% occurrence of $S_4 > 0.15$ at 400 MHz. The data in Fig. 17 show less than 10% occurrence for $S_4 > 0.2$, indicating that equatorial scintillations tend to be more prevalent than auroral zone scintillations.

The seasonal dependence of scintillation is shown for the Millstone midlatitude and auroral region observations in Fig. 18. These data show a spring minimum and a fall maximum for the auroral data at $S_4 > 0.2$. The August peak is caused by a 3-day interval associated with the highest magnetic activity indices observed during the January 1971 to March 1973 time period. The August event biases the observations. The seasonal variation of mean SI at 54 MHz reported by

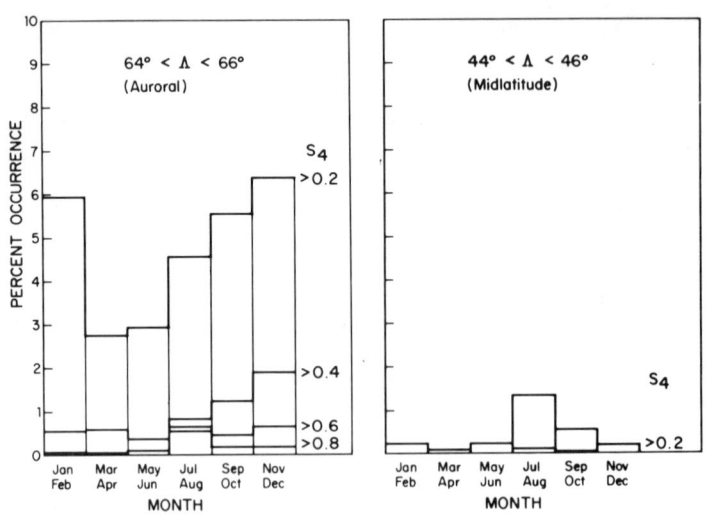

Fig. 18 Seasonal variation of scintillation at 400 MHz for auroral and midlatitudes for 2471 satellite passes observed between January 1971 and March 1973.

Aarons et al.[51] shows a minimum in the number of occurrences in winter as compared to other seasons for all observations to the north and south of the Sagamore Hill Observatory. The 54-MHz data are, however, for the minimum of the sunspot cycle. The data, therefore, show little real seasonal variation. Equatorial region observations tend to show relatively higher percentage occurrences at the times of the equinoxes and a minimum at the northern solstice.

Observations of scintillation in the auroral region generally show a correlation between mean SI values and magnetic activity Kp. Figure 19 shows the increase in the occurrence of scintillation with increasing K_p for the auroral region. Little change in percentage occurrences of $S_c > 0.2$ is noted for the midlatitude region. Aarons et al.[51] noted that the position of the scintillation boundary is correlated with Kp, moving south as Kp increases. In contrast to the apparent dependence of scintillation on magnetic activity in the auroral region, Koster[8] notes that scintillation in the equatorial region is suppressed during periods of enhanced magnetic activity. These results may vary with the sunspot cycle, but insufficient data are available to test dependences on sunspot numbers. Available long-term observations of scintillation over a sig-

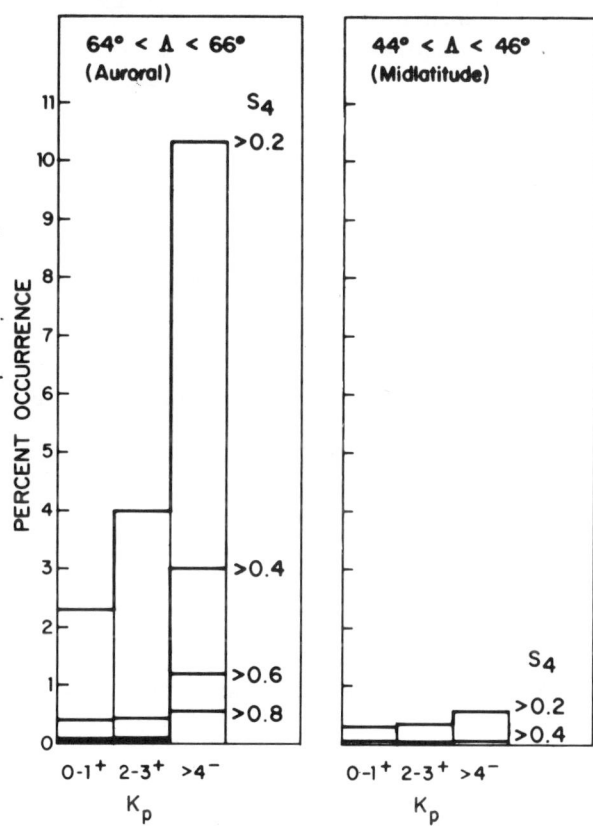

Fig. 19 Magnetic activity dependence of scintillation at 400 MHz for auroral and midlatitude for 2471 satellite passes observed between January 1971 and March 1973.

nificant part of the sunspot cycle have shown only that scintillation tends to be both more prevalent and more intense on the average during years of high sunspot number.

Conclusions

The statistical summary of the Millstone data shows a clear dependence upon magnetic activity, although diurnal and seasonal variations are not well defined. The data are, however, for a relatively short time period when compared with the sunspot cycle and are, therefore, not complete enough to provide an empirical model for the prediction of scintillation statistics for use in communication system design. To provide the required statistics, data from a large number of observations spread over at least half a sunspot cycle are required. Other available observations generally have been made at frequencies below 137 MHz and are useful only for the prediction of scintillation at lower frequencies due to the limiting effect of strong scintillation.

The auroral region observations clearly show that scintillation will affect system design at frequencies up to at least 400 MHz. The equatorial region observations of Taur[9] show that frequencies up to 6 GHz are affected. Since scintillation is a fact of life at uhf and vhf, what methods can be used to mitigate its effects? Although the data are somewhat inadequate, some recommendations can be made. First, we make some negative recommendations. The two-frequency correlation function analysis showed that scintillation is a wideband phenomenon and frequency diversity is not practical. The cross-polarization observations showed that scintillations on the principal and orthoganol circular polarizations were correlated, implying that polarization diversity will not work at uhf. Although these analyses and measurements were performed for auroral region observations, we believe that the results also apply in the equatorial region. For the auroral region, the weak scintillation data showed power law power spectra for electron density irregularities. Preliminary analysis of 254-MHz data reported for the equatorial region[45] also show power law spectra (with a three-dimensional index $p = 4$). This suggests that, although the mechanisms causing equatorial region irregularities may be different from those active in the auroral region, the resultant effects of the irregularities on propagation are the same, and the results obtained from the Millstone data apply in the equatorial region.

One possible solution is space diversity. The Millstone observations were taken using a low orbiting satellite. The temporal variation of the signal level is caused by the motion of the line of sight through the irregularities. The satellite motion is known, and hence the temporal variations may be interpreted in terms of the spatial variation of electron density. The frequency scale on Fig. 13 may be interpreted as a wave-number scale with, for the satellite receiver geometry pertinent to Fig. 13, the 1-Hz value equivalent to 2π km^{-1} (scale size of 1 km). Since the width of each power spectrum is approximately the reciprocal of the correlation distance, the correlation distance is approximately 0.5 km at uhf and 0.8 km at vhf. This implies that the scintillation observed at receivers spaced by more than the correlation distance ($\sim$ Fresnel zone size) should be uncorrelated and could be combined to get diversity. For strong scintillation, the power spectrum broadened, implying that smaller diversity distances may be useful in that limit.

Another possible solution is time diversity. The fade duration statistics reported here reflect the motion of the line of sight through the medium, and, for weak scintillation, the duration values should be inversely proportional to the translation velocity of the ray at the height of the irregularities. For geostationary satellites, the line of sight is fixed, and the irregularities drift by. The fade duration values then should be inversely proportional to the drift rate. Using a 100-m/sec drift rate as typical,[4] the fade durations should be an order of magnitude larger than those shown in Fig. 6. For strong scintillation, the fade durations should be shorter than for weak scintillation, as shown in the discussion of Fig. 6. The power spectra shown in Fig. 13 imply a correlation time the order of 0.5 sec at uhf and 0.8 sec at vhf. As with the fade duration values, the correlation time is dependent upon the velocity of the irregularities perpendicular to the line of sight. Since the signal becomes uncorrelated in time, time diversity also is possible. For geostationary satellites, correlation time depends both on the Fresnel zone size and the drift velocity of the irregularities. The latter is a random variable that has to be observed at many locations over a long period of time to provide an adequate statistical description. Since the fade rate or correlation time is different for strong scintillation, observations must be made in the weak scintillation regime to define drift velocities.

Recommendations

The only solutions we have recommended are the use of space or time diversity. Some guidelines for their application may be obtained from Millstone and other available data.

However, to optimize system design, more information is required. We recommend the following:

1) Additional available weak scintillation data should be processed to determine the structure of the power spectrum to establish if the power law form represents all observations.

2) Available weak scintillation data from geostationary satellite observations should be analyzed to determine drift rate statistics.

3) New observations using both low orbiting and geostationary satellites with several coherently related carrier frequencies widely spread in the uhf, shf band at and above the frequencies of interest should be made to provide adequate statistics of amplitude, phase, and drift velocity. The frequencies should be chosen such that weak scintillation always is observed at one of the frequencies.

References

[1] Little, C. G. and Maxwell, A., "Fluctuations in the Intensity of Radio Waves from Galactic Sources," Philosophical Magazine, Vol.42, No. 338, March 1951, pp. 267-278.

[2] Whitney, H. E., Aarons, J., and Malik, C., "A Proposed Index for Measuring Ionospheric Scintillations," Planetary and Space Science, Vol. 17, No. 5, May 1969, pp. 1069-1073.

[3] Aarons, J., Whitney, E., and Allen, R. S., "Global Morphology of Ionospheric Scintillations," Proceedings of the IEEE, Vol. 59, No. 2, Feb. 1971, pp. 159-172.

[4] Lawrence, R. S., Jespersen, J. L., and Lamb, R. C., "Amplitude and Angular Scintillations of the Radio Source Cygnus-A Observed at Boulder, Colorado," Journal of Research NBS, Vol. 65D, No. 4, July/Aug. 1961, pp. 330-350.

[5] Kent, G. S., "High Frequency Fading of the 108 Mc/s Wave Radiated From an Artificial Earth Satellite as Observed at an Equatorial Station," Journal of Atmospheric and Terrestrial Physics, Vol. 22, No. 3, March 1961, pp. 255-269.

[6] Little, C. G., Reid, G. F., Stiltner, E., and Merritt, R. P., "An Experimental Investigation of the Scintillation of Radio Stars Observed at Frequencies of 223 and 456 Megacycles per Second from a Location Close to the Auroral Zone," Journal of Geophysical Research, Vol. 67, No. 5, May 1962, pp. 1763-1784.

[7] Fremouw, E. J. and Bates, H. F., "Worldwide Behavior of Average VHF-UHF Scintillation," Radio Sciences, Vol. 6, No. 10, Oct. 1971, pp. 863-869.

[8] Koster, J. R., "Equatorial Scintillation," Planetary and Space Science, Vol. 20, No. 11, Nov. 1972, pp. 1999-2014.

[9] Taur, R. R., "Ionospheric Scintillation at 4 and 6 GHz," COMSAT Technical Review, Vol. 3, No. 1, Spring 1973a, pp. 145-163.

[10] Booker, H. G., Ratcliffe, J. A., and Shinn, D. H., "Diffraction from an Ionospheric Screen with Application to Ionospheric Problems," Philosophical Transactions of the Royal Society, Vol. A242, 1950, pp. 579-607.

[11] Mercier, R. P., "Diffraction by a Screen Causing Large Random Phase Fluctuations," Proceedings of the Cambridge Philosophical Society, Vol. 58, 1962, pp. 382-400.

[12] Briggs, B. H. and Parkin, I. A., "On the Variation of Radio Star and Satellite Scintillations with Zenith Angle," Journal of Atmospheric and Terrestrial Physics, Vol. 25, No. 6, June 1963, pp. 339-365.

[13] Koster, J. R., "Ionospheric Studies Using the Tracking Beacon on the 'Early Bird' Synchronous Satellite," Annales de Geophysique, Vol. 22, No. 3, 1966, pp. 435-439.

[14] Pomalaza, J., Woodman, R. F., Tisnado, G., Sandoval, J., and Guillev, A., "A Progress Report on Scintillation Observations at Ancon and Jicamarco Observations," NASA/GSFC X-520-70-398, Oct. 1970, NASA Goddard Space Flight Center, Greenbelt, Md.

[15] Pomalaza, J., Woodman, R. F., Tisnado, J., and Nakasone, E., "Study of Equatorial Scintillations, A Progress Report," NASA/GSFC X-750-73-244, Dec. 1972, NASA Goddard Space Flight Center, Greenbelt, MD.

[16] Budden, K. G., "The Amplitude Fluctuations of the Radio Wave Scattered from a Thick Ionospheric Layer with Weak Irregularities," Journal of Atmospheric and Terrestrial Physics, Vol. 27, No. 2, Feb. 1965, pp. 155-172.

[17] Tatarskii, V. I., The Effects of the Turbulent Atmosphere on Wave Propagation, Nauka, Moscow, 1967; translation available from U.S. Dept. of Commerce, National Technical Information Service, Springfield, Va.

[18] Gilhoni, J. C., ed., "Millstone Radar Propagation Study: Radar Instrumentation," TR-507, 1973, Lincoln Laboratory, Massachusetts Institute of Technology, Lexington, Mass.

[19] Ochs, G. R. and Lawrence, R. S., "Saturation of Laser-Beam Scintillation Under Conditions of Strong Atmospheric Turbulence," Journal of the Optical Society of America, Vol. 59, No. 2, Feb. 1969, pp. 226-227.

[20] Dunphy, J. R. and Kerr, J. R., "Scintillation Measurements for Large Integrated-Path Turbulence," Journal of the Optical Society of America, Vol. 63, No. 8, Aug. 1973, pp. 981-986.

[21] Bischoff, K. and Chytil, B., "A Note on Scintillation Indices," Planetary and Space Science, Vol. 17, No. 5, Nov. 1969, pp. 1059-1066.

[22] Rino, C. L. and Fremouw, E. J., "Statistics for Ionospherically Diffracted UHF/VHF Signals," Radio Science, Vol. 8, No. 3, March 1973, pp. 223-233.

[23] Whitney, H. E., Aarons, J., Allen, R. S., and Seemann, D. R., "Estimation of the Cumulative Amplitude Probability Distribution Function of Ionospheric Scintillation," Radio Science, Vol. 7, No. 12, Dec. 1972, pp. 1095-1104.

[24] Nakagami, M., "The m-distribution--A General Formula of Intensity Distribution of Rapid Fading," Statistical Methods on Radio Wave Propagation, edited by W. G. Hoffman, Pergamon, New York, 1960.

[25] Rufenach, C. L., "Power-Law Wavenumber Spectrum Deduced from Ionospheric Scintillation Observations," Journal of Geophysical Research, Vol. 77, No. 24, Sept. 1972, pp. 4761-4772.

[26] Elkins, T. J. and Papagiannis, M. D., "Measurement and Interpretation of Power Spectrums of Ionospheric Scintillation at a Subauroral Location," Journal of Geophysical Research, Vol. 74, No. 16, Aug. 1969, pp. 4105-4115.

[27] Garriot, O. K. and daRosa, A. V., "Electron Content Obtained from Faraday Rotation with Phase Length Variations," Journal of Atmospheric and Terrestrial Physics, Vol. 32, No. 4, April 1970, pp. 705-727.

[28] Porcello, L. J. and Hughes, L. R., "Observed Fine Structure of a Phase Perturbation Induced During Transauroral Propagation," Journal of Geophysical Research, Vol. 73, No. 19, Oct. 1968, pp. 6337-6346.

[29] Whitney, H. E. and Ring, W. F., "Dependency of Scintillation Fading of Oppositely Polarized VHF Signals," *IEEE Transactions on Antennas and Propagation*, Vol. AP-19, No. 1, Jan. 1971, pp. 151.

[30] McClure, J. P., "Polarization Measurements During Scintillation of Radio Signals from Satellites," *Journal of Geophysical Research*, Vol. 69, No. 7, April 1964, pp. 1445-1447.

[31] Roger, R. S., "The Effect of Scintillations on the Polarization of Satellite Transmissions near 20 Mc/s," *Journal of Atmospheric and Terrestrial Physics*, Vol. 27, No. 3, May 1965, pp. 335-348.

[32] Young, A. T., "Interpretation of Interplanetary Scintillations," *Astrophysical Journal*, Vol. 168, No. 3, Pt. 1, Sept. 1971, pp. 543-562.

[33] Basu, S., Allen, R. S., and Aarons, J., "A Detailed Study of a Brief Period of Radio Star and Satellite Scintillations," *Journal of Atmospheric and Terrestrial Physics*, Vol. 26, No. 8, Aug. 1964, pp. 811-823.

[34] Aarons, J., "Ionospheric Irregularities at Arecibo, Puerto Rico," *Journal of Atmospheric and Terrestrial Physics*, Vol. 29, No. 12, Dec. 1967, pp. 1619-1624.

[35] Chivers, H. J. A., "The Simultaneous Observation of Radio Star Scintillations on Different Radio-Frequencies," *Journal of Atmospheric and Terrestrial Physics*, Vol. 17, No. 3, 1959/1960, pp. 181-187.

[36] Chivers, H. J. A. and Davies, R. D., "A Comparison of Radio Star Scintillations at 1390 and 79 Mc/s at Low Angles of Elevation," *Journal of Atmospheric and Terrestrial Physics*, Vol. 24, No. 7, July 1962, pp. 573-584.

[37] Aarons, J., Allen, R. S., and Elkins, T. J., "Frequency Dependence of Radio Star Scintillations, *Journal of Geophysical Research*, Vol. 72, No. 11, Jan. 1967, pp. 2891-2902.

[38] Allen, R. S., "Comparison of Scintillation Depths of Radio Star and Satellite Scintillations," *Journal of Atmospheric and Terrestrial Physics*, Vol. 31, No. 2, Feb. 1969, pp. 289-297.

[39] Blank, H. A. and Golden, T. S., "Analysis of VHF/UHF Frequency Dependence, Space, and Polarization Properties of Ionospheric Scintillation in the Equatorial Region," 1973 IEEE International Communications Conference Proceedings, June 1973, pp. 17-27 to 17-35.

[40] Taur, R. R., private communication, 1973.

[41] Budden, K. G., "The Theory of the Correlation of Amplitude Fluctuations of Radio Signals at Two Frequencies, Simultaneously Scattered by the Ionosphere," Journal of Atmospheric and Terrestrial Physics, Vol. 27, No. 8, Aug. 1965, pp. 883-897.

[42] Joint Satellite Studies Group, "On the Latitude Variation of Scintillations of Ionospheric Origin in Satellite Signals," Planetary and Space Science, Vol. 16, No. 6, June 1968, pp. 775-781.

[43] Lansinger, J. M. and Fremouw, E. J., "The Scale Size of Scintillation Producing Irregularities in the Auroral Ionosphere," Journal of Atmospheric and Terrestrial Physics, Vol. 29, No. 10, Oct. 1967, pp. 1229-1242.

[44] Burrows, K. and Little, C. G., "Simultaneous Observations of Radio Star Scintillations on Two Widely Spaced Frequencies," Jodrell Bank Annals, Vol. 1, No. 1, Jan. 1952, pp. 29-35.

[45] Paulson, M. R. and Hopkins, V. F., "Effects of Equatorial Scintillation Fading on Satcom Signals," NELC/TR 1875, May 1973, Naval Electronics Laboratory Center, San Diego, Calif.

[46] Frihagen, J. and Liszka, L., "A Study of Auroral Zone Ionospheric Irregularities Made Simultaneously at Tromso, Norway and Kirana, Sweden," Journal of Atmospheric and Terrestrial Physics, Vol. 27, No. 4, April 1965, pp. 513-523.

[47] Jesperson, J. L. and Kamas, G., "Satellite Scintillation Observations at Boulder, Colorado," Journal of Atmospheric and Terrestrial Physics, Vol. 26, No. 4, April 1964, pp. 457-473.

[48] Kelleher, R. F. and Sinclair, J., "Some Properties of Ionospheric Irregularities as Deduced from Recordings of the San Morco II and BE-B Satellites," Journal of Atmospheric and Terrestrial Physics, Vol. 32, No. 7, July 1970, pp. 1259-1271.

[49] Aarons, J., "A Descriptive Model of F-Layer High-Latitude Irregularities as Shown by Scintillation Observations," Journal of Geophysical Research, Vol. 78, No. 31, Nov. 1973, pp. 7441-7450.

[50] Evans, J. V., ed., "Millstone Radar Propagation Study: Scientific Results," TR-509, 1973, Lincoln Laboratory, Massachusetts Institute of Technology, Lexington, Mass.

[51] Aarons, J. A., Mullan, J., and Basu, S., "The Statistics of Satellite Scintillations at a Subauroral Latitude," Journal of Geophysical Research, Vol. 69, No. 9, May 1964, pp. 1785-1794.

GEOPHYSICAL PROPERTIES OF THE IONOSPHERIC
IRREGULARITIES RESPONSIBLE FOR RADIO SCINTILLATION

J. P. McClure[*]

University of Texas, Dallas, Texas

Abstract

The properties of F-region ionospheric irregularities are described based on in situ measurements of the actual waveforms of ion concentration N_i. The spectral properties of the irregularities are discussed. In high, middle, and low latitudes most of the irregularities observed fall into a single "noiselike" category having power spectra that can be approximated by f^{-n} or S^n, where S is the irregularity scale size and n is approximately 2. Thus the spectral components have a maximum gradient that is almost independent of their size. Other categories of irregularities also are observed occasionally. In midlatitudes, a "ground-glass" category exists which, in contrast with "noiselike" irregularities, has little spectral power for S greater than a few kilometers. Near the equator, "sinusoidal" irregularities with wavelengths from 1 to 20 km and amplitudes up to 50% sometimes are seen. The relationship between irregularity behavior and that of other geophysical parameters is discussed. Finally, some recently proposed possible irregularity-producing mechanisms consistent with all available evidence are reviewed.

Presented as Paper 74-53 at the AIAA 12th Aerospace Sciences Meeting, Jan. 30-Feb. 1, 1974, Washington, D. C. This research was supported by the National Science Foundation under Grant GA-31318 and by NASA under Grants NGL-44-004-026 and Contracts NAS 5-9311 and NAS 5-23184.
[*]Research Scientist.

I. Introduction

The purpose of this paper is to review some of the geophysical properties of the irregularities in ionospheric electron concentration N_e responsible for radio scintillation. Most of the data we discuss were obtained from the OGO 6 satellite, although we shall also draw on other sources of information. We shall pay particular attention to the class of irregularities having the largest amplitudes, those responsible for the most intense scintillation. For completeness, we also shall discuss several other categories of ionospheric irregularities which have been observed.

We also shall review some of the geophysical data relevant to the origin of the irregularities. We shall not discuss the complicated physics of the plasma instabilities which undoubtedly are present in the ionosphere when N_e is inhomogeneous, but we shall discuss some of the possible sources of the energy required to generate the inhomogeneities and some of the possible mechanisms that could couple this energy into the ionospheric plasma. There are several recent theories explaining how the largest-amplitude N_e irregularities are generated and why these largest irregularities exist only at night near the equator, whereas they exist at all local times in high latitudes. Perhaps the most interesting result reviewed here is a suggested mechanism that appears to explain successfully all of the observed features of equatorial irregularities, including the at-first-glance paradoxical new data from OGO 6 on the relationship between the presence of irregularities and Fe^+ ions in the equatorial F region.

II. Ionospheric N_e Irregularities

The OGO 6 satellite (perigee $\sim$ 400 km, apogee $\sim$ 1100 km, orbital inclination $\sim$ 82°) carried a set of instruments for ionospheric studies which included a retarding potential analyzer (RPA)[1] capable of measuring many important properties of ionospheric irregularities. The measurements made were of the ion gas only, not the electron gas responsible for propagation effects such as scintillation, but, because of charge neutrality considerations, N_e and the ion concentration N_i are virtually identical in the ionosphere. The instrument was used as a conventional RPA half the time and as an "irregularity detector" half the time. In the RPA mode, the ion temperature, ion composition, and, under favorable conditions (sufficient concentrations of appropriate ions), the "ram"

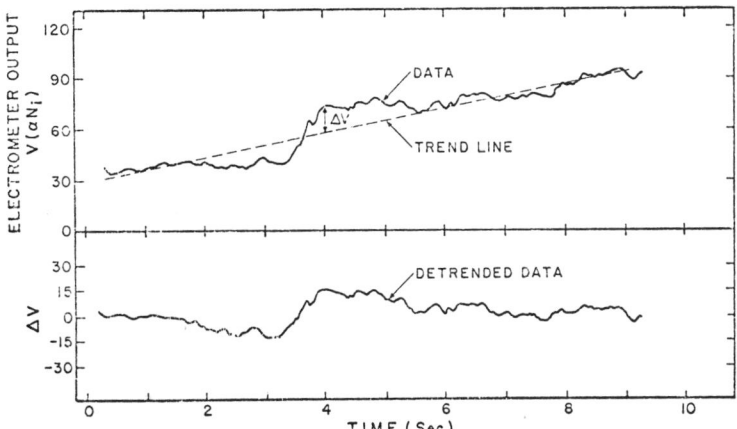

Fig. 1 Typical data on ionospheric irregularities obtained from the RPA on the OGO 6 satellite. By inspection, one can see that the longer wavelengths dominate over the smaller.

velocity v_r of the ions in the satellite frame of reference were obtained. In the irregularity mode, none of the ions intercepted at the forward-facing aperture of the instrument were retarded, and hence the electrometer output voltage was proportional to the ion current intercepted $N_i v_r$. Thus it was possible to correlate, for example, the presence or absence of ionospheric irregularities with changes in the ionic composition. This built-in flexibility of the instrument permitted the discovery of several previously unsuspected properties of the irregularities, as we shall discuss below.

The OGO 6 RPA had a variable telemetry rate. A sample of typical data obtained using the highest available resolution is shown in Fig. 1. One data point was obtained approximately every 35 m along the satellite path; 1 sec corresponds to a distance of approximately 8 km. These "typical" data are characterized by the appearance of their waveform, which, as a visual inspection of the detrended data shown in Fig. 1 reveals, is that of "pink" or "red" noise, where longer wavelengths dominate, rather than that of white noise.

Well over 90% of all F-region ionospheric irregularities have the red noise character of those shown in Fig. 1. Before proceeding in Sec. III to a detailed review of the spectral analysis of these typical irregularities, we shall review

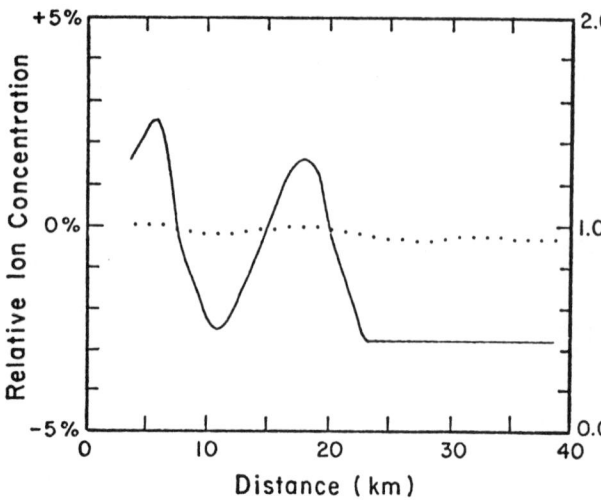

Fig. 2 Quasi-sinusoidal ionospheric irregularity waveforms, an equatorial phenomenon, obtained at 2116 hr LT on day 324, 1969, at -10° dipole lat, -62° long, 431-km alt. The solid line refers to the left-hand and the dots to the right-hand scale.

briefly some of the other categories of ionospheric waveforms which are observed occasionally. In Fig. 2, an example of "sinusoidal" waveforms is shown. The sinusoids have a peak-to-peak amplitude of ∼5% and a wavelength of ∼13 km. Amplitudes ranging from 0.03%, near the threshold of detectability of the instrument, to over 50% and wavelengths ranging from approximately 1 to 20 km have been observed.[2] Sinusoids are strictly an equatorial phenomenon. They never have been observed more than 30° of latitude away from the dip equator. On a very few occasions sinusoids of small but gradually increasing amplitude have been observed as the satellite approaches a region of strong equatorial irregularities having typical "red noise" waveforms.

Just as sinusoidal irregularities are limited to equatorial latitudes, the special subclass of irregularities illustrated in Fig. 3 is limited to middle latitudes. This subclass has been called "ground glass" because, in contrast with the more typical red noise irregularities, these irregularities invariably have small amplitudes (<1%), and almost all of the spectral power is contained in the smaller scale sizes.

In Fig. 4, we show an example of a subclass of irregularities called "breaking waves" because of their resemblance to ocean waves about to break on a beach. This type of irregularity has been observed at all latitudes.[2] The crest-to-trough amplitude for the example shown is approximately 10%, although these waveforms usually have smaller amplitudes. The waves shown in Fig. 4 and others not shown from the same equatorial pass change their sense at the time the satellite crosses the geomagnetic equator. This reversal phenomenon was observed only once.

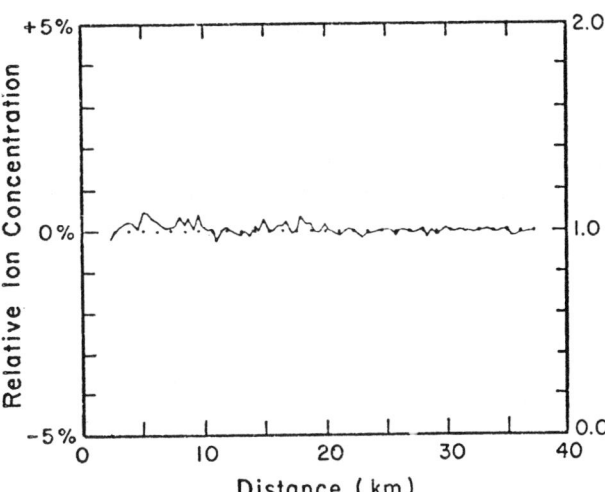

Fig. 3 "Ground-glass" irregularities, a mid-latitude phenomenon, observed at 2230 hr LT on day 322, 1969, at 60° dipole lat, -122° long, 503-km alt.

The most severe scintillation occurs at night near the equator. A sample of data showing large-amplitude N_i structure at night near the equator is shown in Fig. 5. The data points are spaced approximately 1.1 km apart, and at times N_i changes by nearly a factor of 2 in this distance. In other even more disturbed examples, N_i appears to change by an even greater amount in a similar distance. No such example is shown because in these cases the interpretation of the data becomes somewhat ambiguous: the linear automatic range-changing electrometer alters its sensitivity by a factor of $10^{1/2}$ whenever the output signal lies outside of certain

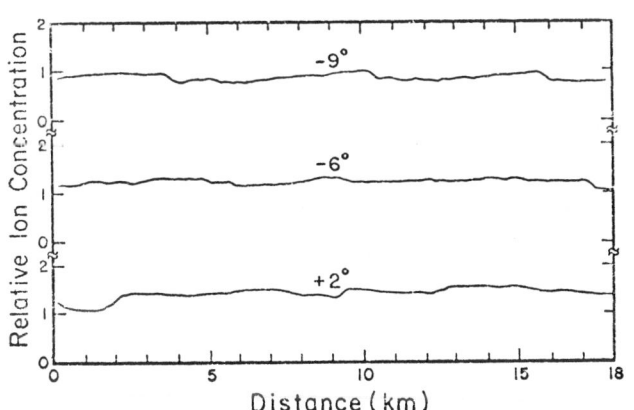

Fig. 4 Breaking wave irregularities, observed at all latitudes. Day 324, 1970, near 455-km alt, -77° long, 2207 hr LT.

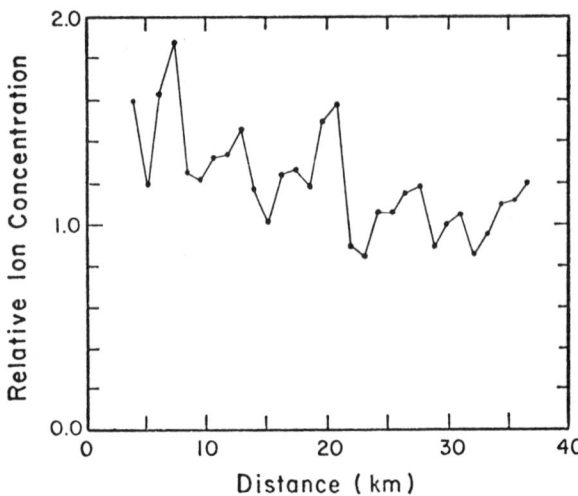

Fig. 5 Large-amplitude equatorial irregularities, observed at 2117 hr LT on day 325, 1969, at 0° dipole lat, -51° long, and 406-km alt.

predetermined limits. However, it is clear from the OGO 6 observations that an often-present feature of the equatorial ionosphere at night is structure having amplitudes as large as or larger than that shown in Fig. 5. In the high-latitude region, structure having similar percentage amplitudes is even more common (but the absolute amplitudes of ΔN_i are smaller because N_i is generally smaller in high than in low latitudes). It occurs almost invariably (occurrence probability ~100%) above a certain latitude regardless of the local time, whereas near the equator the occurrence pattern of the large-amplitude structure is much more complicated in that the occurrence is 1) greatest near the equinoxes; 2) greatest in the "Atlantic" longitude region[3]; 3) greatest in years of high solar activity; 4) inversely dependent on geomagnetic activity; and 5) greatest at night and nearly zero in the daytime. These morphological features are reviewed by Crane[4] and by others[5,6] and will not be discussed further here. We shall, however, examine part of one typical orbit of OGO 6 data here in order to illustrate some of the important features of the morphology of ionospheric irregularities discovered using this satellite.

Figure 6 shows the variations of $\Sigma (=\Delta N_{rms}/N)$ and some of the typical relationships between Σ and other parameters derived from the RPA. For clarity, only Σ values >0.5% are shown, although our threshold for Σ is much lower (0.01-0.03%). The center panel of the figure gives the flux of electrons of energy >10 ev, and the total ion concentration N_i is plotted in the lower panel. The lower scales give the satellite altitude, invariant latitude Λ, and magnetic dipole local time MLT. The MLT is defined as the angle between the magnetic dipole meridian of the satellite and that of the sun; Λ is defined as arc cos $1/\sqrt{L}$, where L is the MacIlwain parameter given approximately by the equatorial geocentric distance in Earth radii of a given geomagnetic field line. These parameters are useful for ordering high-latitude data that might be expected to be related to particle precipi-

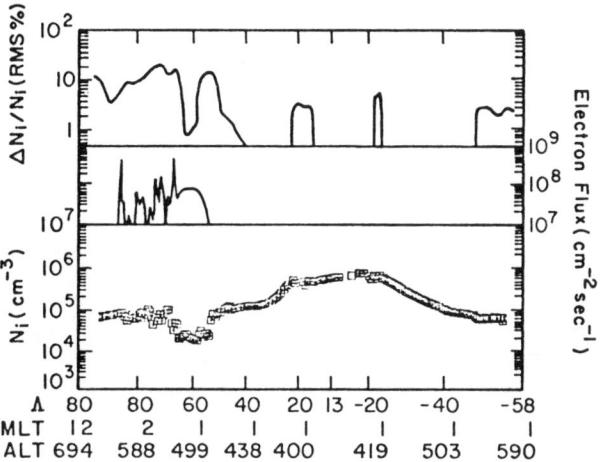

Fig. 6 Typical morphological relationships between the irregularity amplitude, the flux of electrons with energy >10 ev, and the ion concentration N_i (day 23, 1970, orbit 3347).

tation patterns. Near the equator, dip latitude and conventional local time LT are more useful (for geophysical reasons), but in that region LT and MLT are approximately equal, and the dip equator roughly coincides with the minimum value of Λ.

Figure 6 snows several regions of ionospheric irregularities and their boundaries. The equatorward boundary of the midlatitude irregularity region is abrupt in the northern hemisphere and gradual in the southern hemisphere. Stuart[7] presents a statistical study based on satellite scintillation data of the abruptness and other details of midlatitude scintillation boundaries. The two regions of equatorial irregularities shown near $\Lambda = \pm 20°$ (dip latitude near $\pm 15°$) have abrupt boundaries, but equatorial boundaries also can be very gradual. Two patches of Fe^+ ions were observed on this orbit; their boundaries corresponded with those of the equatorial irregularities.

The electron flux is smoothly varying below 65°N and quite irregular above 65°N. No fluxes were observed in the southern hemisphere to 45°S, and no RPA flux data are available beyond 45°S. Conjugate photoelectrons are responsible for the smoothly varying flux between 55° and 65°N; the lower latitude limit of these fluxes is reached when the conjugate ionosphere is no longer sunlit, and we observe that this latitude depends, of course, on longitude, season, local time, etc. The irregular fluxes above 65°N are caused by auroral electrons.[8] The presence or absence of midlatitude irregularities does not seem to be related to the presence or absence or conjugate photoelectrons.

In the example shown in Fig. 6, the "midlatitude trough" in N_i,[9,10] centered near 65°N, is roughly rectangular in shape. Other OGO 6 data show that it can take on a variety of other shapes, as well as other widths and depths. The high-latitude edge of the nightside trough usually is very steep and almost always is associated with the onset of soft auroral electron fluxes and enhanced values of Σ. Usually Σ reaches a maximum value of the order of 10 to 20% in the nighttime auroral zone and tends to decrease somewhat inside the polar cap, as it does in the example shown here. However, on some (rare) occasions Σ may reach very small values (< 1%) inside the polar cap.

When there are irregularities in the midlatitude ionosphere, OGO 6 data often show that there is a minimum in $\Delta N/N$ in the trough region. An extreme example of such a minimum appears just below $\Lambda = 65°$ in Fig. 6. The existence of such minima leads us to speculate that they may be a result of some of the same physical processes responsible for the trough. Since both N and $\Delta N/N$ tend to be smaller in the trough region than in higher or lower latitudes, there should be a pronounced minimum in ionospheric scintillation in that region, although it appears that no such minimum has been reported yet in the literature.

III. Irregularity Spectra

It is important to examine the spectral behavior of ionospheric irregularities, because these properties have a bearing on the origin of the irregularities as well as on the properties of the scintillation they produce. The variation of irregularity amplitude with scale size has been determined by power spectrum analysis techniques.[11] As we have seen, the data from the RPA are in the form of electrometer voltage (which is proportional to ion concentration) as a function of time. Each section of data to be analyzed spectrally was detrended first by removing the linear variation as given by the line of best fit (cf. Fig. 1). Spectral estimates then were calculated by the fast Fourier transform (FFT) technique. Spectral smoothing was achieved by averaging over a number of consecutive spectral estimates.

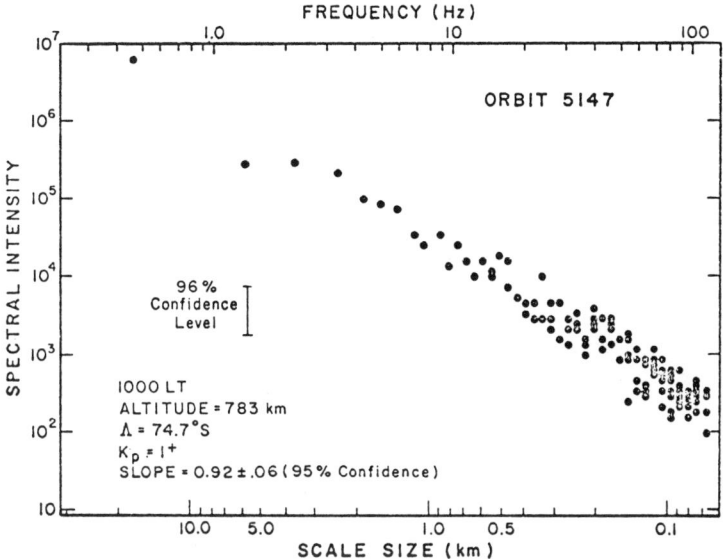

Fig. 7 Irregularity spectrum for 9.6 sec of data (2048 data points) obtained on orbit 5147 on day 147, 1970.

The input to the FFT was a zero-mean time series of values of incremental electrometer voltage ΔV normalized to represent positive and negative percentage fluctuations in the total ion concentration $\Delta N/\overline{N}$, where $\overline{N}$ is obtained from the trend line as illustrated in Fig. 1. The frequency window size was held constant in the calculation of a given frequency spectrum, although it varied from one spectrum to another. Thus the 96% confidence level indicated on the spectral plots shown below applies to all of the points in a given plot.

The examples shown here of the many Fourier spectra examined by Dyson et al.[11] are plots of unnormalized spectral intensity vs frequency; also provided are scales of normalized percentage irregularity amplitude and size, thus emphasizing the relationship between these two parameters. Although the results are discussed below in terms of these geophysically significant parameters, it should be kept in mind that we are discussing frequency spectra, i.e., that there has been no change of variables from frequency to scale size, which of course would change the form of the spectra obtained. Unless otherwise defined, "scale size" means "satellite velocity divided by frequency." We shall discuss mainly relative changes in irregularity amplitude, but the amplitude scales shown in our spectral figures are normalized absolutely as follows: the FFT output in each frequency window was converted to the rms irregularity amplitude $\Delta N_{rms}/\overline{N}(\%)$ equivalent to the total fluctuation power in the given window. Thus the integral of the fluctuation power $\int s(f)df$ over the observed frequency range equals the variance of the original time series of $\Delta N/\overline{N}$ values.

Spectra were calculated at various latitudes using several different record lengths. An example is shown in Fig. 7. This spectrum was derived from 9.6 sec of data (2048 data points). The spectrum is well defined over two orders of magnitude of scale size from 70 m upward and is very linear when plotted with logarithmic scales. The spectrum has a slope of 0.92 ± 0.06. (The slope of the power spectrum would be twice this value.) In calculating the error in slope, the known variance of the data points in the spectrum has been taken into account using the method outlined by Bendat and Piersol.[12] This spectrum is typical of most of those observed for the typical red noise type of irregularities in that it obeys a power law of index very close to unity. The only significant difference between spectra obtained from different record segments is that the absolute value of the amplitude at a given scale size depends on the overall variance of the given data segment; i.e., the slope of the spectrum remains very nearly constant. For two orbits (5147 and 6412), spectra have been

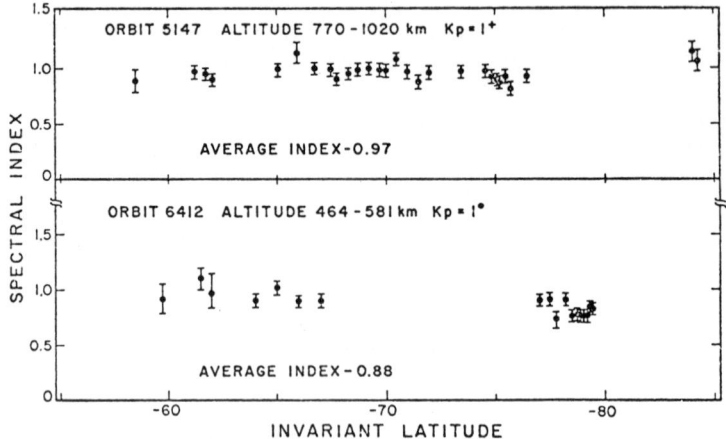

Fig. 8 Spectral index as a function of invariant latitude for orbits 5147 and 6412, which occurred on May 27 and Aug. 24, 1970, respectively.

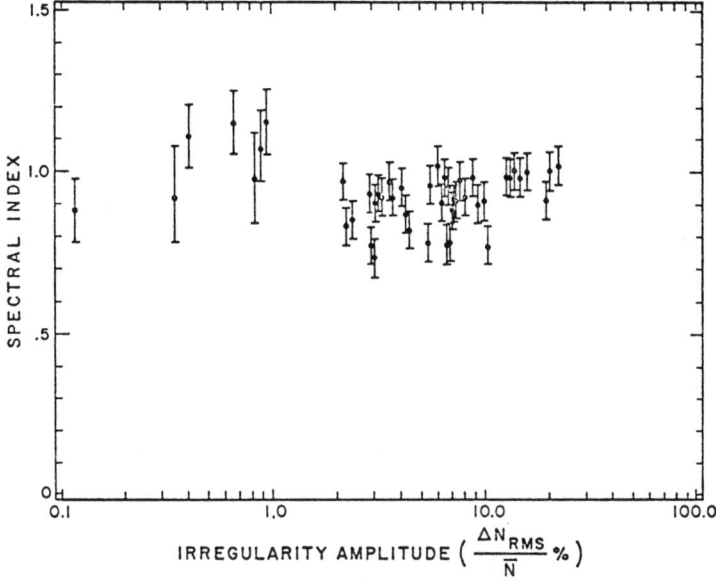

Fig. 9 Spectral index as a function of percentage rms ion concentration for orbits 5147 and 6412.

calculated at different high-latitude locations along the orbital path using record lengths of 4.8 sec (∿35 km) and 9.6 sec (∿70 km). The values obtained for the spectral index are plotted in Fig. 8 as a function of invariant latitude. The points with the larger error bars are for the 4.8-sec segments of data. There are no significant changes in the spectral index as a function of latitude. The average value is different for the two passes but not significantly so. The data were collected at local times between 20 and 10 hr, i.e., between 8 PM and 10 AM. The local time, of course, changes rapidly at high latitudes, where the longitude changes rapidly along the satellite path. A further demonstration of the relative constancy of the power law index is given in Fig. 9, where the spectral index data of Fig. 8 are plotted as a function of the percentage rms amplitude of the ion concentration variations. There is no significant change in the spectral index, even though the rms amplitude of $\Delta N/N$ at a given scale size varies over more than two orders of magnitude. An average value of spectral index of 0.95 is obtained by combining the data of Fig. 9. All low-latitude spectra examined also have yielded spectral indices close to unity.

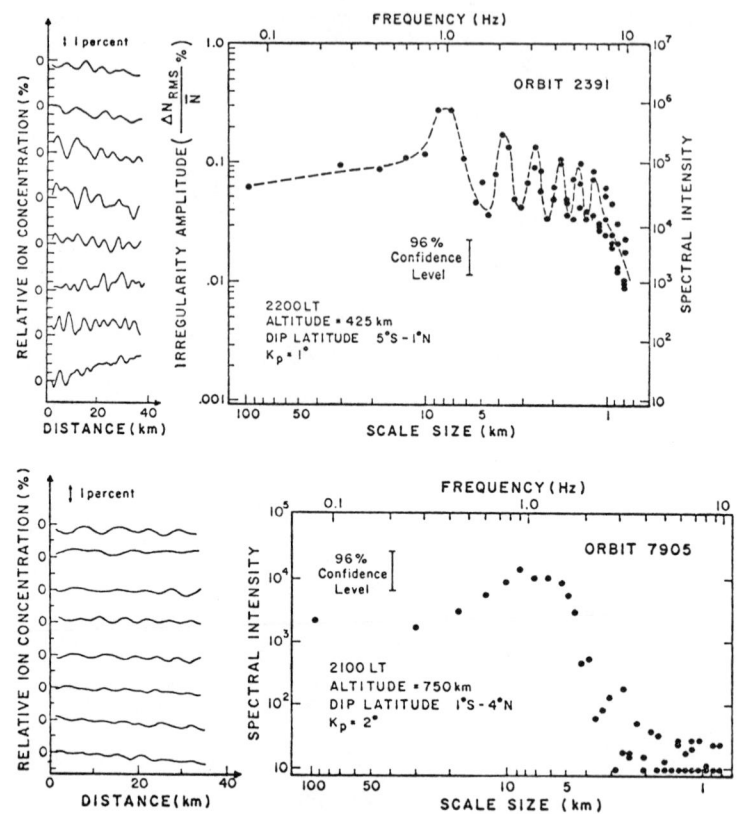

Fig. 10 The consecutive 5-sec samples of sinusoidal irregularity waveforms shown were detrended separately. The average spectrum of the samples is shown. The samples were separated by 5-sec RPA sweep periods.

Two examples of sinusoidal irregularities observed near the equator are shown in Fig. 10, together with their calculated spectra. In this as in the previous and successive spectral figures, the scale sizes are those along the orbital path. The two spectra in Fig. 10 differ in that in one case power is contained within a limited scale size band, whereas in the other example the waveform is somewhat more monochromatic, with evidence that the fluctuations consist of a fundamental plus harmonics. The scale size of the fundamental is 7.5 km.

A typical spectrum of the ground-glass category of irregularities is illustrated in Fig. 11. It is apparent from the

IONOSPHERIC IRREGULARITIES

ion concentration measurements that the irregularities are primarily small scale, although, in this example, some relatively large scale sizes are present, as evidenced by the amplitude modulation of the ground glass which is particularly noticeable in the first two segments of data. The larger scale sizes are not typical of the ground-glass phenomena and have been filtered from the data so that the spectrum in Fig. 11 is for the higher-frequency variations evident in the data. The spectrum shows a random variation of power with scale size; i.e., the "ground glass" has an irregular spectrum for scale sizes smaller than about 3 km.

IV. Auroral Particles and Scintillation

As emphasized in the brief discussion of morphology associated with Fig. 6, the OGO 6 results show that at high latitudes at night the equatorward boundary of the region of large-amplitude irregularities corresponds with that of large fluxes of precipitating auroral electrons. This is not an unexpected result. Suggestions often have been made that precipitation might be responsible for the production of ionospheric irregularities in the auroral region. For example, Petrie[13] and Dyson[14] showed that the occurrence of large-amplitude auroral irregularities is similar to that of various auroral phenomena. However, until recently no direct observations of a one-to-one relationship between irregularities and particles ever had been documented.

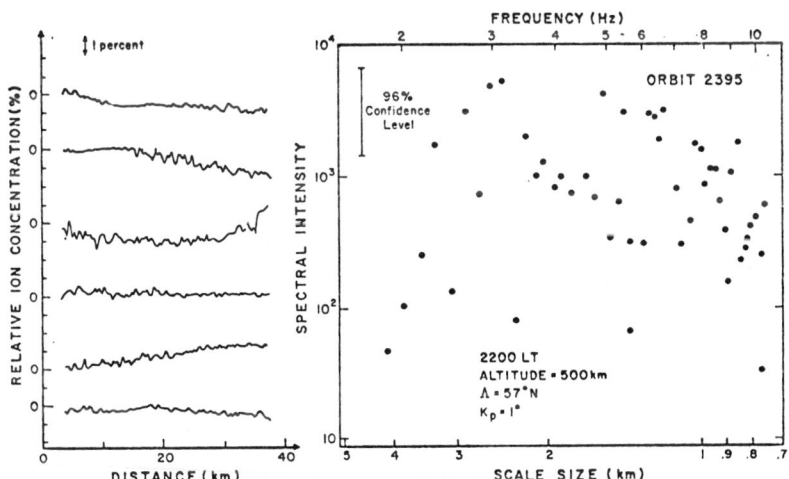

Fig. 11 Consecutive samples of ground-glass irregularity waveforms and their average spectrum.

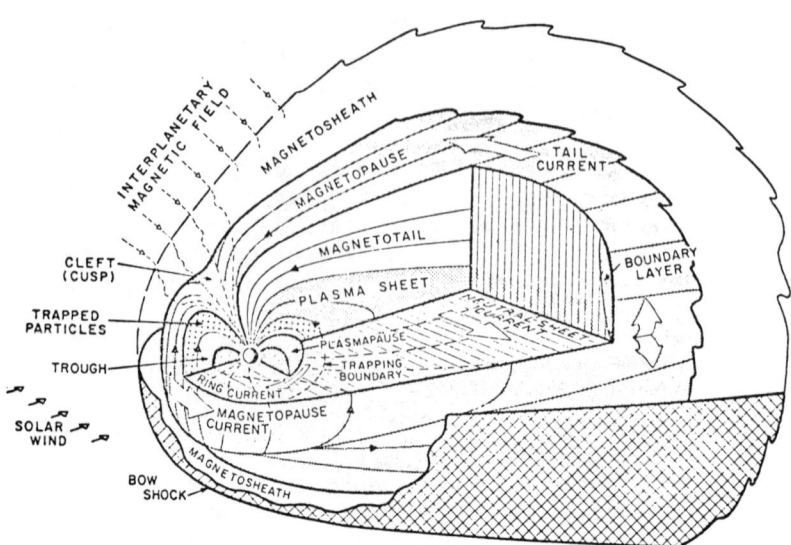

Fig. 12 A model of Earth's magnetosphere according to Heikkila.[15] Strong fluxes of solar wind (magnetosheath) plasma of energy <300 ev, which interact strongly with the F region, have direct access to the ionosphere via the "dayside clefts." Magnetospheric convection will cause irregularities created in the cleft region to drift poleward.

Figure 12, taken from Heikkila's[15] very readable tutorial review of auroral processes, shows schematically a number of features of Earth's magnetosphere which will be useful in the following discussion. Low-energy solar wind electrons and protons flowing past Earth produce a bow shock and a transition region called the magnetosheath outside of the magnetosphere. The OGO 6 data shown in Fig. 6 were obtained on the night side of Earth. The location of the plasmapause and the N_i trough can be seen in both Figs. 6 and 12. Poleward of the trough, irregularly precipitating plasma sheet electrons and large-amplitude N_i irregularities are typically seen in the OGO 6 data, as mentioned previously.

The OGO 6 satellite was not ideally suited for studying further details of the relationship between particles and irregularities because, among other reasons, it measured only the integrated flux of electrons having energy >10 ev. The ISIS-1 satellite carried a soft-particle spectrometer, which produced detailed pictures of energetic particle precipitation. Information about N_i irregularities was obtained from a swept frequency ionospheric sounder. ISIS-1 data from the dayside were examined by Dyson and Winningham.[16] Their final results are shown in Fig. 13. The equatorward boundary of the "strong topside irregularity zone" (STIZ) is identical with the corresponding boundary of the strong precipitating fluxes of solar wind electrons of energy <300 ev which appear to have direct access to the magnetosphere and ionosphere via the "dayside cleft" (cf. Fig. 12) discovered by Heikkila and Winningham[17] and by Frank.[18] Further details of the behavior of the particle precipitation and of the irregularities support the idea that the irregularities result from the precipitation. It was concluded by Dyson and Winningham[16] that the ionizing and heating effects of the precipitation produce ionospheric

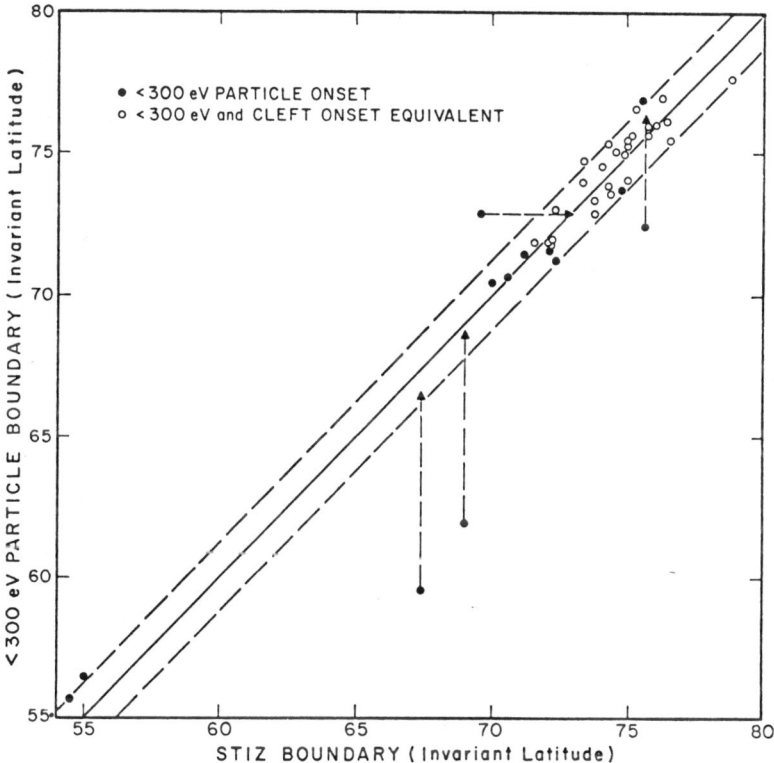

Fig. 13 Comparison of the equatorward boundaries of 300-ev precipitation and stiz based on ISIS-1 data. Arrows indicate the position of the cleft boundary.[16]

irregularities, but that other processes that occur in association with the precipitation also could contribute to the production of the irregularities. Magnetospheric convection will cause irregularities created in the dayside cleft region to drift poleward and into the nightside of the polar cap.

V. Cause of Equatorial Spread-F

Instrumental problems were suspected when OGO 6 data began indicating the presence in the nighttime equatorial F region of small concentrations (usually 5 to 500 cm^{-3}) of an ion roughly twice as heavy as the heaviest expected F-region ions (NO^+, O_2^+, N_2^+, all near 30 atomic mass units). Such data were not returned from the satellite except at equatorial latitudes. The evidence soon became overwhelming[19] that the unknown ion was Fe^+, and that, in addition, the presence of at least a detectable amount (our sensitivity limit for ions of mass >30 AMU was near 5 cm^{-3}) of Fe^+ is a necessary but not a sufficient condition for the existence of equatorial ionospheric irregularities.[3] Such irregularities were observed without Fe^+ only 2.5% of the time, whereas Fe^+ was observed commonly without irregularities. Both this relation and indeed the very presence of Fe^+ above 400 km in the F region were at first very hard to believe in, but at the present time meteoric ions have been detected independently in the equatorial F region by at least two other techniques (optically[20] and via an ion mass spectrometer[21]). Also, the source of these ions is understood[22] (vertical electromagnetic drifts from the source region near 100-km alt into the F region; the initial lifting from 100 km is induced by vertical electric fields, which can occur only within a few hundred kilometers of the geomagnetic dip equator).

It also is believed that we now know why trace amounts of Fe^+ ions virtually always are found on or near equatorial field tubes containing ionospheric irregularities. Time delay considerations show that it is probably not a mechanism associated with the irregularities which somehow transport the Fe^+ into the upper F region (this is incompatible with the aforementioned evidence regarding Fe^+ as a necessary but not sufficient condition for equatorial irregularities). Hanson et al.[23] suggest that the basic cause of equatorial irregularities is electric fields generated in the ionosphere by zonal or meridional neutral winds acting on regions having structure in their field-aligned Pedersen conductivity integrals. The conductivity structure, it is suggested, is caused by large "fingers" of long-lived metallic ions that randomly extend downward to $\sim$150 km, well below the normal nighttime F layer.

The resulting plasma convection leads to large-scale (>10 km) irregularities and sets up gradients in the electric field and/or plasma pressure which drive other instability mechanisms to produce smaller-scale irregularities. The proposed mechanism requires a different neutral wind velocity above and below approximately 200 km for winds in the magnetic east-west direction; but, for winds in the magnetic meridian direction, such a velocity gradient is not required. This mechanism may explain the enhanced occurrence probability for irregularities in the Atlantic zone seen in data from the OGO 6 satellite,[3] since zonal winds have their largest magnetic meridional component in this sector. Also, since both the metallic ion distribution and the irregularities are believed to be approximately field-aligned, the two should be correlated strongly in the upper F region, as observed in the OGO 6 data.[3]

In addition to being consistent with these two new OGO 6 results on equatorial irregularities (the Atlantic maximum and the relation with Fe^+), the proposed mechanism also is consistent with radar observations of equatorial spread-F of McClure and Woodman.[24] These radar observations showed that the vertical drift velocity (or east-west electric field) in a wide region of well-developed spread-F was enhanced and turbulent, reaching positive and negative maximum values several times larger than the steady velocity of the undisturbed ionosphere and changing irregularly every few tens of kilometers in altitude and every few minutes in time. We believe that such turbulent electric fields could be generated by the proposed mechanism.

However, there is another category of weaker spread-F observed at Jicamarca which has characteristics different from those described by McClure and Woodman. It occurs in thin layers on the bottomside of the F region, near the altitude where N_e reaches approximately 1% of its maximum value, and it is not always associated with ionospheric scintillation.[25] It drifts with the background drift velocity of the undisturbed F region; i.e., it is not associated with enhanced and turbulent electric fields. It may be that this type of spread-F is associated with the large N_e gradients that normally exist in this part of the bottomside equatorial F region at night.[2]

VI. Summary and Conclusions

The research described in this paper has provided a great deal of new information of fundamental importance in dealing with the practical problems of radio propagation through an ionosphere containing small-scale inhomogenieties in N_i. It has provided a statistical description of the amplitude vs scale size spectrum of the irregularities, showing that the irregularity amplitude is directly proportional to the scale size in almost all cases, independent of the location or the severity of the irregularities. This result is valuable in several areas, including theoretical modeling of ionospheric radio scintillation and attempts to understand the origin of the irregularities. Other morphological results at high and low latitudes clearly have increased our understanding of irregularity occurrence patterns previously obtained from topside ionospheric sounding and from radio-scintillation observations.

Finally, the recent observations of the detailed relationships between irregularities and other geophysically significant parameters have permitted us to outline some of the details of the source mechanisms for the irregularities. For example, a detailed relation was found, as might have been expected, between strong irregularities and the precipitation of soft electrons. A much more unexpected result was being able to postulate a reasonable mechanism that explains the very puzzling correlation between Fe^+ ions and irregularities in the equatorial F region. At the same time, this mechanism appears to explain the origin of the irregularities themselves in a manner consistent with all of the available observational data about them.

References

[1] Hanson, W. B., Sanatani, S., Zuccaro, D., and Flowerday, T. W., "Plasma Measurements with the Retarding Potential Analyzer on OGO VI," Journal of Geophysical Research, Vol. 75, No. 28, Oct. 1970, pp. 5483-5501.

[2] McClure, J. P. and Hanson, W. B., "A Catalog of Ionospheric F Region Irregularity Behavior Based on OGO VI Retarding Potential Analyzer Data," Journal of Geophysical Research, Vol. 78, No. 31, Nov. 1973, pp. 7431-7440.

[3] Hanson, W. B. and Sanatani, S., "The Relationship Between Fe^+ Ions and Equatorial Spread F," Journal of Geophysical Research, Vol. 76, No. 31, Nov. 1971, pp. 7761-7768.

[4] Crane, R., "Morphology of Ionospheric Scintillation," AIAA Paper 74-52, Jan.-Feb. 1974, Washington, D.C.; also published elsewhere in this volume.

[5] Fremouw, E. J. and Rino, C. L., "An Empirical Model for Behavior of Average F-Layer Scintillation at VHF/UHF," Radio Science, Vol. 8, No. 1, Jan. 1973, pp. 234-255.

[6] Taur, R. R., "Ionospheric Scintillation at 4 and 6 GHz, COMSAT Technical Review, Vol. 3, No. 1, Jan. 1973, pp. 145-159.

[7] Stuart, G. F., "Characteristics of the Abrupt Scintillation Boundary," Journal of Atmospheric and Terrestrial Physics, Vol. 34, No. 3, March 1972, pp. 1455-1471.

[8] Winningham, J. D., Akasofu, S.-I., Yasuhara, F., and Heikkila, W. J., "Simultaneous Observations of Auroras from the South Pole Station and of Precipitating Electrons by ISIS 1," Journal of Geophysical Research, Vol. 78, No. 28, Oct. 1973, pp. 6579-6594.

[9] Muldrew, D. B., "F Layer Ionization Troughs Deduced from Alouette Data," Journal of Geophysical Research, Vol. 70, No. 11, June 1965, pp. 2635-2650.

[10] Sharp, G. W., "Mid-latitude Trough in the Night Ionosphere," Journal of Geophysical Research, Vol. 71, No. 5, March 1966, pp. 1345-1656.

[11] Dyson, P. L., McClure, J. P., and Hanson, W. B., "In Situ Measurements of the Spectral Properties of F-Region Ionospheric Irregularities," Journal of Geophysical Research, Vol. 79, No. 10, April 1974, pp. 1497-1502.

[12] Bendat, J. S. and Piersol, A. G., Measurement and Analysis of Random Data, Wiley, New York, 1966, p. 227.

[13] Petrie, L. E., "Preliminary Results on Mid and High Latitude Topside Spread F," Spread F and Its Effect upon Radiowave Propagation and Communication, Technivision, England, 1966, p. 67.

[14] Dyson, P. L., "Direct Measurements of the Size and Amplitude of Irregularities in the Topside Ionosphere," Journal of Geophysical Research, Vol. 74, No. 26, Dec. 1969, pp. 6291-6304.

[15] Heikkila, W. J., "Aurora," EOS Transactions of the American Geophysical Union, Vol. 54, No. 8, Aug. 1973, pp. 764-768.

[16] Dyson, P. L. and Winningham, J. D., "Topside Ionospheric Spread F and Particle Precipitation in the Dayside Magnetospheric Clefts," Journal of Geophysical Research, Vol. 79, No. 34, Dec. 1974, pp. 5219-5230.

[17] Heikkila, W. J. and Winningham, J. D., "Penetration of Magnetosheath Plasma to Low Altitudes Through the Dayside Magnetospheric Cusps," Journal of Geophysical Research, Vol. 76, No. 4, Feb. 1971, pp. 883-891.

[18] Frank, L. A., "Comments on a Proposed Magnetospheric Model," Journal of Geophysical Research, Vol. 76, No. 10, April 1971, pp. 2512-2515.

[19] Hanson, W. B. and Sanatani, S., "Meteoric Ions Above the F_2 Peak," Journal of Geophysical Research, Vol. 75, No. 28, Oct. 1970, pp. 5503-5509.

[20] Boxenberg, A. and Gerard, J.-C., "Ultraviolet Observations of Equatorial Dayglow Above the F_2 Peak," Journal of Geophysical Research, Vol. 78, No. 22, Aug. 1973, pp. 4641-4649.

[21] Taylor, H. A., private communication, 1973.

[22] Hanson, W. B., Sterling, D. L., and Woodman, R. F., "The Source and Identification of Heavy Ions in the Equatorial F Layer," Journal of Geophysical Research, Vol. 77, No. 28, Oct. 1972, pp. 5530-5541.

[23] Hanson, W. B., McClure, J. P., and Sterling, D. L., "On the Cause of Equatorial Spread F," Journal of Geophysical Research, Vol. 78, No. 13, May 1973, pp. 2353-2356.

[24] McClure, J. P. and Woodman, R. F., "Radar Observations of Equatorial Spread F in a Region of Electrostatic Turbulence," Journal of Geophysical Research, Vol. 77, No. 28, Oct. 1972, pp. 5617-5621.

[25] Farley, D. T., Balsley, B. B., Woodman, R. F., and McClure, J. P., "Equatorial Spread F: Implications of VHF Radar Observations," Journal of Geophysical Research, Vol. 75, No. 34, Dec. 1970, pp. 7199-7216.

MODELING AND PREDICTION OF IONOSPHERIC SCINTILLATION

Edward J. Fremouw[*]

Stanford Research Institute, Menlo Park, Calif.

Abstract

There are two major aspects of scintillation modeling: description of the scattering irregularities, and statistical characterization of the communication channel. Parameters of concern are the width of the irregularities' spatial spectrum, its falloff with increasing wavenumber, its anisotrophy (e.g., relative to the geomagnetic field), and the height and thickness of the irregular region. Scintillation varies with geomagnetic latitude, time of day, season, and epoch of the solar cycle, with considerable day-to-day variation. Modeling to date has taken into account only the trends and therefore is predictive only in a statistical sense. The sole signal statistic so far modeled is scintillation index, a measure of the rms fluctuation in received signal strength. This paper describes an existing scintillation model and prospects for improvements within the next few years.

Presented as Paper 74-54 at the AIAA 12th Aerospace Sciences Meeting, Jan. 30-Feb. 1, 1974, Washington, D.C. The author thanks C. L. Rino of the Stanford Research Institute for many illucidating discussions and gratefully acknowledges the cooperation of experimenters at the Air Force Cambridge Research Laboratories and University of California at San Diego who provided unpublished data for that analysis. The scintillation modeling efforts reviewed in this paper were supported by NASA and by the Advanced Research Projects Agency, with strongly contributing studies supported by the Defense Nuclear Agency.
[*]Senior Physicist.

I. Introduction

Ionospheric researchers have studied scintillation for two reasons: 1) to glean information about structure in the ionosphere, and 2) to provide engineering descriptions of signal perturbations encountered in transionospheric propagation. These same two reasons underlie the particular research endeavor of scintillation modeling. On the one hand, empirical modeling is a means of ordering data into a form that, hopefully, will provide insight into the characteristics and production mechanisms of ionospheric irregularities. On the other, it is a mechanism for collating knowledge into a format convenient for applications, namely, representation in terms of equations and a manageable number of parameters that can be used for computation, given a set of conditions.

In principal, these two objectives require opposite procedures for their accomplishment, but in practice they become almost irretrievably merged. Ideally, the basic researcher would like to deduce irregularity parameters from scintillation observations by means of straightforward calculation. On the other hand, the applied scientist would like to calculate signal parameters from descriptions of electron-density irregularities. As a matter of fact, these two workers turn out to be the same person (actually a community of individuals), who performs both tasks in an iterative fashion.

The second of the jobs is the only one that can be attacked directly. Thanks fundamentally to Maxwell's equations, there is an abundance of radiowave scattering theories that permit calculation of signal parameters if properties of the medium are known. (See Ref. 1 for a recent review of theories.) The process of inferring parameters of the medium, however, is inductive. It typically involves postulation of a parameter set, calculation of corresponding signal parameters, comparison with observation, and iterative adjustment of the postulate. Uniqueness can never be established by this procedure. Fortunately, in situ (satellite) measurements of electron number density with sufficient spatial resolution to provide guidance in describing ionospheric irregularities are now becoming available, as described by McClure.[2] But this is a relatively recent development, and the scintillation modeling to be described in this paper is mainly the result of iterative propagation calculation and result testing.

Whatever propagation theory is used for the calculations, the geometry considered is similar to that illustrated in Fig. 1, which shows a vertical plane containing the line of sight between the transmitter above the irregular ionospheric layer and the receiver on Earth's surface. The most important geometric quantities are the distances of the layer's penetration point from the receiver and transmitter, z_1 and z_2, respectively, and the angle of incidence i of the propagation vector on the layer. The incidence angle, of course, is related to the zenith angle θ of the transmitter from the receiver by means of spherical-Earth geometry, and it turns out that the significance of the distances z_1 and z_2 is their control of Fresnel-zone size, which depends also on the radio wavelength. The geometry is specified in terms of the heights of the transmitter and of the layer, H and h, respectively, and it is not surprising that the thickness of the layer Δh also plays a roll in determining signal parameters.

There are two major aspects to scintillation modeling. First, one asks just how the parameters of the irregular layer and the characteristics of the irregularities in it affect transionospheric radio signals; this is the domain of signal-statistical channel modeling. Second, how do the irregularity parameters and, therefore, the signal characteristics vary from time to time, place to place, and with changing geophysical conditions? This is the domain of morphology, as described by Crane.[3] A broadly useful scintillation model must answer both questions.

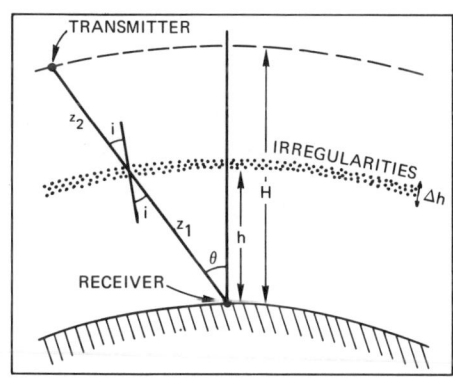

FIGURE 1 GEOMETRY CONSIDERED FOR SCINTILLATION CALCULATIONS.

In Sec. II of this paper, we shall review the signal-statistically rudimentary modeling that has been accomplished thus far, concentrating on its morphological aspects. In Sec. III, we shall describe the theoretical and empirical underpinnings for true signal-statistical channel modeling. Prospects for refining the existing model morphologically and for extending its signal-statistical domain will be assessed in Sec. IV.

II. Modeling of Scintillation Index

Scintillation modeling performed thus far is based on the theory of diffraction by a weakly modulating phase screen developed by Briggs and Parkin.[4] From this theory is calculated a single signal-statistical parameter, the fractional rms fluctuation of signal intensity, defined as

$$S_4 = \langle (A^2 - \langle A^2 \rangle)^2 \rangle^{\frac{1}{2}} / \langle A^2 \rangle \tag{1}$$

where A is the real amplitude of the signal, and the angular brackets indicate averaging.

The S_4 "scintillation index" is one of four quantitative indices that are commonly used to describe the strength of amplitude scintillation, the others being as follows:

$$S_1 = |A - \langle A \rangle| / \langle A \rangle \tag{2}$$

$$S_2 = \langle (A - \langle A \rangle)^2 \rangle^{\frac{1}{2}} / \langle A \rangle \tag{3}$$

$$S_3 = \langle |A^2 - \langle A^2 \rangle| \rangle / \langle A^2 \rangle \tag{4}$$

The relationship between the four indices depends on the shape of the probability density function for signal amplitude. For scintillation modeling performed to date, they have been assumed to be proportional to one another, with the constants of proportionality being those obtained for a Rayleigh distribution.

The Briggs and Parkin theory allows calculation of S_4 in terms of ionospheric-irregularity parameters. In this theory, the irregular layer is thought of as a thin screen that spatially modulates the phase of the radio wave propagating

PREDICTION OF IONOSPHERIC SCINTILLATION

through it. Diffraction effects within the layer are ignored, which is acceptable for sufficiently narrow-angle scatter. The strength of scattering is described in terms of a phase-modulation index ϕ_o (in radians), at the output of the screen, which is given by the following equation:

$$\phi_o = K \cdot \lambda [(a \, \xi_o \, \sec i)^{\frac{1}{2}}/\beta^{\frac{1}{2}}] (\Delta h)^{\frac{1}{2}} (\Delta N) \tag{5}$$

where K is a constant of proportionality which depends on the detailed shape of the irregularities' spatial spectrum (which controls the sharpness of gradients in the medium).

The primary ionospheric parameters in (5) are the rms fluctuation ΔN in electron density, the thickness Δh of the irregular layer, the transverse irregularity scale size ξ, and the irregularity axial ratio a. In addition, ϕ_o depends on the incidence angle i and an irregularity projection factor β, where $\beta = (a^2 \sin^2 \psi + \cos^2 \psi)$, and ψ is the angle between the geomagnetic field and the radio line of sight. The radio wavelength is denoted by λ.

The particular form taken by Eq. (5) depends on assumption of a spatial spectrum; similar forms would arise for different assumptions, with slight modifications in parameter definition. Briggs and Parkin assumed that the spectrum is gaussian-shaped, by specifying its Fourier transform (the spatial autocorrelation function) to be a gaussian of e^{-1} width ξ_o across the geomagnetic field and width $a\xi_o$ along the field.

In the phase-screen approach to scintillation, only the wave's phase is supposed to vary spatially across the screen, with amplitude being constant. Amplitude variations build during postscattering propagation in the manner of a diffraction pattern.[5] Thus, the scintillation index is very dependent upon distance from the screen as related to wavelength by means of the Fresnel-zone radius λz, resulting in the following relation for S_4:

$$S_4 = \sqrt{2} \, \phi_o \, [1 - (\cos u_1 \cos u_2)^{\frac{1}{2}} \cos (u_1 + u_2)/2]^{\frac{1}{2}} \tag{6}$$

where u_1 and u_2 are Fresnel-distance parameters defined as

$$u_1 = \tan^{-1}(2\lambda z/\pi\xi_o^2) \qquad (7)$$

$$u_2 = \tan^{-1}(2\lambda z/\pi\beta^2\xi_o^2) \qquad (8)$$

The arguments in Eqs. (7) and (8) are seen to be essentially the ratio of Fresnel-zone size to irregularity scale size, with z being the reduced distance $z = z_1 z_2/(z_1 + z_2)$, where z_1 and z_2 are as shown in Fig. 1.

Inspection of Eqs. (5) and (6) reveals that S_4 depends upon both ionospheric parameters and quantities that are essentially geometric ones relating to the diffraction process. Thus, scintillation modeling cannot proceed totally empirically on the basis of index data only. Application of a model will require the ability to calculate scintillation index for an arbitrary geometry to be specified by the user on the basis of an ionospheric-parameter model supplied by the researcher.

To achieve this end, the first scintillation modeling presented in the literature[6] was performed by coding Eqs. (5-8), along with a number of auxiliary geometric expressions, to permit calculation of S_4 as a function of ionospheric parameters and of various satellite and radio-star observing conditions. The results then were compared with published observations performed under those conditions and the model iteratively improved. The main endeavor was to provide proper parameter values for calculating ϕ_o, especially to select the appropriate behavior of irregularity strength ΔN.

It was necessary first to select values for the other geophysical quantities involved in the calculations. The simplest to handle was the layer thickness Δh, which was treated easily as a constant. Doing so means that model testing was actually of the product $\Delta N(\Delta h)^{\frac{1}{2}}$; separating the effects of the two variables is unnecessary for application. In order to be as realistic as possible geophysically, a value was taken for Δh from measurements reported in the literature, namely, 100 km.[7-9] The same measurements indicate an average value of 350 km for h, without systematic trends, and this value was used. For the axial ratio a, the constant value 10 was used, based on observations performed under a variety of conditions.[10-12]

The essence of the procedure was to postulate models for ΔN and ξ_o, to insert the model values in Eq. (5) along with the other parameters needed, and then to employ Eqs. (5) and (6) to calculate the value of S_4 expected for a given set of published observations. In this manner, the model was tested and improved, using 12 data sets from a variety of observational circumstances. A thirteenth set, not used in model development, was included in final testing.

The procedure was designed to account for dissimilar experimental circumstances and data-reduction procedures. For each data set, the transmitter and receiver locations used in calculation were chosen to be representative of the actual ones, and the magnetic-field geometry was accounted for on the basis of an Earth-centered, but axially tipped, dipole model. After the scintillation index was calculated, averages were performed in a manner similar to those performed by the observer. The final result then was compared with the reduced data presented in the literature.

The index calculated was S_4, which was converted if necessary to S_1, S_2, or S_3 by means of the proportionality constants suggested by Briggs and Parkin[4] and tested by Bischoff and Chytil.[13] The papers used gave scintillation magnitude either as one of these four indices or as some other index calibrated in terms of one of them. The model initially postulated treated ξ_o as a constant (1 km) and described ΔN by means of three additive terms specifying, respectively, equatorial, midlatitude, and high-latitude contributions to the rms fluctuation of electron density. The forms of the terms were chosen to be consistent with scintillation morphology, essentially the characteristics described by Crane.[3]

In the initial postulate, the expression for ΔN was defined quantitatively by 14 numerical constants, to be evaluated by comparison of model-based calculations of scintillation index with observed values. For the most part, changes in the initial postulate that came about through iterative testing were in the nature of evaluating the constants. Some changes in form were made, however, most notably the addition of a fourth term to account for aurorally associated scintillation and introduction of a latitudinal variation for ξ_o.

As a result of the foregoing procedure, the following model for scintillation-producing irregularities in the F layer was put forth: h = 350 km, Δh = 100 km, a = 10, and ξ_o and ΔN given by empirical equations that appear in Ref. 6. The transverse scale size was modeled as a function of geomagnetic latitude only, and the rms fluctuation of electron number density was given as

$$\Delta N = \Delta N_{eq}(R,D,t,\ell) + \Delta N_{mid}(t,\ell) + \Delta N_{hi}(R,t,\ell) + \Delta N_{aur}(R,t,\ell) \qquad (9)$$

where the independent variables are the following: mean sunspot number R, day of the year D, time of day t, and geomagnetic latitude ℓ. The terms of Eq. (9) appear in explicit form in Eqs. (9-12), respectively, of Ref. 6.

The form of the first term in Eq. (9) describes the well-known peaking of equatorial scintillation in the midnight hours and the decay of activity through the early morning hours, a simple harmonic seasonal dependence with peaks at the equinoxes, a linear dependence on sunspot number, and a gaussian latitudinal dependence that drops to e^{-1} at 12° on either side of the geomagnetic equator. The second term describes simple diurnal and latitudinal variations of scintillation at middle latitudes.[14] The behavior of high-latitude scintillation other than that directly associated with auroral disturbance is described in the third term of Eq. (9). This behavior is attributed to diurnal and solar-cycle migrations of the scintillation boundary. The final term of Eq. (9) describes what is believed to be aurorally associated scintillation arising in a region, near the auroral oval, the latitudinal extent of which is proportional to sunspot number, as is the strength of the irregularities it contains.

Comparisons of scintillation index calculated from the preceding model with observations are shown in Figs. 2-8. The calculated curves are solid where the assumption of weak

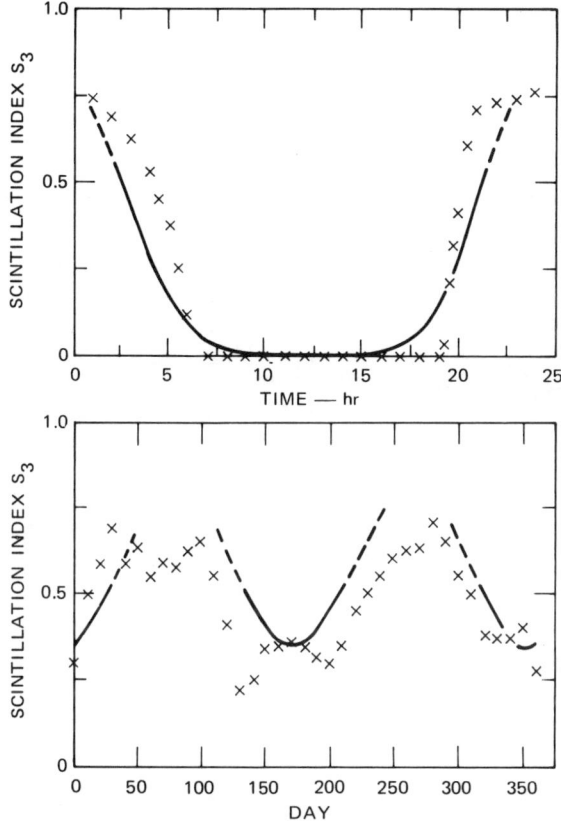

FIGURE 2 COMPARISON OF MODEL CALCULATIONS WITH GEOSTATIONARY-SATELLITE OBSERVATIONS FROM GHANA.[15] The top is diurnal variation: frequency = 136 MHz, sunspot number = 107, day number = 31. The bottom is seasonal variation: frequency = 136 MHz, sunspot number = 97, time is 0200. The observations are shown as discrete points, and the calculations as a curve. The curve is solid where the weak-scatter assumption is valid and dashed where it is questionable. Where it is invalid, no calculated results are given.

modulation is satisfied ($\phi_o < 0.7$) and dashed where the assumption is questionable ($0.7 \leq \phi_o \leq 1.0$). Where the assumption is invalid ($\phi_o > 1.0$), no calculated value is given. Comparison of results with equatorial observations[15] appears in Fig. 2; the fits are reasonably close where the weak-scatter assumption holds. The rise of the observed values in the evening hours, which is more abrupt than those calculated, could be accounted for by a change in form of the equatorial term of the ΔN model, and parameter adjustments could reduce other discrepancies. This hardly seem justified, however, in light of two more serious limitations of the model at equatorial latitudes.

The first limitation stems from lack of an opportunity to test the predicted sunspot-number dependence of scintillation. There appear to be no long-term equatorial data available in terms of quantitative indices, although there remains the possibility of calibrating some earlier observational results in such terms.[16] The equatorial term of the scintillation model may be considered relatively reliable at vhf and perhaps uhf under average ionospheric conditions for sunspot numbers on the order of 100 (typical of solar maximum). For other sunspot numbers, however, it is only an untested estimate, and more experimental work is needed.

The second limitation may be inherent in the average nature of the model but is of some practical concern and of a good deal of scientific interest. In the past few years, instances of significant scintillation on surprisingly high frequencies (as high as 6 GHz) have been reported by equatorial observers.[17-19] Modeling performed to date does not account for these important observations.

At middle latitudes, the model just described produced quite acceptable fits to the data of Preddey et al.[20] and of Preddey.[21] Figure 3 shows the comparison of calculated diurnal variation at middle latitudes with the former data. Figure 4 compares calculated and observed latitudinal dependence for daytime and for nighttime, using the latter data. The bars shown on the data points indicate the range of day-to-day variations observed in scintillation index. (They are not measurement uncertainties.)

Figure 4 also shows latitudinal dependence in the southern-hemisphere scintillation-boundary region, under essentially solar minimum conditions (sunspot number = 30). The fit is seen to be quite good at night, the time of most practical concern. The match is less satisfactory in the daytime, reflecting the dictates of data sets from other stations, notably the northern-hemisphere observations of Aarons et al.[22] and of Fremouw.[23]

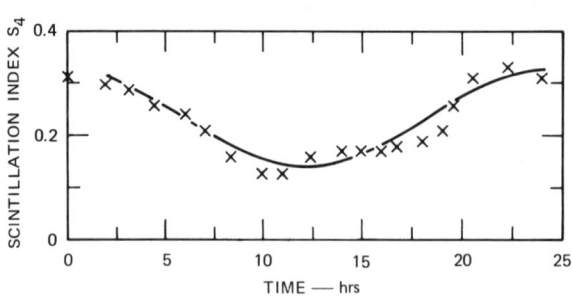

FIGURE 3 COMPARISON OF MODEL CALCULATIONS WITH HIGH-INCLINATION-SATELLITE OBSERVATIONS OF THE DIURNAL VARIATION OF SCINTILLATION FROM BRISBANE, AUSTRALIA.[20] Frequency = 40 MHz; sunspot number = 13.

Similar observations were conducted by Preddey in the boundary region near solar maximum (sunspot number = 103) and the corresponding

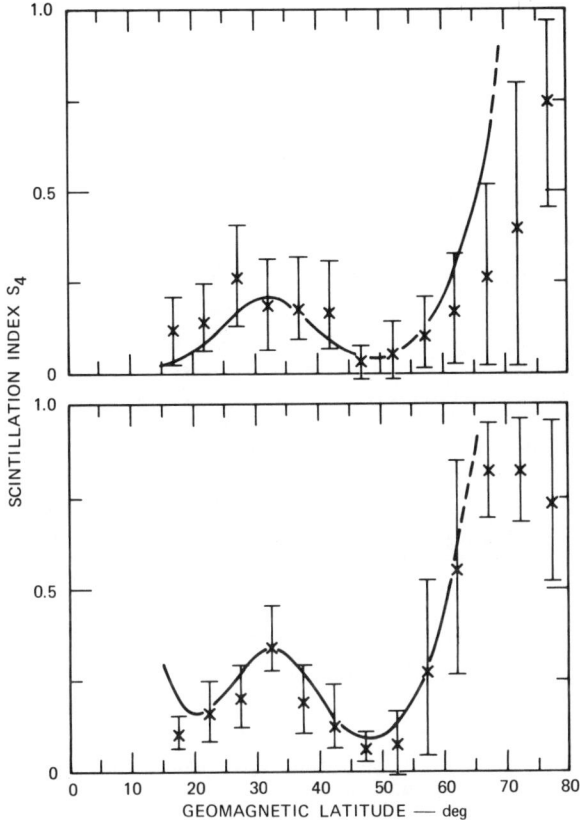

FIGURE 4 COMPARISON OF MODEL CALCULATIONS WITH HIGH-INCLINATION-SATELLITE OBSERVATIONS IN THE MIDDLE-LATITUDE AND SCINTILLATION-BOUNDARY REGIONS OF THE SOUTH PACIFIC.[21] The top is daytime: frequency = 40 MHz, sunspot number = 30. The bottom is night: frequency = 40 MHz, sunspot number = 30.

calculations were performed. In general, the fit is not so satisfactory as for solar-minimum conditions, with the nighttime results again being better than those for the daytime. The mismatch is due largely to the dictates of the extreme solar-maximum (sunspot number = 184) data obtained by Lawrence et al.[24] at Boulder during the International Geophysical Year (IGY).

Comparison of the calculated diurnal variation of 108-MHz scintillation with the Boulder data is given in Fig. 5. In general, the fit is seen to be reasonably close, although there is some discrepancy near both noon and midnight. The midday discrepancy is due to at least two causes. First, most midday scintillations at Boulder were ascribed by Lawrence et al. to E-layer irregularities, on the basis of ionosonde data, whereas the model is for F-layer irregularities only. Second, the influence of the southern-hemisphere data[21] was to depress the calculated daytime index in the latitude region of the Boulder observations. Regarding the midnight discrepancy, little can be said because the calculations indicate breakdown of the weak-scatter assumption even at 108 MHz in the Boulder IGY data.

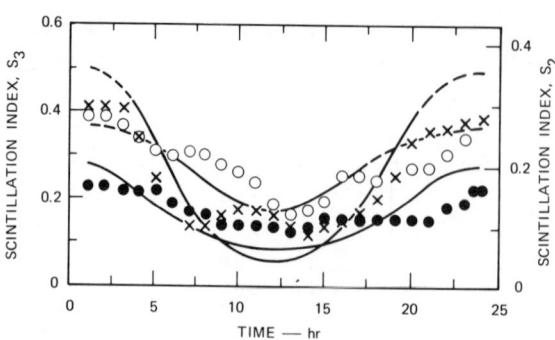

FIGURE 5 COMPARISON OF MODEL CALCULATIONS WITH RADIO-STAR OBSERVATIONS OF THE DIURNAL VARIATION OF SCINTILLATION FROM BOULDER, COLORADO[24] (X: frequency = 108 MHz, sunspot number = 184) AND COLLEGE, ALASKA[23,25] (O: frequency = 223 MHz, sunspot number = 200; ●: frequency = 68 MHz, sunspot number = 15).

Turning to auroral-zone observations, Fig. 5 also shows the diurnal variation of the observed and calculated scintillation index for the Alaska radiostar data of Little et al.[25] and of Fremouw.[23] The fits are considered quite good, although the calculations produced a slightly stronger diurnal variation near solar minimum (sunspot number = 15) than was observed. For the solar-maximum (sunspot number = 200) data, the calculated values are heavily dependent on the fourth term of the ΔN model.

Results of the only direct test of frequency dependence made in the modeling are presented in Fig. 6, comparing calculations against the two-frequency, scintillation-ratio observations of Lansinger and Fremouw.[26] The fit is quite good but is a test of frequency dependence only at vhf in the auroral zone near solar minimum.

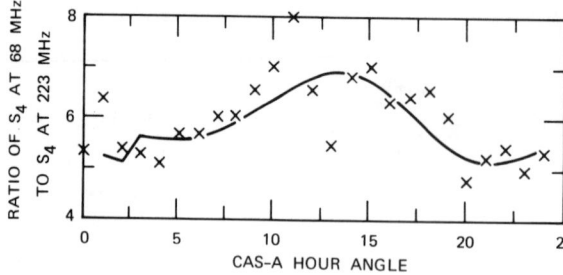

FIGURE 6 COMPARISON OF MODEL CALCULATIONS WITH RADIO-STAR OBSERVATIONS OF THE RATIO-OF-SCINTILLATION INDEX AT TWO FREQUENCIES FROM COLLEGE, ALASKA.[26] The apparent discontinuity in the calculated curve is a magnetic-field effect and would be smooth for a denser calculation grid. Sunspot number = 15.

When the model just described was published,[6] several significant limitations and shortcomings were known. First, it was recognized that testing was incomplete, especially for very high latitudes (i.e., the polar caps). At other latitudes, there was, and still is, a paucity of data extending over sufficiently long periods to test sunspot-number dependence. There is a special need for long-term data in the form of a statistically quantitative index from near the geomagnetic equator. In addition to long-term observations, data still are needed for detailed evaluation of the latitudinal dependence of scintillation there. Furthermore, any longitudinal dependence that may exist is not addressed by the model.

Probably the most distressing shortcoming of the model was its inability to maintain consistently good fits with data from the subauroral scintillation-boundary regions, for which a good collection of data was available. The problem is illustrated in Fig. 7, where scintillation index calculated from the described model (solid curves) is compared with two sets of data from the boundary region in the northern hemisphere, one from eastern North America[22] (X's) and one from western Europe[27] (O's). The American data were used in developing the model, and a good fit was obtained in the boundary region at an early stage in the development. Incorporation of southern-hemisphere data[21] caused a deterioration that was not surmounted. When the European data were employed as an independent check after model development had been completed, the discrepancy was found to persist.

It was thought significant that both sets of observations shown in Fig. 7 were in terms of one simplified scaling index,[28] whereas the southern hemisphere data shown in Fig. 4 were in terms of another.[21] This suggested the possibility of error in relating one or both of these empirical indices to

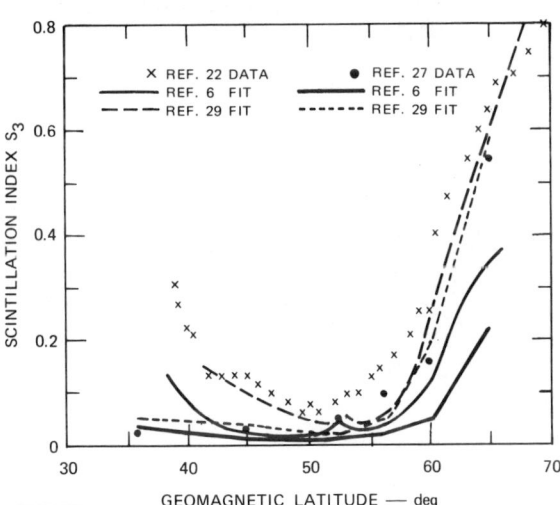

FIGURE 7 COMPARISON OF CALCULATIONS FROM TWO MODELS[6,29] WITH 54-MHz. HIGH-INCLINATION-SATELLITE DATA FROM NORTH AMERICA[22] (sunspot number = 47) AND FROM EUROPE[27] (sunspot number = 30).

the statistical ones defined in Eqs. (1-4). Such an error still may be involved, but recently Pope[29] has suggested that the discrepancy results from a true hemispheric asymmetry in the latitudinal dependence of ionospheric-irregularity strength in the scintillation-boundary regions. This suggestion is based on in situ measurements of electron-density irregularity.[30]

Subsequent to its publication, the model's shortcoming in the boundary region became even more evident as new North American observations were performed.[31] This is illustrated in Fig. 8, which also shows that the discrepancy increases with increasing geomagnetic activity. To improve the model in the boundary region, Pope[29] has proposed a modification to the third term of Eq. (9). The modification would be twofold. First, a 4° increase would be introduced in the scintillation boundary's mean latitude in the southern hemisphere as compared with the northern hemisphere. Second, sunspot number dependence of the boundary latitude would be replaced with a dependence on planetary magnetic index K_p.

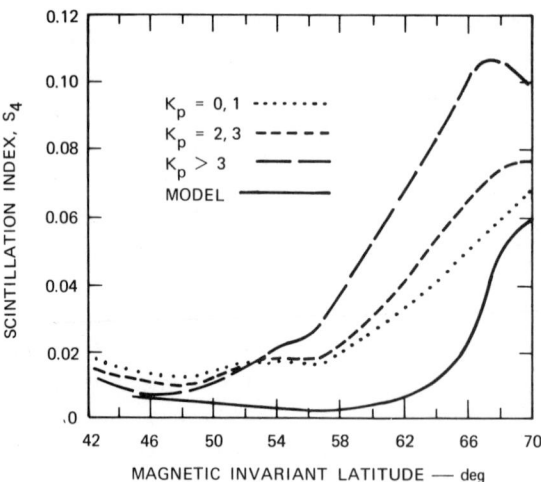

FIGURE 8 COMPARISON OF CALCULATIONS FROM FREMOUW-RINO MODEL (solid curve) WITH 400-MHz OBSERVATIONS OF HIGH-INCLINATION SATELLITES PERFORMED IN MASSACHUSETTS BY WAND et al.[31] UNDER VARIOUS GEOMAGNETIC CONDITIONS. Pope has suggested a model revision,[29] not yet fully tested, to account for magnetic (K_p) dependence.

The improvement that would be obtained in model behavior for describing the latitude of the northern scintillation boundary at low vhf for moderate magnetic activity is shown by the dashed curves in Fig. 7. The modification has not been tested against southern-hemisphere scintillation data, but its hemispheric asymmetry is in the right sense to maintain or improve the model's fit with the data of Fig. 4.

Similarly, the K_p dependence suggested would improve the model's fit with the data shown in Fig. 8. The effect on long-term behavior of the model brought about by removing the sunspot-number dependence in favor of a geomagnetic dependence, however, has not been calculated.

Another contribution to the failure of the original model to predict the location of the northern scintillation boundary for uhf observations, as shown in Fig. 8, may be the spatial-spectrum assumption inherent in the Briggs-Parkin scattering formulas employed. Incorporation of a gaussian spatial autocorrelation function is equivalent to assuming a gaussian shape for the spectrum, specified entirely by a single parameter reciprocally related to ξ_o. Thus, both the width of the spectrum and its falloff rate with increasing spatial frequency are set by ξ_o.

Now, the effect of postscattering propagation is such that the amplitude diffraction pattern develops first in the higher spatial frequencies. As regards amplitude scintillation, the diffraction process is equivalent to a "Fresnel filter" that suppresses the effects of ionospheric structure having wavelengths large compared with the Fresnel-zone radius for the radiofrequency involved.[32] If the ionospheric spatial spectrum itself is falling off with increasing spatial frequency, as indeed it does, the net result is that the amplitude fluctuations observed at the ground are produced predominantly by that portion of the ionospheric spatial spectrum whose wavelengths approximate the Fresnel-zone radius.

Thus, the amplitude scintillation observed at a given frequency (low vhf, say) is controlled by a fairly narrow range of the ionospheric spatial spectrum. It is on such observations that the values for ξ_o used in the model were based. Application of the model to higher frequencies then involves an assumption that the spatial spectrum falls off at a rate similar to that of the gaussian dictated by the ξ_o value employed. If the true spatial spectrum is richer than the gaussian in high-frequency components, for instance, application of a model based on vhf observations to uhf predications would result in underestimating the scintillation index. This could have contributed to the discrepancy illustrated in Fig. 8.

If one adheres to scattering equations based on the gaussian-spectrum assumption, it appears that satisfactory data fits could be achieved only over large frequency ranges by employing a frequency-dependent scale size, $\xi_o(f)$, which leads to use of such terms as "a spectrum of scale sizes." Since ξ_o after all is itself a parameter meant to characterize the spatial spectrum, this is tantamount to talking of a "spectrum of spectrums," a cumbersome concept even though it arises from a possibly useful ad hoc means of ordering scintillation data. What really is needed, of course, is a mathematical form more realistic than the gaussian to describe the underlying spatial spectrum of ionospheric structure, a point to which we shall return in Sec. III.

III. Channel Modeling

In spite of the shortcomings of the existing empirical model for scintillation index, it is felt to represent a useful means for collating data. Furthermore, its limitations are thought to be well identified; the boundary-latitude shortcoming in particular appears amendable to solidly based alleviation in the near future. Even so, a reliable model for calculating scintillation index still would be rather rudimentary from the communication engineer's viewpoint; it is not the only channel parameter of concern to him.

Indeed, the engineer may not be interested in scintillation index per se. Although it does provide a measure of the rms fluctuation of signal strength (directly so in the case of S_4), the engineer usually is more interested in the probability of signal fading below some threshhold value. To estimate this probability, the applied scientist must know the cumulative distribution function (CDF) of amplitude and/or of intensity. This, in turn, depends upon the underlying statistics of the complex-voltage phasor at the antenna terminals.

Propagation theories provide means for calculating some moments of complex-signal distributions, but the underlying distributions themselves result from a postulate. The postulate usually stems from application to the central limit theorem (often tacitly) to some statistical quantity that appears as a spatial integral in the theoretical development.

Depending on whether Maxwell's equations are solved by means of the first Born approximation or by the Rytov expansion, it may seem more "natural" to suppose that the signal's quadrature components are gaussian random variates, or that its phase and log-amplitude are gaussian variates. The former postulate leads to so-called gaussian statistics and the latter to so-called log-normal statistics; in either case, the two variates may be correlated.

With the gaussian and the log-normal hypotheses equally satisfying intuitively and based on equally weak theoretical foundations, the dilemma between them can be resolved (and the possibility that neither is correct investigated) only by means of experiment. Ideally, one would like to observe the complex phasor that represents the output of an antenna receiving a scintillating continuous wave (CW) signal. The ideal phase reference would be the signal that would have been received in the absence of scintillation.

It is expected that future satellite-beacon observations will permit an approximation to this ideal experiment by means of coherent transmission of several frequencies, including one sufficiently high as to provide a reference that is essentially undisturbed by the ionosphere.[33] Figure 9 illustrates schematically the scintillating phasor to be investigated in the complex plane by this technique. The question of the underlying signal statistics may be probed by investigating the shape of contours of equal probability for the phasor's tip in the complex plane. The representative contour shown in the figure is elliptical, it being well known that this is the shape expected for gaussian signal

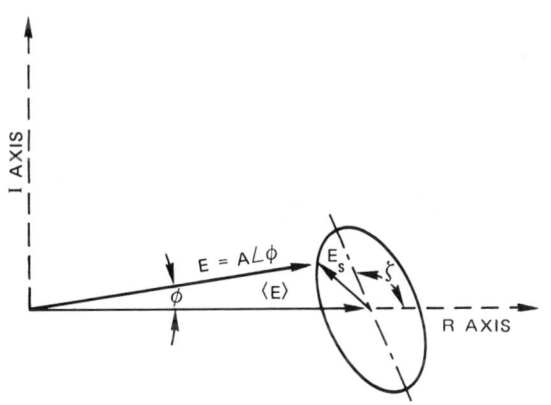

FIGURE 9 COMPLEX-PLANE REPRESENTATION OF THE PHASOR OF INTEREST FOR STUDIES OF FIRST-ORDER SIGNAL STATISTICS, INCLUDING A REPRESENTATIVE EQUIPROBABILITY CONTOUR FOR GAUSSIAN STATISTICS.

statistics.[34] For log-normal statistics, a representative contour would be rather "sausage"-shaped.

In Fig. 9, E represents the resultant received phasor whose amplitude is A and whose phase ϕ is referenced to that of the long-term average (complex) voltage $\langle E \rangle$. The latter phasor is the "nondeviated component" to which a closed-loop receiver with an extremely narrow-loop bandwidth would "lock." This mean phasor is identical to that which would exist in the absence of scintillation, except that some of its power has been transferred into the randomly varying component phasor E_S by the ionospheric scattering process. The first-order complex signal statistics are defined in terms of parameters of the equiprobability contour.

For the gaussian-statistics case illustrated in Fig. 9, the probability ellipse can be specified fully in terms of its mean radius, eccentricity, and orientation angle ζ. The mean radius is proportional to the percentage of received power which resides in the scattered component phasor. Normalizing so that the power in the nondeviated component is unity, this percentage (which may be regarded as the reciprocal of a "coherence ratio") is equal to the power σ^2 in the scattered component, which (for unit resistance) is well known to equal the sum of the voltage variances along the major and minor axes of the ellipse, σ_1^2 and σ_2^2, respectively.[34]

The eccentricity and orientation of the ellipse can be specified by a single complex quantity B, whose magnitude is $\sigma_1^2 - \sigma_2^2$ and whose phase angle is 2ζ. Now suppose we implement our ideal experiment by utilizing two quadrature detectors that deliver (real) voltages proportional to the components of E which are in phase with $\langle E \rangle$ and in phase-quadrature with it, respectively. We are interested in the variances, σ_R^2 and σ_I^2, respectively, of these voltages and their covariance C_{RI}. Our interest in these quantities lies in the fact that they are measurable, summarizing parameters that can be used to characterize the equiprobability ellipses (i.e., the first-order complex-signal statistics) by means of the following relationships[35]:

$$\sigma^2 = \sigma_R^2 + \sigma_I^2 \tag{10}$$

$$R_e(B) = \sigma_R^2 - \sigma_I^2 \qquad (11)$$

$$I_m(B) = 2C_{RI} \qquad (12)$$

Now, the ideal experiment that would allow complete determination of the complex-signal statistics has not been performed yet. Indeed, most scintillation observations have been of amplitude or intensity only. However, for gaussian statistics, the probability density function (PDF) for amplitude can be calculated, by means of a Bessel function series, in terms of the complex-ellipse parameters.[34] Alternatively, it can be computed by numerical integration of the bivariate gaussian PDF for complex voltage, as can the PDF for intensity. Once the PDF for amplitude or for intensity is obtained, it is a simple matter to integrate it to obtain the corresponding CDF.

Considerations similar to the foregoing for gaussian statistics can be made on the basis of the log-normal hypothesis for complex-signal statistics, with the variances of phase and of log-amplitude and their covariance playing the roll of the corresponding quadrature-component parameters. The calculation of the amplitude and intensity PDF's and CDF's in this case is straightforward. Rino[35] has made comparative calculations and data tests for intensity PDF's based on gaussian and on log-normal statistics.

Figure 10 shows calculated intensity PDF's for two different values of scintillation index S_4. There is only one log-normal-based PDF for each value of S_4, a potentially significant advantage for scintillation modeling. In contrast, Fig. 10 shows three gaussian-based PDF's for each value of S_4 parameterized on a quantity Z that is related to the ellipticity and orientation of the equiprobability ellipse shown in Fig. 9. The relation is through the complex quantity B, whose components are expressed in Eqs. (11) and (12).

Also shown in Fig. 10 are PDF's based on the Nakagami m-distribution for amplitude, which shares with the log-normal-based PDF the advantage of being unique for a given S_4. The theoretical basis for the Nakagami distribution is somewhat obscure, but it has enjoyed some empirical success in

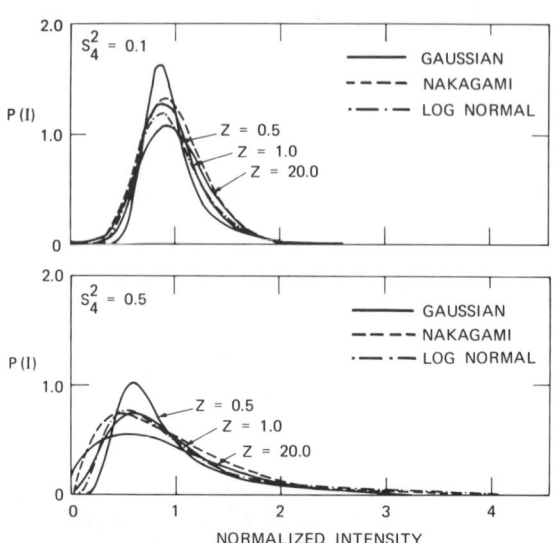

FIGURE 10 FAMILIES OF GAUSSIAN-BASED PDFs PARAMETERIZED ON Z, COMPARED WITH THE CORRESPONDING LOG-NORMAL-BASED AND NAKAGAMI-BASED PDFs FOR TWO VALUES OF SCINTILLATION INDEX.

describing amplitude scintillation.[13,36] Over some range of parameter values, the Nakagami amplitude PDF is an approximation for that based on gaussian complex-signal statistics. One such range is for very large values of the Z parameter, when gaussian statistics reduce to the well-known Rice statistics for a signal in noise [i.e., equal power in the in-phase and phase-quadrature noise components and zero correlation between them, which from Eqs. (10) and (11) corresponds to $B = 0$].

In order to test the relative merits of the gaussian and log-normal hypotheses for signal statistics, Rino[35] has compared intensity historgrams with calculated intensity PDF's for three different scintillation conditions: 1) microwave scatter by a laboratory plasma, 2) radio-star scintillation through the solar wind, and 3) satellite scintillation through the ionosphere. Rather consistently, a better fit (as determined either from the χ^2 or the least-squares test) is obtained with the gaussian-based PDF (for optimally chosen Z) than for the log-normal-based PDF.

Figure 11 shows three such comparisons with solar-wind data collected by researchers at the University of California at San Diego (UCSD).[37] As expected from theory, the differences between the PDF's are slight for small S_4 values and increase for increasing scintillation index. It is clear both by inspection and from the square-error values that the

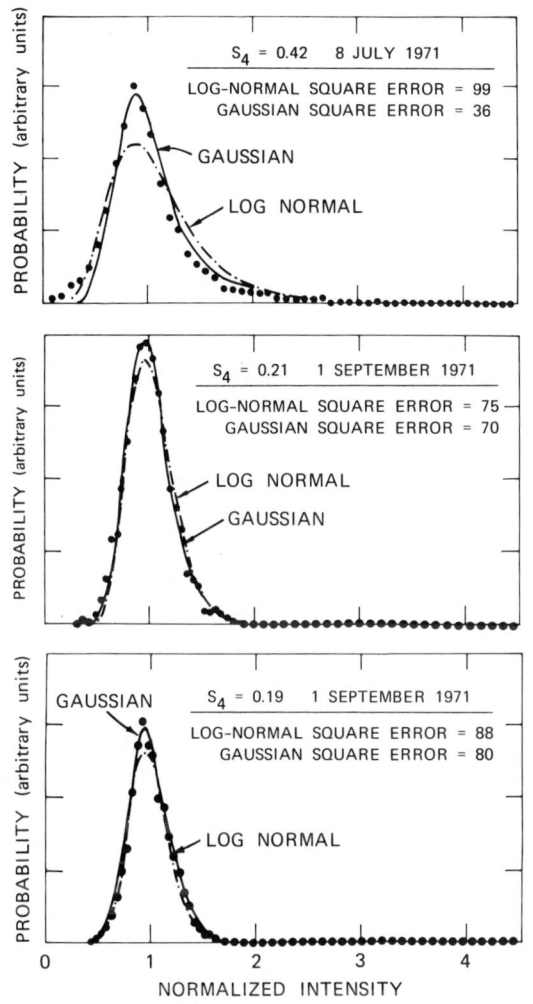

FIGURE 11 BEST-FIT GAUSSIAN AND LOG-NORMAL PDFs FOR 75-MHz INTERPLANETARY SCINTILLATION DATA PROVIDED BY COLES.[37]

gaussian hypothesis provides a better approximation to reality than does the log-normal hypothesis.

A similar comparison is shown in Fig. 12 (top) for ionospheric scintillation data collected by workers at Air Force Cambridge Research Laboratories (AFCRL)[38] from observations of the vhf transmissions from ATS-5. The gaussian hypothesis provides a smaller χ^2 error than does the log-normal, although the difference between the two is apparent to the eye only near the peak of the distribution. Also shown (bottom of Fig. 12) are the calculated CDF's and the corresponding histogram for the ATS-5 data. For practical purposes, the differences between the two calculated curves are insignificant, both of them providing excellent CDF fits to the data.

Only a few data sets have been compared with PDF's calculated from the gaussian and the log-normal hypotheses. Consistently, however, the gaussian hypothesis provides the better of the two, and the best-fit PDF's obtained so far display a considerable degree of similarity. Independent of

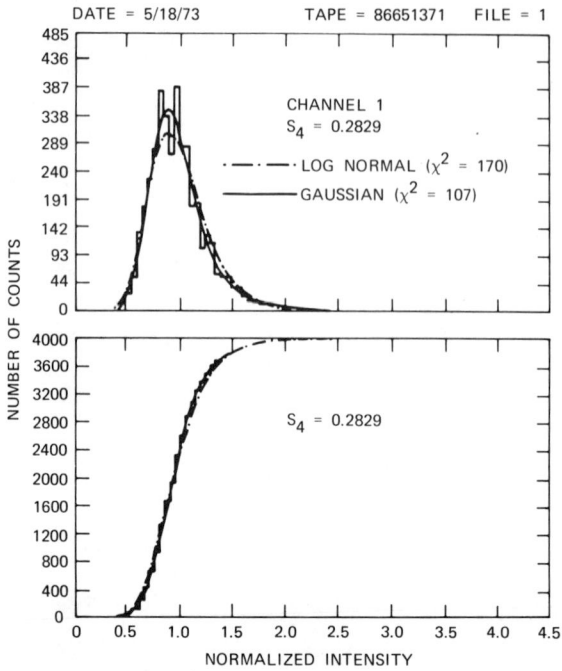

FIGURE 12 BEST-FIT GAUSSIAN AND LOG-NORMAL PDFs (top) AND CDFs (bottom) FOR IONOSPHERIC SCINTILLATION DATA COLLECTED BY WHITNEY AND AARONS[38] AT 136 MHz.

the value of S_4, the magnitude of B is generally larger than 0.8 and its angle less than $30°$. This implies that more than 80% of the scattered power is in phase quadrature with the undeviated signal component and that the in-phase and phase-quadrature components of the scattered signal are partially correlated.

These results readily explain why past data comparisons[39] with predictions based on Rician statistics generally have provided a poor fit.

The best-fit gaussian PDF, in general, tends to be more peaked about its mean than either the corresponding Rice, Nakagami, or log-normal PDF's. As a practical engineering matter, however, the differences may well be insignificant for amplitude statistics; the Nakagami PDF gives more conservative estimates of the true CDF than does the more accurate gaussian PDF. As an example, Fig. 13 shows fade margins for two values of S_4 and different PDF's by setting $B = 0.85 \angle 20°$, which seems to be a representative value.

IV. Prospects for Improved Scintillation Modeling

The consistency found in the gaussian-based PDF's is encouraging for scintillation modeling. If borne out by additional data comparisons, first-order channel modeling will be considerably facilitated, especially if phase-related statistical quantities share the consistency found for amplitude-related ones. The reason is that the B-related parameter Z

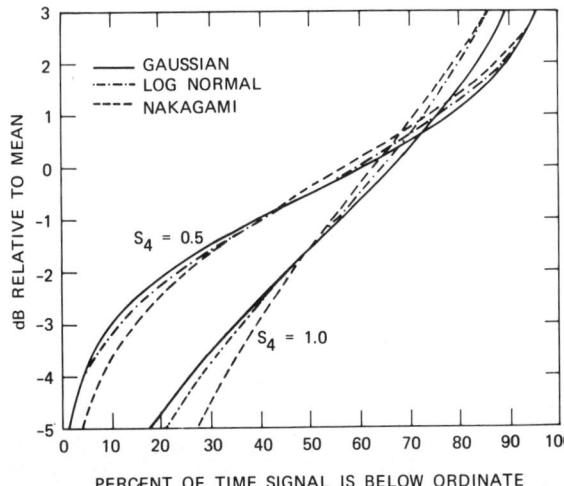

FIGURE 13 FADE-MARGIN COMPARISON FOR CDFs BASED ON GAUSSIAN, LOG-NORMAL, AND NAKAGAMI STATISTICS.

is essentially the Fresnel-distance measure that appears as the argument in Eq. (7). Recent theoretical work by Rino[35] has related Z to ionospheric parameters for the power-law spatial spectrum described by McClure.[2] Consistency in Z may mean that the irregularity characteristics that control scintillation, except for their strength, vary only through a rather small range. If so, this will permit modeling of complex-signal statistics with a manageable number of parameters.

In the meantime, existing and forthcoming in situ measurements of the spatial spectrum should permit improved prediction of the frequency dependence of S_4 as compared with that in the presently available scintillation-index model. This, coupled with recent morphological findings regarding the irregularity boundary, means that a much improved model for high-latitude scintillation index should be forthcoming within a year or so.

Hopefully, observations of signals from the beacons and transponders aboard ATS-F will improve the morphological description of equatorial scintillation, also. An equally pressing matter in this regard is detailed measurement of the irregularity spatial spectrum near the equator, where it may well be nonmonotonic. If this is the case, challenging theoretical work will be required to relate the results to observable signal characteristics in time for experimental testing against data from the forthcoming DNA-002 satellite beacon. This latter instrument, which is scheduled for launch in August 1974, will permit measurements closely approximating the ideal complex-signal observations described in Sec. III.

V. Conclusion

It appears, then, that much improved scintillation index models may be available in a matter of a year or so and that a model of first-order amplitude statistics is on the horizon. Pending successful collection of complex-signal data for testing theoretical predictions that now exist, the same model should be able to provide estimates of first-order phase-related quantities.

Second-order statistics, having to do with the spatial and temporal structure (e.g., spaced-receiver correlations, and temporal spectra) of scintillating signals as well as with coherence over broad bands of frequency, may soon equal first-order statistics in importance for engineering application, as discussed by Massey.[40] If, indeed, the gaussian hypothesis for complex-signal statistics proves reliable as more complete tests are made, then, in principal, higher-order statistical quantities should be calculable from models of ionospheric irregularities. Thus, the foundation for comprehensive channel modeling for application to scintillation-prone communication links appears to be forming.

Foreseeable models are statistical in nature, both as regards signal structure and scintillation morphology. Deterministically predictive models will depend on fundamental understanding of the production of ionospheric irregularities. The production mechanisms may be similar under different conditions, but the causative agents may well differ. Ionospheric research by means of in situ probes and radar backscatter methods at both low and high latitudes may shed light on the causes in both the equatorial and high-latitude scintillation zones. Questions of magnetospheric proportions may need to be answered, however, before scintillation conditions are truly predictable, especially in the high-latitude case.

Meanwhile, statistical scintillation modeling is in a period of comparatively rapid growth, because of theoretical advances, data-processing improvements, and impending experimental opportunities. For application of these scientific results to engineering enterprises, professional organizations can play a decidedly catalytic role.

References

[1] Barabanenkov, N., Kravtsov, A., Rytov, S. M., and Tamarski, V. I., "Status of the Theory of Propagation of Waves in a Randomly Inhomogeneous Medium," Soviet Physics (Elpsala), Vol. 13, No. 5, May 1971, pp. 551-575.

[2] McClure, P., "Geophysical Properties of the Ionospheric Irregularities Responsible for Radio Scintillation," AIAA Paper 74-53, Jan.-Feb. 1974, Washington, D.C.; also published elsewhere in this volume.

[3] Crane, R., "Morphology of Ionospheric Scintillation," AIAA Paper 74-52, Jan.-Feb. 1974, Washington, D.C.; also published elsewhere in this volume.

[4] Briggs, B. H. and Parkin, I. A., "On the Variation of Radio Star and Satellite Scintillations with Zenith Angle," Journal of Atmospheric and Terrestrial Physics, Vol. 25, No. 6, June 1963, pp. 339-366.

[5] Ratcliffe, J. A., "Some Aspects of Diffraction Theory and Their Application to the Ionosphere," Reports on Progress in Physics, Vol. XIV, 1956, p. 188.

[6] Fremouw, E. J. and Rino, C. L., "An Empirical Model for Behavior of Average F-Layer Scintillation at VHF/UHF," Radio Science, Vol. 8, No. 3, March 1973, pp. 213-222.

[7] Liszka, L., "An Investigation of the Height of Scintillation-Producing Irregularities," Arkiv For Geofysik, Vol. 4, December 1964, pp. 523-528.

[8] Yeh, K. C. and Swenson, G. W., Jr., "F-Region Irregularities Studies by Scintillation of Signals from Satellites," Journal of Research National Bureau of Standards, Sec. D, Vol. 68D, No. 8, August 1964, pp. 881-894.

[9] Kent, G. S. and Koster, J. F., "Some Studies of Nighttime F-Layer Irregularities at the Equator Using Very High Frequency Signals Radiated from Earth Satellites," Spread-F and Its Effects Upon Radiowave Propagation and Communications, edited by P. Newman, Technivision, Maidenhead, England, 1966, pp. 333-356.

[10] Jones, I. L., "Further Observations of Radio Stellar Scintillation," Journal of Atmospheric and Terrestrial Physics, Vol. 19, No. 1, January 1960, pp. 26-36.

[11] Liszka, L., "A Study of Ionospheric Irregularities Using Transmissions at 54 Mc/s," Arkiv For Geofysik, Vol. 4, 1963, pp. 227-246.

[12] Koster, J. R., "Some Measurements of the Irregularities Giving Rise to Radio-Star Scintillations at the Equator," Journal of Geophysical Research, Vol. 68, No. 9, May 1963, pp. 2579-2590.

[13] Bischoff, K. and Chytil, B., "A Note on Scintillation Indices," Planetary and Space Science, Vol. 17, No. 5, May 1969, pp. 1059-1066.

[14] Fremouw, E. J. and Bates, H. F., "Worldwide Behavior of Average VHU-UHF Scintillation," Radio Science, Vol. 6, No. 10, October 1971, pp. 863-869.

[15] Koster, J. R., "Equatorial Studies of the VHF Signal Radiated by Intelsat II, F-3, 1, Ionospheric Scintillation," Progress Rept. 3, Contract F61052-67-C-0027, Sept. 1968, University of Ghana-Legon, Accra, Ghana.

[16] Koster, J. R., private communication, 1971.

[17] Christiansen, R. M., "Preliminary Report of S-Band Propagation Disturbance During Alsep Mission Support (November 19, 1969-June 30, 1970)," Rept. X-861-71-236, 1971, NASA Goddard Space Flight Center, Greenbelt, Md.

[18] Skinner, N. J., Kelleher, R. F., Hacking, J. B., and Benson, C.W., "Scintillation Fading of Signals in the SHF Band," *Nature, Physical Science*, Vol. 232, 1971, pp. 19-21.

[19] Craft, H. D., Jr. and Westerlund, L. H., "Scintillations at 4 and 6 GHz Caused by the Ionosphere," AIAA Paper 72-179, January 1972, San Diego, Calif.

[20] Preddey, G. F., Mawdsley, J., and Ireland, W., "Midlatitude Radio-Satellite Scintillation: The Morphology Near Sunspot Minimum," *Planetary and Space Science*, Vol. 17, May 1969, pp. 1161-1171.

[21] Preddey, G. F., "Midlatitude Radio-Satellite Scintillation: The Variation with Latitude," *Planetary and Space Science*, Vol. 17, No. 8, August 1969, pp. 1557-1561.

[22] Aarons, J., Mullen, J. P., and Basu, S., "The Statistics of Satellite Scintillations at a Subauroral Latitude," *Journal of Geophysical Research*, Vol. 69, No. 9, May 1964, pp. 1785-1794.

[23] Fremouw, E. J., "Aberrations of VHF-UHF Signals Traversing the Auroral Ionosphere," Final Rept., Contract NAS5-3904, Rept. UAG R-181, 1966, Geophysical Institute of the University of Alaska, College, Alaska.

[24] Lawrence, R. A., Jespersen, J. L., and Lamb, R. C., "Amplitude and Angular Scintillations of the Radio Source Cygnus-A Observed at Boulder, Colorado," *Journal of Research National Bureau of Standards*, Sec. D, Vol. 65(D), No. 4, July-August 1961, pp. 333-350.

[25] Little, C. G., Reid, G. C., Stiltner, E., and Merritt, R. P., "An Experimental Investigation of Radio Stars Observed at Frequencies of 223 and 456 Megacycles per Second from a Location Close to the Auroral Zone," *Journal of Geophysical Research*, Vol. 67, No. 5, May 1962, pp. 1763-1784.

[26] Lansinger, J. M. and Fremouw, E. J., "The Scale Size of Scintillation-Producing Irregularities in the Auroral Ionosphere," Journal of Atmospheric and Terrestrial Physics, Vol. 29, May 1967, pp. 1229-1242.

[27] Joint Satellite Studies Group, "On the Latitude Variation of Scintillations of Ionosphere Origin in Satellite Signals," Planetary and Space Science, Vol. 16, No. 6, June 1968, pp. 775-781.

[28] Whitney, H. E., Aarons, J., and Malik, C., "A Proposed Index for Measuring Ionospheric Scintillations," Planetary and Space Science, Vol. 17, 1969, pp. 1069-1073.

[29] Pope, J., private communication, 1973.

[30] Sagalyn, R., Smiddy, M., and Ahmad, M., "High-Latitude Irregularities in the Top Side Ionosphere Based on ISIS I Thermal Ion Probe Data," Journal of Geophysical Research, Vol. 79, No. 28, October 1974, pp. 4252-4261.

[31] Wand, R. H., Evans, J. V., Ghiloni, J. C., and Power, R. A., "Morphology of Auroral and Sub-Auroral Zone Scintillation," URSI Meeting, August 1973, Boulder, Colorado.

[32] Rufenach, C. L., "A Radio Scintillation Method of Estimating the Small-Scale Structure in the Ionosphere," Journal of Atmospheric and Terrestrial Physics, Vol. 38, July 1971, pp. 1941-1951.

[33] Fremouw, E. J., "A Planned Polar-Orbiting Wideband Beacon," Symposium on Future Application of Satellite Beacon Measurements, May-June 1972, Graz, Austria.

[34] Beckman, P. and Spizzichino, A., The Scattering of Electromagnetic Waves from Rough Surfaces, Pergamon, New York, 1963, pp. 119-136.

[35] Rino, C. L. and Fremouw, E. J., "Ionospheric Scintillation Studies," Final Rept., Contract NAS5-21891, SRI Project 2273, 1973b, Stanford Research Institute, Menlo Park, Calif.

[36] Whitney, H. E., Aarons, J., Allen, R. S., and Seemann, D. R., "Estimation of the Cumulative Amplitude Probability Distribution Function of Ionospheric Scintillations," Radio Science, Vol. 7, No. 12, December 1972, pp. 1095-1104.

[37] Coles, W., private communication to C. Rino, 1973.

[38] Whitney, H. and Aarons, J., private communication, 1973.

[39] Armstrong, J. W., Coles, W. A., and Rickett, B. J., "Observations of Strong Interplanetary Scintillation at 75 MHz," Journal of Geophysical Research, Vol. 77, No. 16, June 1, 1972, pp. 2739-2743.

[40] Massey, J., "Methods of Alleviation of Ionospheric Scintillation Effects on Digital Communication," AIAA Paper 74-55, Jan.-Feb. 1974, Washington, D.C.; also published elsewhere in this volume.

IMPACT OF SCINTILLATION
ON TRANSIONOSPHERIC
COMMUNICATIONS

Howard A. Blank[*]
Computer Sciences Corporation, Falls Church, Va.

and

Thomas S. Golden[+]
NASA Goddard Space Flight Center, Greenbelt, Md.

Abstract

Analyses of automatic gain control (AGC) analog data from the small astronomy satellite (SAS) at vhf (136 MHz) have shown amplitude fluctuations in the signals received at several equatorial tracking sites operated by NASA. Coincident digital telemetry data, transmitted from SAS at 1 kbit/sec, have also been analyzed. Errors in data are shown to increase markedly for night-time satellite passes around the periods of the two equinoxes as compared to the errors observed during other periods. This paper establishes the qualitative and quantitative correlations between the ionospheric scintillation phenomenon and the occurrence of errors. Strong time and geographic cause-and-effect relationships are established by determining the statistical correlation between scintillation indices (SI) and bit error rates (P_e). Detailed analyses of the AGC fading statistics and the fine-grain error statistics, such as the error-free gap distribution $G(j)$ and the bit-error autocorrelation function

---

Presented as Paper 74-51 at the AIAA 12th Aerospace Sciences Meeting, Jan. 30-Feb. 1, 1974, Washington, D. C. The work reported in this paper has been supported by NASA Goddard Space Flight Center under Contract NAS5-24011.
[*]Manager, Advanced Systems Section.
[+]Director of Technology and Transfer Project.

a(j), establish the strong correlation between deep fades and their effects on the burst error structure associated with the transionospheric channel and the communication receiver being employed. The results presented herein serve to characterize the degradation potentiality of the equatorial transionospheric channel during the equinox periods.

1. Introduction

A large amount of theoretical and empirical knowledge has been obtained during the past two decades concerning electromagnetic wave propagation in the ionosphere. Almost all of the investigations of transionospheric propagation have concerned themselves with the study of either the physical propagation mechanisms or the fluctuating statistical characteristics of the received rf signal. Little attention has been devoted to the impact of the random characteristics of the ionosphere on the quality of communications. It has been observed[1] that digital satellite telemetry returns have shown severe degradation in data quality during periods of ionospheric activity.

In order to ascertain in a quantitative, as well as a qualitative, sense the impact of ionospheric disturbances on digital data transmissions, a study has been conducted whose purpose was to determine the characteristics of digital errors caused by the ionospheric scintillation phenomenon. Specifically, empirical telemetry and AGC signal data have been collected from the SAS-1 satellite at six equatorially located ground stations during the spring and fall equinox periods of 1971 and 1972. The satellite stations and time periods were selected in order to obtain conditions of worst-case ionospheric behavior. It was hoped that the resulting communication transmission degradations would be most contrasting as compared to those nominally encountered under nonscintillation conditions.

The telemetry and AGC data are being converted, reduced, and analyzed by a variety of techniques to be described herein. The analyses include a comparison of the error statistics caused by ionospheric scintillation with regard to time and geography, a determination of data dropout and gross burst-error characteristics, and a detailed comparison of the fading and fine-grain error characteristics of individual sets of data. This paper presents some preliminary results from the study.

IMPACT OF SCINTILLATION ON COMMUNICATIONS

2. Equipment Description

The empirical equipment for the telemetry and AGC measurements includes the SAS-1 satellite and NASA tracking stations at Quito, Equador; Ascension Island; Kourou, French Guiana; Brazzaville, Congo; Ouagadougou, Upper Volta; and the Seychelles Islands. The equipment descriptions are given in the following paragraphs.

2.1. SAS-1

The SAS-1 spacecraft was launched in late 1970 and is currently in a 550-km circular-equatorial orbit, with a 3° inclination. The telemetry is transmitted from the satellite at 136.680 MHz via a circularly polarized turnstile antenna. The spacecraft antenna has a gain of 0 db, and the ERP of the spacecraft is 0.25 w. Station passes are 10 to 15 min in length, which yields from 6×10^5 to 9×10^5 bits of telemetry at a 1-kbit/sec data rate. The telemetry data are modulated as 1-bit PCM/NBPM utilizing a ± 1-rad split phase technique. This method leaves approximately 58.5% of the signal power available in the carrier and approximately 38.8% of the signal power available in the first set of modulated sidebands.

2.2. Ground Receiver

The basic equipment configuration at each of the Earth terminals is shown in Fig. 1. A station has either a pair of 40-ft parabolic antennas with 55% efficiency or a pair of 16-element yagis. In either case, the antenna system has 22 db of gain at 136 MHz. Each output drives preamplifiers at vhf, which are then down-converted to IF, filtered through a 10-kHz bandpass filter, and acquired and detected. The receiver carrier tracking loops are third order and have bandwidths of 30 Hz. The detection loss of this receiver is approximately 4 db. The detected PCM output, as well as the AGC error signal, is fed through module recorders (i.e., operational and/or logarithmic amplifiers) and on to an Ampex FR600 magnetic tape recorder.

3. Data Reduction

The data conversion and reduction was accomplished via three different techniques. Two were employed for the digital telemetry data, and one was used for the AGC data. Each technique is described below.

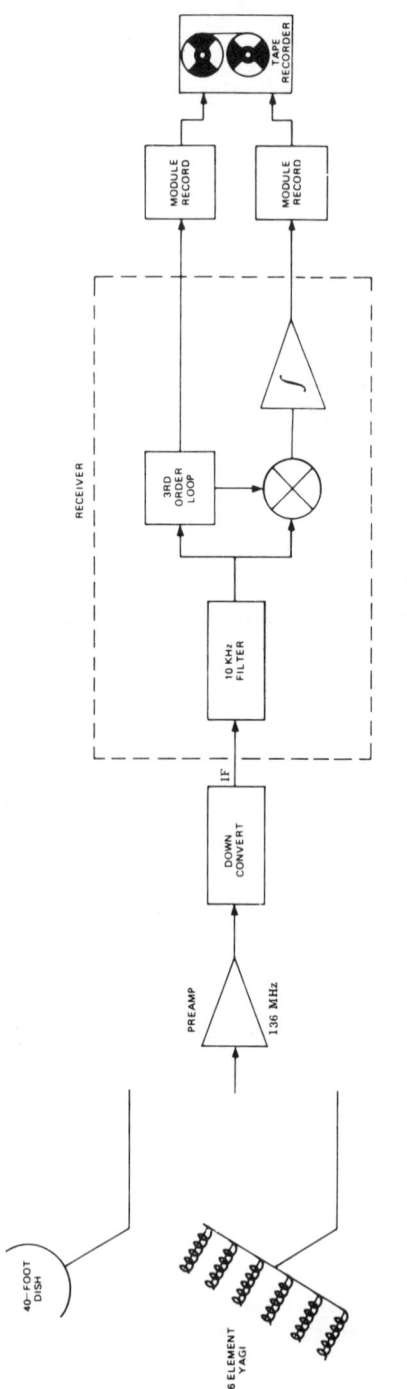

Fig. 1 Receiver configuration.

3.1. Ionospheric Distortion Analysis Program (IDAP) System

IDAP was developed to provide a reliable technique for reducing AGC fading data in order to determine the associated statistical characteristics of the transionospheric scintillation channel. The data conversion and reduction procedure of IDAP is shown in Fig. 2. All of the equipment involved is located at Goddard Space Flight Center (GSFC), and all AGC data processing was conducted using this equipment.

Three basic steps are used in the computer processing of AGC data. The first step is to digitize the analog recording of the AGC collected in the field. The second step is to interpret and translate the digitized levels to obtain a time series of signal strength measurements. The third step is to extract the desired statistics from the translated time series. The following list of analysis programs was written for the reduction of the field data: 1) probability density function vs signal level (dbm or mw); 2) probability distribution function vs signal level (dbm or mw);

IMPACT OF SCINTILLATION ON COMMUNICATIONS

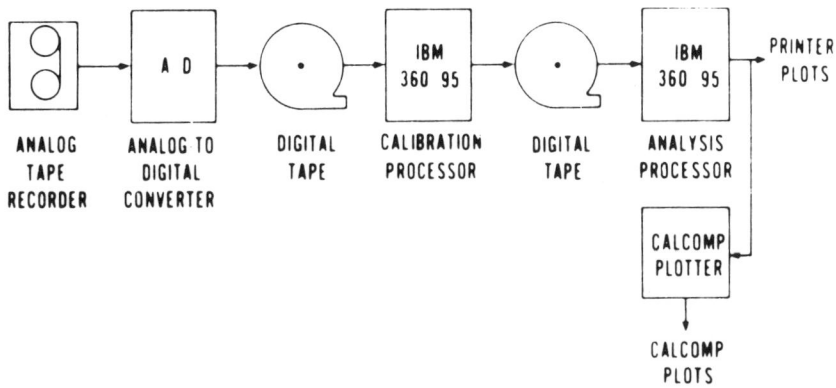

Fig. 2 IDAP system.

3) autocorrelation function vs time lag (sec); 4) cross-correlation function vs time lag (sec); and 5) spectral-density function vs frequency (Hz).

3.2. Quality Control Monitoring System (QCMS)

The SAS-1 PCM telemetry data, along with ground time and other pertinent information, are recorded on analog-magnetic tapes, which are sent to the Data Processing Branch of NASA/GSFC for digitization. The data are processed and placed on digital-formatted computer tapes for further processing. The Data Processing Branch performs quality analysis on each digital file generated from the analog tape by examining the following: 1) the number of recovered telemetry frames, 2) the percentage of data recovery based on frames recovered and frames expected, 3) the data quality based on the telemetry errors occurring in the minor frame sync pattern, and 4) others. The quality indices are placed on 80-column cards during the pre-edit phase of the input processing. These quality cards then were processed on the quality control monitoring system (QCMS). The SAS-1 preprocessing system is shown in Fig. 3.

The QCMS located at NASA/GSFC is used routinely to obtain the quality-quantity statistics and time history of SAS-1 telemetry. The QCMS (see Fig. 4) consists of a cathode-ray tube (CRT) for visual display, with controls and a memory (Idiiom 620 computer) for manipulating the displayed data. The 620 computer system operates as an adjunct to a CDC 3200 computer, through which information on data quality-quantity of processed (station-contact) passes

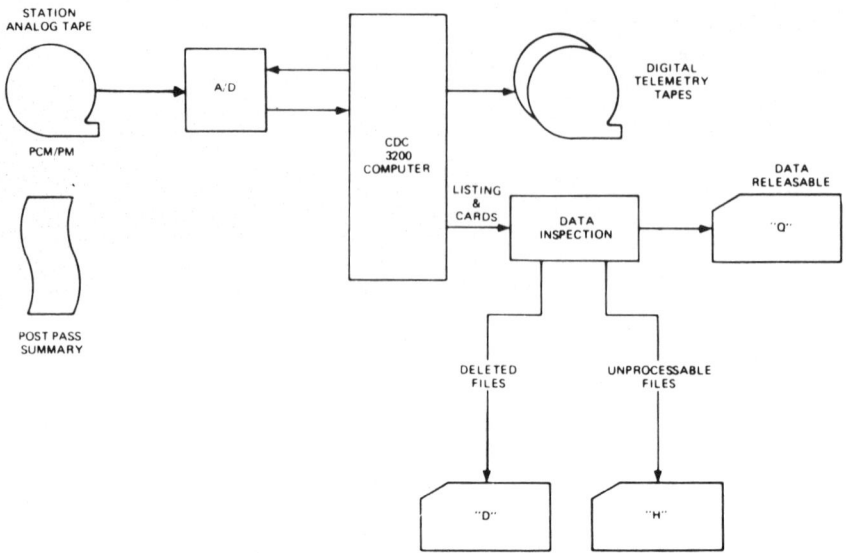

Fig. 3 Simplified SAS-1 preprocessing system.

is entered into the system. The inputs to the QCMS are either 80-column punched cards or digital tape containing card images, both of which contain data quality-quantity information resulting from the analog-to-digital (A/D) and pre-edit conversion processes performed. Plots of the time-history of data recovery, data quality, etc., are presented on the CRT, along with a plot tape from the QCMS; the plot tape is processed on a CALCOMP plotter.

The QCMS utilizes the CDC 3200 for 1) parameter interpretation-graph building, 2) statistical calculations, and 3) generating of

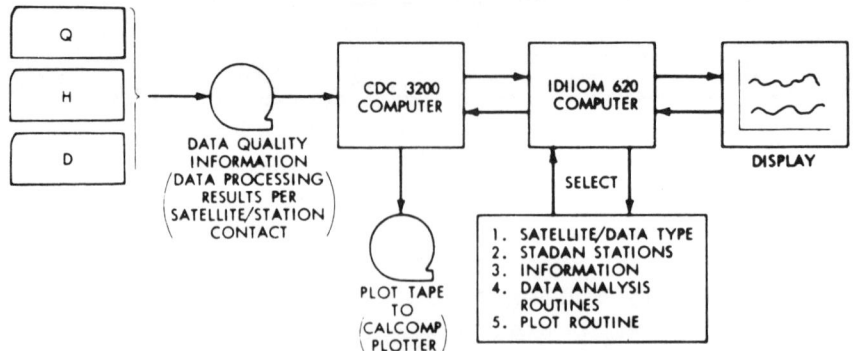

Fig. 4 Quality control monitoring system (QCMS).

IMPACT OF SCINTILLATION ON COMMUNICATIONS

hard-copy outputs. The Idiiom 620 computer is responsible for 1) input of constraints (or parameters) and 2) graph display and control.

3.3. Fine-Grain Error Reduction

In addition to the generalized QCMS analysis, the SAS-1 telemetry data were required to be analyzed for fine-grain error characteristics in order to evaluate the burst-error behavior of the scintillation channel. The data reduction procedure is shown in Fig. 5. The PCM/PM station tapes are A/D converted, timed, and detected, producing digital data tapes with all of the acquirable telemetry data. The digital telemetry then is error-detected against known transmitted bit patterns, and a digital error tape is produced. The digital error tape then is analyzed to obtain useful fine-grain error statistics.

Among the potentially available fine-grain error statistics are three that are most useful in ascertaining the error behavior of the transionospheric scintillation channel, as follows:

Error-free gap distribution:

$$G(j) \triangleq \Pr\left\{ 0^{j-1} \mid 1 \right\}$$

Bit-error autocorrelation function:

$$a(j) \triangleq \Pr\left\{ x^{j-1} 1 \mid 1 \right\}$$

Block-error distribution:

$$P(m,n) \triangleq \Pr\left\{ m \text{ errors in a block of n bits} \right\}$$

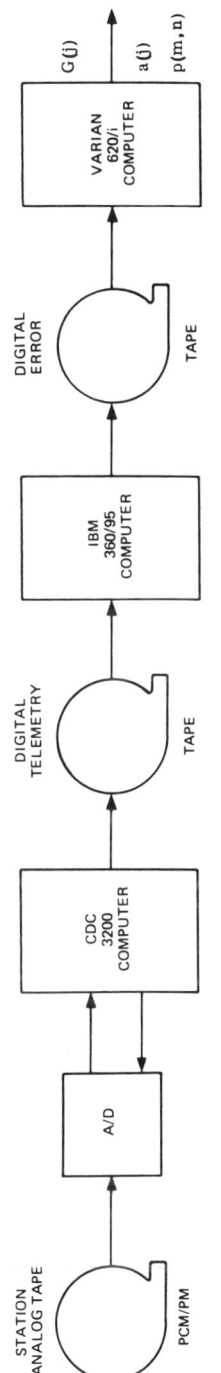

Fig. 5 Fine-grain error reduction procedure.

(Throughout this paper, 1 denotes a bit in error, 0 denotes a bit not in error, x denotes either. Thus, 0^{j-1} denotes j-1 consecutive error-free bits.) Each telemetry tape reduced and reported herein was analyzed to determine G(j), a(j), and P(m,n).

4. Analysis of Data

This section presents the results of the data collection, reduction, and analysis of AGC and telemetry data obtained from SAS-1 transmissions to six Earth stations located in the geomagnetic equatorial region. The results are presented in three paragraphs segregated in order to indicate transmission error behavior during nonscintillation periods, as well as error behavior during those periods of expected ionospheric scintillation activities. The latter is presented in two parts dealing with gross and fine-grain error behavior, respectively. The fine-grain error analysis includes the results of the corresponding channel signal statistics, as well as a direct comparison of simultaneous burst error and signal fading statistics.

4.1. Nonscintillation Error Behavior

The SAS-1 system was designed to provide a low bit-error probability (P_e) in order to insure high-quality telemetry transmission. With the equipment described in Sec. 2, processed data from January-June 1971 for all stations indicate[2] an average energy-per-bit to gaussian noise density ratio (E_b/N_o) of 8.2 db, with an average block data recovery probability (P_{BR}) of 0.98. During periods of nonscintillation activity, the satellite-Earth transmission channel can be considered to be an independent, random error channel. For the modulation technique employed, an $E_b/N_o = 8.2$ db will yield an $P_e = 1.5 \times 10^{-4}$. Processed data from September-December 1971 indicate[2] an $E_b/N_o = 9.0$ db, which results in an $P_e = 3.3 \times 10^{-5}$. During this period, $P_{BR} \to 1$. The improved performance of the telemetry system during the latter part of 1971 is due to increased ERP in the spacecraft. This went from 0.25 to 2.00 w in November 1971. All subsequent transmissions used the higher ERP.

Both of the preceding time frames include periods of scintillation activity that may bias the error rate results. Consequently, error behavior analyses were conducted for solstice periods. The first of these was conducted for December 1970-January 1971 at

Quito, Equador. The results indicate that $\overline{E_b/N_o} = 9.60$ db, with a standard deviation ($\sigma[E_b/N_o]$) of 2.85 db. The results were obtained from a sample of 1203 satellite passes, representing a total of approximately 7.26×10^8 transmitted telemetry bits. For the statistics of E_b/N_o just cited, Table 1 shows the corresponding values of P_e. Figure 6 gives the raw data from which the statistics of Table 1 were obtained. Plotted are E_b/N_o in decibels vs time of day at Quito for the winter solstice of 1970. Each triangle represents a full satellite pass, and the curve is a least-squares fit to the data. Note the degradation due to ionospheric scintillation during the 0-500-hr GMT period.

An even more detailed example of SAS-1 system behavior during periods of quiet ionospheric activity is illustrated by the results shown in Table 2. Error rate results are given for midday hours at Quito, Equador and Ascension Island for the summer solstice of 1971, and for Ouagadougou, Upper Volta and the Seychelles Islands during the winter solstice of 1972. These results show favorable agreement with those previously quoted.

Two other statistics of interest to be used for future comparison are the average block data recovery probability (P_{BR}) and the average zero block error probability (P_{BO}). The value of P_{BR} is of interest because it tells us the equipment "dropout" probability. That is, P_{BR} is a combination of station receiver synchronization errors and computer equipment processing errors. Knowing P_{BR} for nonscintillation periods will provide a control for receiver synchronization error probabilities during severe scintillation periods. The value of P_{BO} is of interest because it yields

Table 1 Error rates for Quito, Equador during December 1970-January 1971

Statistic	E_b/N_o, db	P_e
$\overline{E_b/N_o} + \sigma[E_b/N_o]$	12.45	1.4×10^{-9}
$\overline{E_b/N_o}$	9.60	1.0×10^{-5}
$\overline{E_b/N_o} - \sigma[E_b/N_o]$	6.76	1.1×10^{-3}

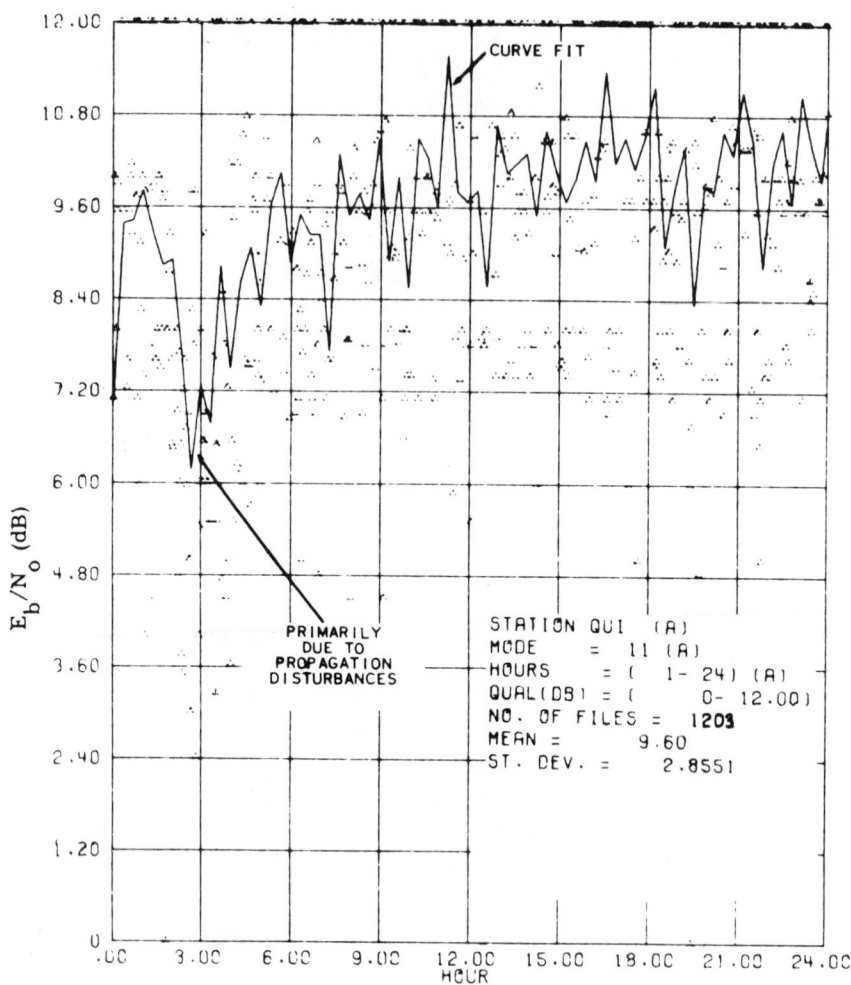

Fig. 6 E_b/N_0 vs time for winter solstice, 1970 at Quito.

Table 2 Error rates for solstice periods

Station	E_b/N_o, db	P_e
Quito	8.00	2×10^{-4}
Ascension	7.70	3×10^{-4}
Ouagadougou	8.20	1.5×10^{-4}
Seychelles	8.00	2×10^{-4}

IMPACT OF SCINTILLATION ON COMMUNICATIONS 407

information as to the "burstiness" of channel errors. The smaller the value of P_{BO}, the more clustered are the channel errors. As for P_{BR}, knowledge of P_{BO} for solstice periods provides a control for equinox periods.

Table 3 gives the results obtained for P_{BR} and P_{BO} for the same stations and solstice periods shown in Table 2. These results indicate that equipment limitations keep $0.90 \lesssim P_{BR} \lesssim 0.95$ and that the relative independence of error occurrences keeps $P_{BO} < 0.90$. The importance of the statistics quoted in Table 3, as well as those quoted in Table 2, will become more apparent when we discuss error behavior during periods of strong scintillation activity.

Table 3 Block recovery and error-free probabilities for solstice periods

Station	P_{BR}	P_{BO}
Quito	0.9036	0.8954
Ascension	0.8970	0.8590
Ouagadougou	0.9549	0.8875
Seychelles	0.9468	0.8987

For future reference, note that the following error statistics appear to provide a set of control values for the equatorial region:

$$P_r \cong 2 \times 10^{-4} \quad (i.e., E_b/N_o = 8.00 \text{ db})$$

$$P_{BR} \cong 0.9200$$

$$P_{BO} \cong 0.8800$$

4.2. Error Behavior for Equinox Periods

Results of error statistics have been obtained for the six stations for the equinox periods of 1971 and 1972, during the early evening hours. It is expected that ionospheric scintillation activity will be at a peak during these periods, leading to poor error performance. The results of error rate performance for Quito, Equador are shown in Table 4 for the equinox periods of 1971-1972. The average error rate over these four equinox periods is $P_e = 1.38 \times 10^{-2}$, which corresponds to an $E_b/N_o = 3.80$ db. A comparison of this result to that given in Table 2 (i.e., $P_e = 2 \times 10^{-4}$, $E_b/N_o = 8.00$ db) shows a two-order-of-magnitude error rate degradation

Table 4 Error results for Quito

Period	P_e	P_{BR}	P_{BO}
Spring 1971	3.25×10^{-2}	0.7564	0.9546
Fall 1971	9.50×10^{-3}	0.5010	0.6573
Spring 1972	6.50×10^{-3}	0.7987	0.9505
Fall 1972	4.50×10^{-3}	0.8021	0.9554

between the solstice and equinox periods. The corresponding power loss is 4.20 db, leading to a preliminary conclusion of scintillation fading power of $\sigma = 4.20$ db.

Table 4 also shows the "dropout" probabilities P_{BR} for the equinox periods at Quito. The four results lead to an average of $P_{BR} = 0.7835$ (i.e., Fall 1971 is not included in this calculation for equipment failure reasons). Comparing this value to that of Table 3 shows that 12.01% (i.e., 0.9036-0.7835) of the transmitted blocks were not recoverable because of severe scintillation fading. That is, for approximately 12% of the blocks transmitted from SAS-1 to Quito, the ground receiver was unable to acquire because of poor signal-to-noise ratios. Figure 7 shows P_{BR} vs time (i.e., 0-500 hr GMT) for Quito during the Fall 1972 equinox period. Each triangle represents one entire satellite pass, and there were a total of 131 passes during this period. The curve represents a least-squares fit to the points. The curve of Fig. 7 clearly illustrates the effect of equatorial ionospheric scintillation on receiver acquisition difficulties.

Table 4 also shows the error-free block probabilities P_{BO} for the equinox periods at Quito. The results average out to $P_{BO} = 0.9530$ (i.e., Fall 1971 is omitted). The corresponding value of Table 3 is 0.8954, showing approximately a 6% decrease in the number of blocks in error between the equinox and solstice periods. This is due to the burst nature of the scintillation fading. Figure 8, from the Spring equinox of 1971, illustrates this point. Plotted is P_{BO} vs time in days, with each triangle representing a satellite pass. Note that, with the exception of a few passes, most values of $P_{BO} > 0.90$, with the large majority being $P_{BO} > 0.95$. The fact that so few blocks of data contain all of the errors is most

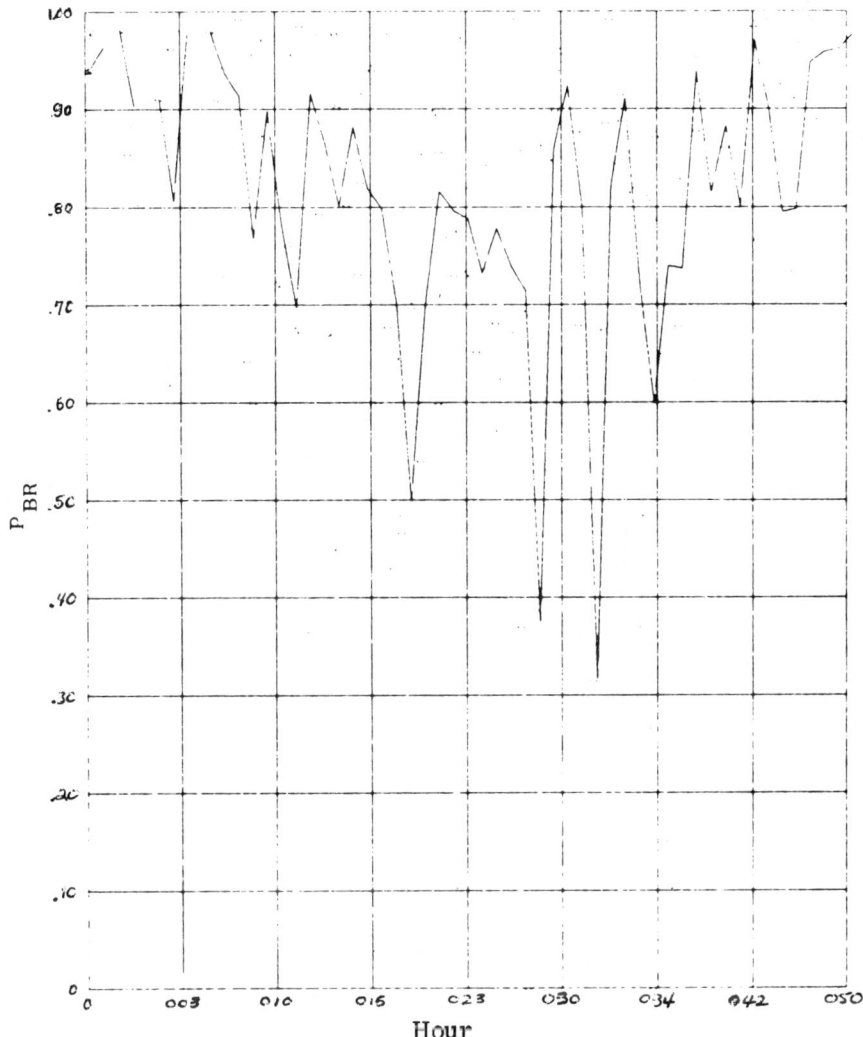

Fig. 7 "Dropout" rate for Quito for August-October 1972.

significant. If one were to recalculate the error rate based only on those blocks actually containing errors, one would obtain the so-called burst error rate P_{eB}. This statistic localizes the effects of deep fades and serves to determine the scintillation fading power for those short periods where the fading is severe enough to cause errors. Table 5 gives the values of P_{eB} for Quito. The average value of P_{eB} over these four equinox periods is $P_{eB} = 0.3149$, which implies an $E_b/N_o = -3$ db. Comparing this to the results of Table 3

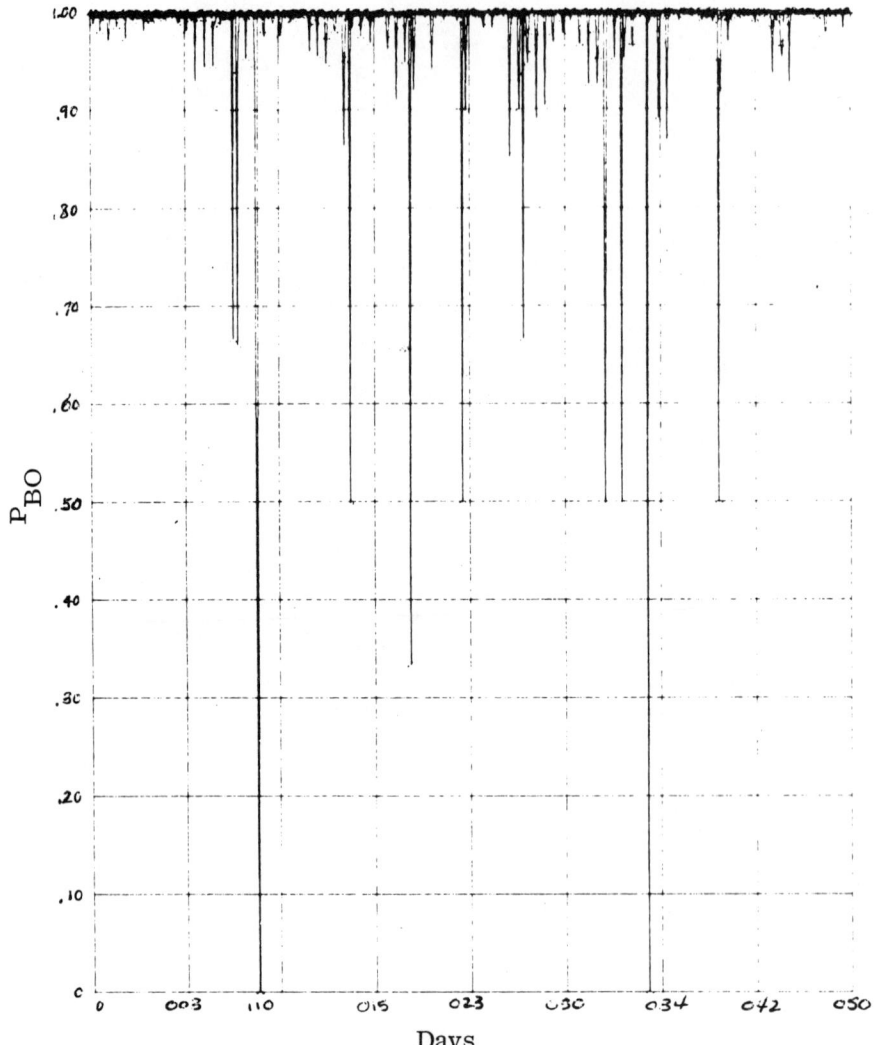

Fig. 8 Error-free block rate for Quito.

(i.e., $E_b/N_0 = 8$ db) shows the fading power $\sigma_B = 11$ db during deep fades.

In order to illustrate the geographic effect of equatorial scintillation behavior, we have selected P_e results from all six Earth stations for the Fall equinox of 1972. The results for both P_e and P_{eB} are shown in Table 6. The average value of error rate over the six stations is $P_e = 3.75 \times 10^{-3}$, implying an average $E_b/N_0 = 5.50$ db.

IMPACT OF SCINTILLATION ON COMMUNICATIONS   411

Table 5 Burst-error probabilities for Quito

Period	P_{eB}
Spring 1971	0.7126
Fall 1971	0.1275
Spring 1972	0.1313
Fall 1972	0.1008

Table 6 Average error and burst-error rates for fall equinox, 1972

Station	P_e	P_{eB}
Quito	4.5×10^{-3}	0.1008
Ascension	3.5×10^{-3}	0.0652
Kourou	3.0×10^{-3}	0.1182
Brazzaville	4.5×10^{-3}	0.0978
Ouagadougou	3.5×10^{-3}	0.1316
Seychelles	3.5×10^{-3}	0.1245

The corresponding average burst-error probability is P_{eB} = 0.1063, yielding an average E_b/N_0 = -1.50 db. These results lead to a calculation of fading power of σ = 2.5 db for all equinox periods, and σ_B = 9.7 db for the short-term, deep fade, burst periods. Note that there appears to be little variation in either P_e or P_{eB} for the longitudinally distributed equatorial stations represented in Table 6. This implies little longitudinal independence during equinox periods.

4.3. Fine-Grain Error Statistics

This section discusses the error characteristics of the trans-ionospheric scintillation channel with regard to the averages of error rate, burst error rate, dropout probability, and error-free block probability. These statistics were obtained from equatorial stations during time frames when scintillation activity is anticipated. A more definitive estimate of scintillation-caused error behavior can be obtained by observing AGC and error pattern characteristics on an individual satellite pass basis. Although this analysis is still in progress, we have included the results of the analysis for the passes shown in Table 7. All passes shown were over Quito.

The values of scintillation fading power σ coefficient of variance scintillation index S_4,[3] and m parameter of the Nakagami distribution[4] are shown in Table 8 for the satellite passes listed in Table 7. As can be seen, the fading power σ varies by a significant amount from pass to pass. The values of S_4 appear to vary quite a bit. The corresponding values of the m parameter indicate that a

Table 7 Sample passes over Quito

Tape number	Date	Time, GMT
679	2/3/71	0205-0218
700	2/7/71	0451-0504
732	2/13/71	0403-0415
757	2/18/71	0445-0458
772	2/21/71	0223-0236
820	3/02/71	0419-0432
839	3/07/71	0339-0352

Nakagami distribution fit to the data will identify tape 700 as nearly one-sided gaussian (i.e., m = 0.5 theoretically), whereas tape 757 is an approximation to a two-sided gaussian distribution. Note that none of the tapes shows a Rayleigh characteristic (i.e., m = 1 theoretically). Tapes 732, 820, and 839 show highly asymmetric probability densities analogous to that of the Rayleigh variate, whereas tapes 679, 757, 772, and 820 indicate somewhat symmetric density functions.

Assuming that the m distribution is valid for this phenomenon, the probability that the received signal power is less than σ, $P_r\{x \le \sigma\}$, is shown in the last column of Table 8. Note that for tape 700 $P_r\{x \le \sigma\} = 0.331$, whereas $P_r\{x \le \sigma\} = 0.317$ for a theoretical one-sided gaussian distribution. Tapes 757, 772, and 820, which are close to two-sided gaussian distributions, show $P_r\{x \le \sigma\} =$

Table 8 AGC statistics at Quito

Tape number	σ, db	S_4	m	$P_r\{x \le \sigma\}$
679	4.226	0.600	2.779	0.110
700	6.812	1.298	0.596	0.331
732	1.130	0.805	1.541	0.442
757	2.348	0.436	5.255	0.141
772	3.690	0.540	3.430	0.149
820	1.985	0.430	5.440	0.145
839	4.876	0.826	1.462	0.151

0.141, 0.149, and 0.145, respectively. A theoretical two-sided gaussian distribution function will show $P_r\{x \leq \sigma\} = 0.158$.

The fading power results of Table 8 show that the average value of σ for the sample passes was $\bar{\sigma} = 3.581$ db. This yields a theoretical value of $P_e = 9.5 \times 10^{-3}$. (This results from an $E_b/N_o = 4.42$ db, which in turn is obtained by subtracting 3.58 db from the nominal $E_b/N_o = 8.00$ db quoted for nonscintillation periods.) Table 9 shows a comparison of theoretical and actual error rates for the test passes. As can be seen, the comparisons are favorable. The actual average P_e, $\bar{P}_e$, is $\bar{P}_e = 1.79 \times 10^{-2}$.

The statistic that is most useful when comparing fading and burst error characteristics is the duration-of-fade, or level-crossing, probability distribution. That is, by comparing the amount of time (in bits) which the channel spends below a given power level with the bit-error autocorrelation function $a(j)$, one potentially is able to ascertain the similarity between the two measures of channel behavior. With this in mind, the duration-of-fade statistics of the satellite passes shown in Table 7 have been calculated. These are shown in Table 10. For each satellite pass, Table 10 shows the median duration of fade at the crossing level corresponding to the median received power level, and the median duration of fade at the crossing level corresponding to the median-σ received power level. The median crossing level statistic is a measure of the scintillation channel fading rate, and the median-σ

Table 9 Actual and theoretical error rates at Quito

Tape number	σ, db	P_e (theor.)	P_e (actual)
679	4.226	1.45×10^{-2}	1.60×10^{-2}
700	6.812	4.80×10^{-2}	4.20×10^{-2}
732	1.130	9.80×10^{-3}	1.15×10^{-2}
757	2.348	3.20×10^{-3}	5.85×10^{-3}
772	3.690	9.90×10^{-3}	1.50×10^{-2}
820	1.985	2.39×10^{-3}	3.50×10^{-3}
839	4.876	2.20×10^{-2}	3.20×10^{-2}

Table 10 Fade duration statistics at Quito

Tape number	Median duration, sec	Median-σ duration, sec
679	3.330	0.0375
700	4.100	0.0200
732	1.650	0.0575
757	2.400	0.0713
772	2.850	0.0700
820	1.830	0.0500
839	3.560	0.0250

crossing level statistic is a measure of the signal duration below the average received signal power level. From Table 10, we see that the median fade duration at the median crossing level was always greater than 1.650 sec (i.e., tape 732), with the largest value being 4.100 sec (i.e., tape 700). The average median fade duration at the median crossing level, for the passes shown in Table 10, is 2.82 sec. This implies an average fading rate of 0.355 fades/sec.

The median duration-of-fade results at the fading power crossing level range from a low of 0.0200 sec (i.e., tape 700) to a high of 0.0713 sec (i.e., tape 757). The average median fade duration at the median-σ crossing level, for the results of Table 10, is 0.0466 sec. Thus, since the data rate of the SAS-1 telemetry system is 1 kbit/sec, one would expect an average of about 47 bits to be transmitted during an average fade of power σ db. In other words, one would expect the bit-error memory of this channel to be on the order of 47 bits in duration. The bit-error memory of any given channel is described by $a(j)$, where $a(0) = 1$ and $a(\infty) \rightarrow P_e$. So, by observing plots of $a(j)$ vs j, and by relative comparison with P_e, one can ascertain the burst-error memory of a channel. Figure 9 shows the average $a(j)$ function for the seven test passes just quoted, as well as the average value of P_e. Two significant observations can be made with respect to the $a(j)$ curve: 1) the equatorial scintillation channel is a classic burst-error channel with short, dense bursts of errors [note that $a(10) = 0.190$, which is a large value of correlation]; and 2) note also that $a(47) = 0.0135$, which is only 1.42 times larger than P_e [i.e., $a(47) = 1.42\ P_e$]. This shows that there is a quantitative one-to-one correlation between the power fading statistics of this channel and the corresponding burst-error statistics.

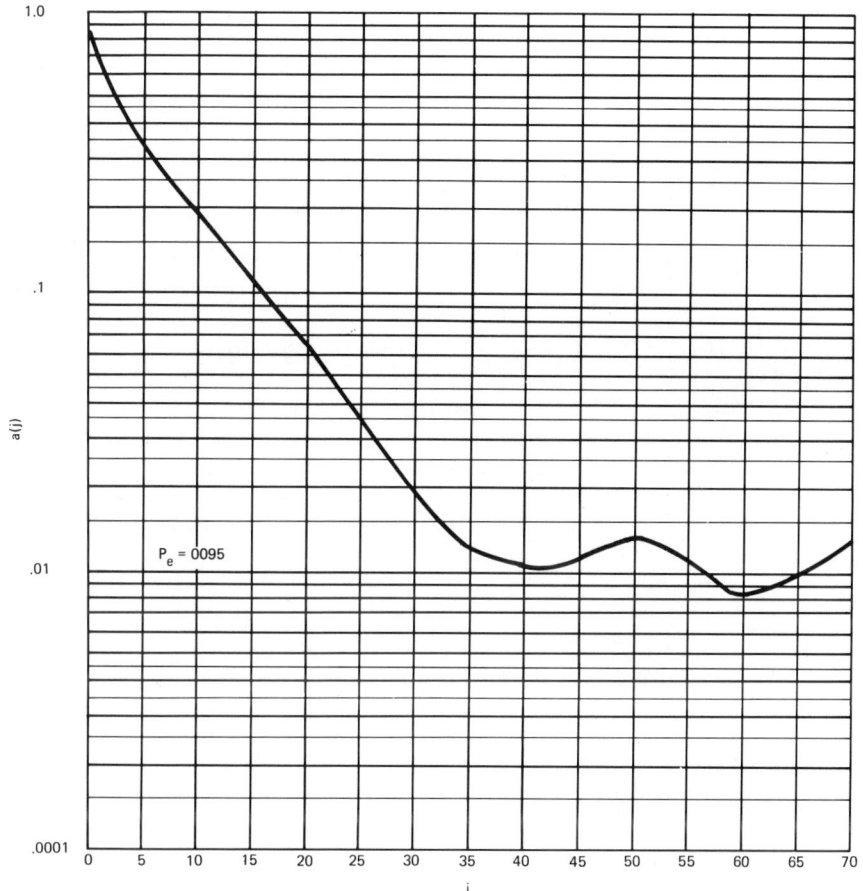

Fig. 9 Bit-error autocorrelation function for Quito.

Another burst-error statistic that is of value in describing the error behavior of the equatorial scintillation channel is the block-error probability $P(m,n)$. The averaged results of $P(m,n)$ vs m for values of n = 10 and 24 are given in Fig. 10. These results were obtained from the aforementioned set of seven Quito tapes. Also shown in Fig. 10 are the values of $1-P_{BO}$ for each block size. From these results, we can observe that the burst effects of the channel cause relatively low block error rates, even for blocks of modest size.

5. Summary

The results presented in this paper establish a number of relationships between the occurrence of errors in a communications

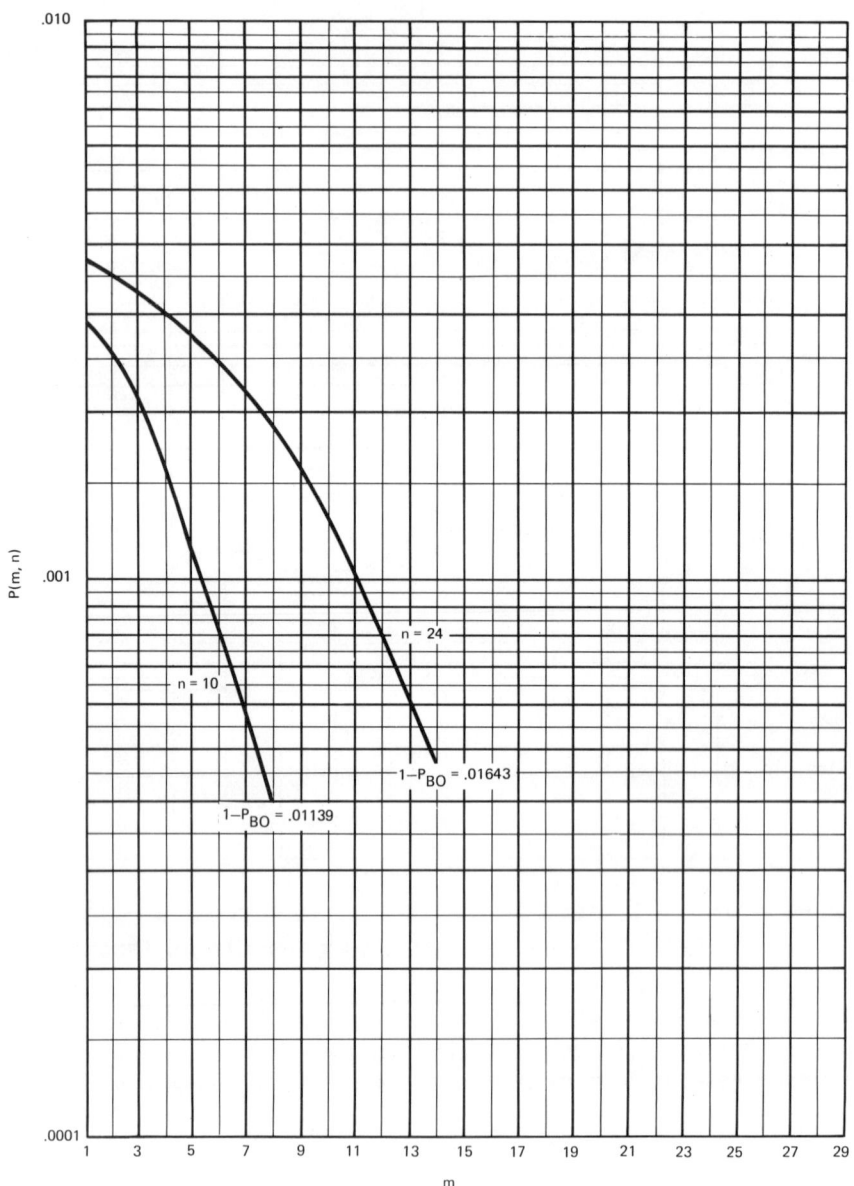

Fig. 10 Block-error distribution function for Quito.

link and the ionospheric scintillation phenomenon in the geomagnetic equatorial region. During nonscintillation time periods, the SAS-1 system performed at an E_b/N_o = 8 db, which is a P_e = 2 x 10^{-4}. For the four equinox periods of 1971-1972 at Quito, Equador, the average fading power of $\sigma = 4.2$ db caused a two-order-of-magnitude error rate degradation to $P_e = 1.38 \times 10^{-2}$, or an $E_b/N_0 = 3.8$ db. The processing-equipment-limited block recovery rate, $P_{BR} = 0.90$, during nonscintillation periods was degraded by 12%, at Quito during equinox periods, to $P_{BR} = 0.78$ by scintillation-caused receiver "dropouts." The error-free block probability at Quito was increased from $P_{BO} = 0.88$, during nonscintillation periods, to $P_{BO} = 0.95$ during equinox periods. This implies a significant increase in error clustering during periods of high scintillation activity. In fact, for Quito during the equinox periods, the burst error rate $P_{eB} = 0.32$, or $E_b/N_0 = -3$ db. This means that the severe scintillation fading that occurred during the transmission of those blocks with at least one error averaged $\sigma_B = 11$ db.

The results of error statistics from six equatorial Earth stations during the Fall 1972 equinox period show an $P_e = 3.75 \times 10^{-3}$, or an $E_b/N_0 = 5.5$ db. This is an order-of-magnitude degradation in P_e over nonscintillation performance. The fading power is $\sigma = 2.5$ db. The burst error rate is $P_{eB} = 0.11$, with $E_b/N_0 = -1.5$ db. This implies the burst fading power of $\sigma_B = 9.7$ db. Most important, there was little noticeable difference in performance among the six longitudinally separated stations.

The analysis of the fine-grain AGC and error statistics for seven satellite passes at Quito, during the Spring 1971 equinox period, reveals some very interesting characteristics of scintillation fading effects on communication performance. For one, it appears quite likely that the Nakagami probability distribution is well qualified to describe the first-order fading statistics of the received signal. For the seven tapes analyzed, the values of $P_e = 9.5 \times 10^{-3}$, and $E_b/N_0 = 4.42$ db, yielding a $\sigma = 3.58$ db. These values are typical of those found for data received at Quito during equinox periods. The fade duration statistics reveal relatively slow fading with regard to the data rate at the median received signal level but show short-duration fades at the σ crossing level (i.e., about 0.047 sec). The corresponding results for a(j) from the data error patterns show that $a(47) = 1.42\ P_e$, which indicates that the channel error memory is on the order of the median duration-of-fade interval at the σ crossing level. For the same set of tapes, the results of P(0, 10) and P(0, 24) (i.e., P_{BO} for the two block sizes, respectively) show that all channel errors are clustered into a small percentage of the transmitted blocks. The preliminary results reported herein show that the vhf equatorial ionospheric scintillation channel is a classic burst-error channel with short, dense bursts.

References

[1] Karas, T. J. and Vincent, G. B., "SAS-A Data Processing Results From Launch Through the Anomaly of the Spacecraft Tape Recorder System," X-564-71-126, March 1971, NASA.

[2] Reeder, R. R., "Small Astronomy Satellite Data Processing Results from February 1, 1971 Through December 31, 1971," X-564-72-304, Aug. 1972, NASA.

[3] Briggs, B. H. and Parkin, I. A., "On The Variation of Radio Star and Satellite Scintillations With Zenith Angle," Journal of Atmospheric and Terrestrial Physics, Vol. 25, No. 2, March 1963, pp. 339-365.

[4] Nakagami, M., "The m-Distribution - A General Formula Of Intensity Of Rapid Fading," Statistical Methods of Radio Wave Propagation, edited by W. C. Hoffman, Pergamon, New York, 1960.

Index to Contributors to Volume 42

Abrahamson, C. M., *HARRIS Electronic Systems Division* 145
Blank, Howard A., *Computer Sciences Corporation* 397
Brown, D. W., *SHAPE Technical Centre, The Netherlands* 217
Browning, J. J., *IBM Federal Systems Division* 35
Crane, R. K., *Massachusetts Institute of Technology* 311
Fremouw, Edward, J., *Stanford Research Institute* 367
Golden, Thomas, S., *NASA Goddard Space Flight Center* 397
Gribbin, W. J., *COMSAT General Corporation* 201
Hockenberry, J. H., *Lockheed Missiles & Space Company* 3
Hooper, W. P., *HARRIS Electronic Systems Division* 115
Ince, A. N., *SHAPE Technical Centre, The Netherlands* 217
Jarett, D., *Bell Telephone Laboratories, Inc.* 291
Kalley, J. J., Jr., *TRW Systems* ... 49
Karlin, Jay J., *Fairchild Space and Electronics Company*........... 83
Klocksiem, J. P., *Lockheed Missiles & Space Company*................ 3
Koenig, W. A., *Lockheed Missiles & Space Company*................ 3
Kowalik, H., *Telesat Canada*.... 183
Krejci, D. W., *Lockheed Missiles & Space Company*................. 3
Kullstam, P. A., *Computer Sciences Corporation* 263
Lu, H. S., *Aeronutronic-Ford Corporation* 25

Ludwig, Lloyd G., *Hughes Aircraft Company*................... 243
Marek, F. L., *IBM Federal Systems Division*..................... 35
Massey, J. L., *University of Notre Dame*...................... 279
McClure, J. P., *University of Texas* 347
Midgley, J. A., *SHAPE Technical Centre, The Netherlands* 217
Miller, A. D. D., *Telesat Canada*. 167
Mork, H. L., *TRW Systems*...... 49
Petrick, G. P., *HARRIS Electronic Systems Division* 145
Raab, Bernard, *Fairchild Space and Electronics Company*........... 83
Rogers, W. M., *HARRIS Electronic Systems Division* 115
Scott, W. G., *Aeronutronic-Ford Corporation* 25
Parr, A. K., *IBM Federal Systems Division*..................... 35
Sanderson, Charles C., *Hughes Aircraft Company*............ 243
Smith, T., *Aeronutronic-Ford Corporation* 25
Smoll, A., *Aeronutronic-Ford Corporation* 25
Spilman, L. D., *Bell Telephone Laboratories, Inc.* 291
Teichman, M. A., *IBM Federal Systems Division* 35
van Hover, A. J. E., *COMSAT General Corporation* 201
Whitman, J. G., Jr., *HARRIS Electronic Systems Division*..... 115

TL
507
P75
v.42

SEP 23 1976